The Cambridge Handbook of
Evolutionary Ethics

Evolutionary ethics – the application of evolutionary ideas to moral thinking and justification – began in the nineteenth century with the work of Charles Darwin and Herbert Spencer, but was subsequently criticized as an example of the naturalistic fallacy. In recent decades, however, evolutionary ethics has found new support among both the Darwinian and the Spencerian traditions. This accessible volume looks at the history of thought about evolutionary ethics as well as current debates in the subject, examining first the claims of supporters and then the responses of their critics. Topics covered include Social Darwinism, moral realism, and debunking arguments. Clearly written and structured, the book guides readers through the arguments on both sides and emphasizes the continuing relevance of evolutionary theory to our understanding of ethics today.

Michael Ruse is Director of the History and Philosophy of Science Program at Florida State University. His publications include *The Philosophy of Human Evolution* (Cambridge 2012), *The Cambridge Encyclopedia of Darwin and Evolutionary Thought* (Cambridge 2013), and *Darwinism as Religion* (2016).

Robert J. Richards is the Morris Fishbein Distinguished Service Professor in the History of Science at the University of Chicago. His publications include *Darwin and the Emergence of Evolutionary Theories of Mind and Behavior* (1987), *Was Hitler a Darwinian? Disputed Questions in the History of Evolutionary Theory* (2013), and *Debating Darwin* (2016).

CAMBRIDGE HANDBOOKS IN PHILOSOPHY

Cambridge Handbooks in Philosophy are explorations of philosophical topics for both students and specialists. They offer accessible new essays by a range of contributors, as well as a substantial introduction and bibliography.

Titles published in this series

The Cambridge Handbook of
Evolutionary Ethics

EDITED BY

Michael Ruse
Florida State University

Robert J. Richards
University of Chicago

CAMBRIDGE
UNIVERSITY PRESS

CAMBRIDGE
UNIVERSITY PRESS

University Printing House, Cambridge CB2 8BS, United Kingdom

One Liberty Plaza, 20th Floor, New York, NY 10006, USA

477 Williamstown Road, Port Melbourne, VIC 3207, Australia

4843/24, 2nd Floor, Ansari Road, Daryaganj, Delhi – 110002, India

79 Anson Road, #06-04/06, Singapore 079906

Cambridge University Press is part of the University of Cambridge.

It furthers the University's mission by disseminating knowledge in the pursuit of
education, learning, and research at the highest international levels of excellence.

www.cambridge.org
Information on this title: www.cambridge.org/9781107589605
DOI: 10.1017/9781316459409

© Cambridge University Press 2017

First published 2017

Printed in the United Kingdom by Clays, St Ives plc

A catalogue record for this publication is available from the British Library.

Library of Congress Cataloging-in-Publication Data
Names: Ruse, Michael, editor. | Richards, Robert J. (Robert John), 1942– editor.
Title: The Cambridge handbook of evolutionary ethics / edited by Michael Ruse,
Florida State University; Robert J. Richards, University of Chicago.
Description: New York: Cambridge University Press, 2017. |
Includes bibliographical references and index.
Identifiers: LCCN 2017022401 | ISBN 9781107132955 (hardback) |
ISBN 9781107589605 (paperback)
Subjects: LCSH: Ethics, Evolutionary. | BISAC: SCIENCE / Philosophy & Social Aspects.
Classification: LCC BJ1311.C36 2017 | DDC 171/.7–dc23
LC record available at https://lccn.loc.gov/2017022401

ISBN 978-1-107-13295-5 Hardback
ISBN 978-1-107-58960-5 Paperback

Contents

Contributors

Naomi Beck studies the use of evolutionary concepts in economic and political thought. She has held research fellowships at various institutions including the University of Chicago, the Max Planck Institute for Economics, and the European University Institute. She is currently finishing a book on F. A. Hayek's theory of cultural group selection.

Frédéric Bouchard is Inaugural ÉSOPE Chair of Philosophy and a professor of philosophy of science at the Université de Montréal. He recently co-edited with Philippe Huneman *From Groups to Individuals: Evolution and Emerging Individuality* (2013).

Justin Clarke-Doane is assistant professor of philosophy at Columbia University. His work centers on metaphysical and epistemological problems surrounding apparently a priori domains, such as morality, modality, mathematics, and logic.

William J. FitzPatrick is Gideon Webster Burbank Professor of Intellectual and Moral Philosophy at the University of Rochester. His work in metaethics has appeared in such journals as *Ethics*, *Mind*, *Philosophical Studies*, *Analysis*, and *Oxford Studies in Metaethics*.

Ben Fraser is postdoctoral researcher in the School of Philosophy at the Australian National University. His research interests lie in metaethics and the philosophy of biology.

Abraham H. Gibson is an NSF Postdoctoral Fellow in the Center for Biology and Society at Arizona State University. He writes on the history of evolutionary theory.

Justin Horn is assistant professor of philosophy at Oklahoma State University. He specializes in metaethics, focusing on issues in moral ontology, moral epistemology, and moral semantics.

Richard Joyce is professor of philosophy at Victoria University in Wellington, New Zealand. He is the author of *The Myth of Morality* (2001) and *The Evolution of Morality* (2006).

Uri D. Leibowitz is assistant professor of philosophy at the University of Nottingham. His main research interests are in metaethics and normative ethics, but his work also covers issues in the philosophy of science, epistemology, and ancient philosophy.

Hallvard Lillehammer is professor of philosophy and assistant dean in the School of Social Sciences at Birkbeck College, University of London. He is the author of *Companions in Guilt: Arguments for Ethical Objectivity* (2007).

Lynn Hankinson Nelson is professor of philosophy at the University of Washington. She is the author of *Biology and Feminism: A Philosophical Introduction* (2015).

Jeffrey O'Connell is a Ph.D. student in the Department of Philosophy at the Florida State University. His research interests are in nineteenth-century philosophy and the history and philosophy of science.

Trevor Pearce is assistant professor of philosophy at the University of North Carolina at Charlotte. He is currently writing a book on pragmatism and biology.

Michael L. Peterson is professor of philosophy at Asbury Theological Seminary. He is the author of *Reason and Religious Belief* (2012) and the co-author, with Michael Ruse, of *Science, Evolution, and Religion: A Debate about Atheism and Theism* (2017).

Richard A. Richards is professor of philosophy at the University of Alabama. He is the author of *The Species Problem: A Philosophical Analysis* (2010) and *Biological Classification: A Philosophical Introduction* (2015).

Robert J. Richards is Morris Fishbein Distinguished Service Professor of History of Science and Medicine at the University of Chicago. His publications include *Darwin and the Emergence of Evolutionary Theories of Mind and Behavior* (1987), *The Romantic Conception of Life: Science and Philosophy in the Age of Goethe* (2002), and *Was Hitler a Darwinian? Disputed Questions in the History of Evolutionary Theory* (2013).

Michael Ruse is Lucyle T. Werkmeister Professor of Philosophy at Florida State University. His many publications include *Can a Darwinian Be a Christian? The Relationship between Science and Religion* (2000), *The Philosophy of Human Evolution* (2012), and most recently *Darwinism as Religion: What Literature Tells Us about Evolution* (2017).

Russ Shafer-Landau is professor of philosophy at University of Wisconsin – Madison. He is the editor of *Oxford Studies in Metaethics*.

Neil Sinclair is associate professor of philosophy at the University of Nottingham. His primary research interest is in expressivist accounts of moral discourse.

Michael Vlerick is assistant professor of philosophy of science at Tilburg University and affiliated researcher at the University of Johannesburg. His work is situated in the areas of evolutionary epistemology, evolutionary morality, and cultural evolution.

Acknowledgments

The two editors would like to thank each other for their long friendship, as shown in their joint labors on this volume. They have never exchanged a harsh word nor have they ever found a claim by the other even remotely plausible. Michael Ruse would like to thank William and Lucyle Werkmeister whose generous legacy supports his professorship and made possible the conference on evolutionary ethics, in the spring of 2015 at Tallahassee, Florida, on which this volume is based. The editors are very much in debt to Jeffrey O'Connell, student and friend, for the huge amount work done in running the conference smoothly and in bringing this volume to publication.

Introduction

Michael Ruse and Robert J. Richards

After the death of the great English novelist George Eliot, a perceptive friend wrote of a meeting with her:

I remember how, at Cambridge, I walked with her once in the Fellows' Garden of Trinity, on an evening of rainy May; and she, stirred somewhat beyond her wont, and taking as her text the three words which have been used so often as the inspiring trumpet-calls of men – the words God, Immortality, Duty – pronounced, with terrible earnestness, how inconceivable was the first, how unbelievable the second, and yet how peremptory and absolute the third. Never perhaps have sterner accents affirmed the sovereignty of impersonal and unrecompensing Law. (Myers 1881)

This was the great dilemma of so many Victorians. Thanks to Germanic higher criticism, thanks to a changing society from rural to urban, thanks to the advances of science, the old religious verities were crumbling and decaying. Matthew Arnold heard the dying echo:

> The Sea of Faith
> Was once, too, at the full, and round earth's shore
> Lay like the folds of a bright girdle furled.
> But now I only hear
> Its melancholy, long, withdrawing roar,
> Retreating, to the breath
> Of the night-wind, down the vast edges drear
> And naked shingles of the world.

If there is no God, then what is to control the savage within us? Where are we to find morality and its foundations? Even the great nonbelievers of the age were not quite sure that they could go it alone. Thomas Henry Huxley, he who invented the term "agnostic," sat on the first London School Board and pushed for compulsory Bible studies. Only then could we be sure, he argued, that the young of the country were receiving proper moral guidance.

Yet, nature abhors a vacuum, in culture as in physics. Many in the mid- and late-nineteenth century thought that there was hope of a secular guide to behavior. It would be supplied by the sensational theory of the age – evolution. If, as Huxley said, we are modified monkeys rather than modified dirt, then this should tell us about ourselves and, most particularly, should tell us about our behavior, including above all our moral behavior: what we should do and why we should do it. In other words, evolutionary theory might tell us about "normative ethics," that is, principles of duty, and about "metaethics," that is, the character of ethical theory itself. Paralleling the Christian normative directives – love your neighbor as yourself – there will be evolved, altruistic instincts. Paralleling Christian metaethics – God's commands determine what is right – there will be evolutionary metaethics – the evolved structure of ethical judgment determines what is right.

From the very beginning, the evolutionary approach to ethics has been bathed in controversy, both about what exactly is being claimed and about whether the claims are true. In the opening issue of the journal *Mind* in 1876, the eminent philosopher Henry Sidgwick, though he had lost his faith, wrote vigorously against efforts by both Charles Darwin and Herbert Spencer to provide evolutionary-based analyses of moral thought and behavior. During the next century and a half Sidgwick's denunciation set the tone for the dismissal by most philosophers of any attempt at an evolutionary ethics. They felt that such thinking committed some of the crudest sins in the catechism of ethical thought. Famously, Sidgwick's student G. E. Moore picked explicitly on the worst exemplification, what he labeled the "naturalistic fallacy." Subsequent philosophers condemned evolutionary ethics for advancing vile social prescriptions and finding advantage in war and proposing extreme forms of laissez-faire economics. Although the term "Social Darwinism" was mainly applied retroactively, it became shorthand for this kind of thinking.

There were always those, usually biologists, who argued for some kind of evolutionary approach to ethical behavior, but these efforts were dismissed as untutored. Things started to change however in the 1960s, when evolutionary biologists became more secure in their science and increasingly turned their attentions to social behavior. At the same time, paleoanthropologists laboring in Africa began to work out the implications of their discoveries for modern-day members of the species Homo sapiens. These two strands of thought gradually came together and a more confident evolutionary ethics started to rise into view. Credit should be given to the Harvard ant specialist Edward O. Wilson, who opened a magisterial survey of modern thinking about the evolutionary basis of social behavior with an existential link to the avant-garde:

Camus said that the only serious philosophical question is suicide. That is wrong even in the strict sense intended. A biologist, who is concerned with questions of physiology and evolutionary history, realizes that self-knowledge is constrained and shaped by the emotional control centers in the hypothalamus and limbic system of the brain. These centers flood our consciousness with all the emotions – hate, love, guilt, fear, and others – that are consulted by ethical philosophers who wish to intuit the standards of good and evil. What, we are then compelled to ask, made the hypothalamus and limbic system? They evolved by natural selection. That simple biological statement must be pursued to explain ethics and ethical philosophers, if not epistemology and epistemologists, at all depths. (Wilson 1975, 3)

Expectedly, voices rose in opposition. Philosophers (including at that point one of the editors of this volume) trotted out all of the tried and true objections. Biologists also chipped in. The dispute was bitter and nasty, even by academic standards. In part this was because many academics had been radicalized by opposition to the Vietnam War, and this spilt over into opposition to a system that was seen as supportive of capitalism and the status quo. The burgeoning of Holocaust studies during the period stood as a warning that biologically based approaches to ethics might lead us toward the path of German thought in the 1930s. However, calmer heads started to prevail and a small but increasing number began to explore the possibilities of bringing evolution fruitfully to ethical thinking. This number included the two editors of this book, one who had always been calmly sympathetic to such an approach and the other, who had somewhat of a Pauline conversion and, like the great apostle, has been writing frenetic epistles ever since.

There is still much work to be done. Empirical scientists, so-called evolutionary psychologists, are moving ahead, rapidly uncovering the bases of moral thought and behavior; historians are working to discern the true links between our thinking today and the thinking of the past; and philosophers vigorously explore the potentialities and limitations of evolutionary ethics. It is in this spirit, looking forward as much as looking back, that we offer this handbook. It is not definitive. No such book at this stage could be definitive. But it is comprehensive, it does offer new insights historically and philosophically, and it shows the points of agreement and of difference, and we very much hope that it can and will act as a stimulus to others. We will now run briefly over the contributions to this volume, dividing them into four subsections.

Historical

The chapter by Hallvard Lillehammer is a wonderfully appropriate start to the collection, because it shows fully how strong and consistent was the opposition of regular philosophers to any kind of evolutionary approach to ethics. Focusing on the work of the important British philosopher W. D. Ross,

Lillehammer shows not only how his critique of evolutionary ethics parallels and draws on Moore but also how the critique broadens out to encompass all naturalistic accounts, including those of French sociologists like Émile Durkheim. Lillehammer argues that Ross's own position, regarding moral truths as directly intuited as necessary, anticipates some of today's critics of evolutionary ethics, including critics included in this collection.

The higher criticism of Biblical texts, which began in Germany and quickly spread through Europe, England, and America, helped spark the crisis of Christianity during the second half of the nineteenth century. One of the apostles of the crisis was Friedrich Nietzsche, whom Jeffrey O'Connell discusses in his chapter. Nietzsche, as O'Connell shows, was critical of Darwinian and Spencerian attempts to form an evolutionary ethic, in part because he understood their emphasis on the importance of utility and selflessness as a holdout of Christian values. For Nietzsche, the important question was not whether we are in fact altruistic, but rather why we value altruism in the first place, and he offered his naturalistic theory of the will to power as an alternative to the English shopkeeper's pallid version of Christian morality.

Trevor Pearce focuses on the pragmatists, whose philosophy Nietzsche would have dismissed. John Dewey, William James, and Charles Sanders Peirce originally adopted evolutionary principles in their thinking. Pearce specifically examines the application of evolutionary theory to ethical thinking that is found around the beginning of the twentieth century in the writings of Dewey and Jane Addams, two Chicagoans at the time. Pearce takes up both Dewey's "evolutionary method" in ethics, a kind of functionalist approach, and Addams's social ethics, something clearly formulated in response to her work at Hull House, the settlement house serving newly arrived European immigrants. Pearce argues that their thinking is of more than historical interest and that it still influences philosophers today.

Economic Darwinism has a shrill ring to sensitive ears. It usually means a kind of individualistic struggle that would leave much of the population poor and destitute. Naomi Beck shows one should be very wary of assuming that all who fall under this label endorse precisely the same proposals. She looks at the work of the Austrian Nobel Laureate in economics Friedrich August von Hayek, finding that he argued for a form of cultural group selection that supposedly leads to a socially benign state, one far better than anything consciously planned.

Finally, in this subsection we have the contribution of Abraham Gibson, who focuses first on Harvard University, showing how Edward O. Wilson is very much the end point of a tradition trailing back to the early twentieth century, back to the work of the ant specialist William Morton Wheeler and the zoologist George Howard Parker. Their influence spread out from Harvard to Chicago and further west, with Spencerian views about dynamic equilibrium much appreciated and a reliance on ideas of group selection. Perhaps, then, it is no real surprise to find that group-selectionist views have recently been taken up by Wilson, the intellectual grandson of Wheeler (who was the supervisor of the supervisor of Wilson).

For Evolutionary Ethics

Michael Ruse opens this section, part historical and part philosophical, showing how he was led in the early 1980s from conventional opposition to evolutionary ethics to enthusiastic support. He distinguishes a more Darwinian approach, which he endorses, from a more Spencerian approach that he finds in the thinking of Robert J. Richards. (Richards, of course, finds that delirious characterization to be the product of a long day in the sun.) Ruse is at pains to insist that he does not think he is saying anything particularly radical, describing his position as that of the philosophy of David Hume brought up to date by the science of Charles Darwin.

It was Richard Joyce in his *The Evolution of Morality* (2006a) who pushed forward the modern discussion, offering a full exposition of the "debunking argument." Those advancing that argument hold that once you have given a selection-based explanation of ethical behavior, you have drained that behavior of moral significance. In his contribution to this volume, Joyce gives a précis of this argument. One particularly valuable aspect to his chapter is the way in which he situates his thinking in the context of the various theories and divisions that one can find upheld (or attacked) by today's philosophers of morality.

In a paper written in this tradition, Justin Horn raises a challenge for moral realism, the view that morality is objective in a very strong sense. In particular, Horn considers whether we ever have justified beliefs about such objective moral truths given that our moral sensibilities have been shaped by evolution. He distinguishes several different ways that evolution could have influenced the content of our moral judgments and suggests that, at best, evolutionary influences track objective moral truths only "indirectly." He then presents a thought experiment akin to Descartes's evil demon and argues that the potentially distorting influence of evolution on our moral judgments undermines the justification of our moral beliefs, so long as we assume those beliefs are about objective moral facts.

Richard Richards is sympathetic to evolutionary naturalism, including the attempt to explain morality as something that evolved through natural selection. He thinks, however, that the success of this means that we can sidestep arid discussions about whether this now means that morality can remain objectively true or whether it now must be seen as subjective in some definite sense. He argues that we should regard moral properties as relational, on the model of our understanding of secondary qualities. Instead of thinking of secondary qualities as properties, we should think of them as relationships – not this apple is red, but this apple appears red to me in such and such conditions. So instead of thinking of values as a property we should think of them as a relationship – not killing babies is wrong, but killing babies is wrong for me in such and such conditions. There is both a subjective element, how subjects are affected and feel, and an objective element, the facts about how subjects are affected.

Robert J. Richards is a longtime contributor to the debate about evolutionary ethics, as enthusiastic about the strategy as he was at the beginning of his career. If Ruse and Joyce represent one approach, which Ruse thinks is more Darwinian than Spencerian, Richards represents the other approach, which he believes a more accurately Darwinian one; after all, Darwin presumed he offered a naturalistic account of the origins and development of conscience and an explanation of judgments that make valid claims to universality and objectivity. For Ruse (and Joyce) the Humean is/ought distinction is the starting point of the discussion, leading to a subjectivist view of morality. In a tradition that Richards acknowledges goes back to Aristotle, for him, the is/ought distinction is no barrier to objectivity on Darwinian terms. He perceives human behavior as having value, as having meaning – morally good, bad, or neutral – allowing for universal and objective judgment. Edward O. Wilson would feel very comfortable about this.

Finally, Ben Fraser takes a much more radical approach to the whole issue. He likens his position to that of the New Atheists. They think that religion is false, but more importantly they think that religion is dangerous. It leads people to do bad things. Fraser thinks the same of morality. It may perhaps have had an adaptive function, but we can no longer assume that this is the case. He includes here the point (made forcibly by Ruse and others) that we objectivize morality, thinking that it is more than mere emotion. He thinks that this leads to rigidity and lack of concern for others. As with the New Atheists, one senses that the wholescale cleansing of the temple is not aimed at some kind of moral anarchy, but rather an escape from the shackles of our evolutionary past.

Against Debunking Arguments

Russ Shafer-Landau simply does not buy into the debunking argument. He is not insensitive to the pressures of evolutionary biology or to the influences of culture. But he thinks we are simply begging the question by saying that this is all there is to the matter or that this is all that there could be to the matter. If one argues that in some sense one is simply aware of moral truths as one is aware of other necessary truths, then evolutionary biology is simply irrelevant. The moral realist emerges unscathed.

William FitzPatrick also argues that evolutionary debunking arguments beg the question against moral realism. He distinguishes two perspectives from which we might attempt to explain moral beliefs: the internal perspective that we occupy as engaged moral agents who take ourselves to have good reasons for believing what we do, and the external perspective that we can occupy as scientists looking on moral beliefs as just more empirical phenomena to be explained through biological or sociological causes having nothing to do with their truth. Those who focus on the external perspective assume that it provides the "best explanation" of our moral beliefs. FitzPatrick argues instead that, if we take the internal perspective seriously and don't just beg the question against realism from the start, then the best explanation of at least some of our moral beliefs may well be one that appeals to our flexible use of intelligence in recognizing good reasons for believing their contents to be true.

Justin Clark-Doane argues that sauce for the ethical goose is also sauce for the mathematical gander. If debunking arguments undermine the objectivity of moral beliefs, then they are no less effective in undermining the objectivity of mathematical beliefs. As he rightly points out, a lot of the discussion thus far about debunking arguments has taken a fairly unreflective attitude toward mathematics, not to mention the extent to which evolutionary discussions are always going to pick out basic truths (such as $1 + 1 = 2$) and the extent to which such discussions give warrant for the necessity that we find in such truths. Clark-Doane is inclined to think that thus far, on the basis of evolution, no definite wedge has been thrust between any or all of the truths we tend to think basic, and, hence, we might pause before we go too far down the debunking road.

Uri D. Leibowitz and Neil Sinclair take seriously the fact that evolutionary accounts of ethics claim that one can explain morality in terms of evolution, and they take even more seriously the fact that there is considerable ambiguity in this claim about whether it is tendencies or abilities that are supposed to be explained. Too often there is a gap between the actual science and the supposed object of explanation. In any case, they point out that because evolution may point to the irrelevance of the truth of morality, it does not follow that morality is unjustified. It may be that independently we can rely on the justified nature of morality. In making a claim like this, Leibowitz and Sinclair link up with some of the other critics of debunking arguments. They are not against the interest and worth of evolutionary explanations, just wary about their full implications.

Michael Vlerick takes evolutionary explanations of morality seriously, arguing that traditional moral realism is untenable given the evolutionary origin of our "moral sense." Nevertheless, in contrast to Joyce's evolutionary anti-realism, he does not hand the same fate to normative justification. In fact, he argues that it is precisely in their biological roots that we find the key to justifying our moral norms. Emotions forged in the Pleistocene, he claims, cannot adequately explain moral urges felt today, for instance, that we ought to donate money to help hungry children in Ethiopia. This leads him, in common with some of the other contributors, to highlight human reason as a significant factor in the ethical broth. He believes that this gives us reason to think of morality as objective for us and allows us to entertain the notion of moral progress.

Elaborations

The final set of chapters in this collection pick up and elaborate the implications of evolution for our understanding of morality. Frédéric Bouchard is in favor of naturalistic explanations of ethics; it is just (in line with some others in the volume) that he is uncomfortable with putting everything about human nature down to the genes. He is also in favor (in line with some and against others in this volume) of a more holistic approach to human individuality and an understanding of mechanisms of change that encompass groups and not just individuals. Indeed, he goes so far as to suggest that the microorganisms inhabiting our bodies might have a significant causal input on some mental states and hence, although he eschews overt discussion of foundations, finds himself directed toward a somewhat relativistic view of morality. It all depends on what is in your intestines.

With good reason, feminists have critiqued both traditional evolutionary thinking and contemporary human sociobiology. Lynn Hankinson Nelson, although seeing much of worth in feminist criticisms, is not willing to abandon the whole of Darwinian theory. She does take issue with the claim of evolutionary psychology that it does not commit the naturalistic fallacy. She argues that both evolutionary psychologists and their feminist critics derive normative conclusions from their claims but that those claims are "evaluatively thick," that is, they carry both empirical and normative content. This does not mean that one cannot usefully apply evolution to ethics; but it does mean one must take care about ignoring ethical implications that, even if subtly or initially unrecognized, are implicated in one's analysis.

Michael Peterson is a theist and this informs his approach to evolutionary ethics. He is not a Creationist and is happy not just with evolution but also with the Darwinian mechanism of natural selection. He argues nevertheless that evolutionary ethics involves an epistemological and metaphysical position that is not strictly scientific but tends to be underpinned by a nontheistic naturalism. There is little surprise therefore that moral realism does not fare well. Peterson argues that if one takes a theistic position, one's epistemology and metaphysics can be more realist in nature and that this reflects more accurately the way philosophy should account for our deepest commonsense moral intuitions. Obviously this does not make theism true but is a mark in its favor.

Part I

Historical

1 Ethics, Evolution, and the A Priori: Ross on Spencer and the French Sociologists

Hallvard Lillehammer

Introduction

For large parts of the twentieth century much moral philosophy in the Anglophone tradition took place in isolation from the study of the natural history of human morality, including evolutionary biology.[1] To a large extent, this isolationist tendency remains an important part of the philosophical landscape today.[2] In this chapter, I raise three questions about this tendency. First, what explains the detachment of philosophical attention from empirical questions about the natural history of morality during this period? It may be tempting to think that the philosophers of the time simply ignored the main developments of the emerging human sciences of their time, but a closer reading of the seminal works of the period shows that they did not. On the contrary, they explicitly responded to these developments and turned away from them on the basis of what they thought of as decisive arguments. So, why did these philosophers turn away from these developments? It may be tempting to think that the explanation is closely connected to the alleged diagnosis of the "naturalistic fallacy" in G. E. Moore's *Principia Ethica* (Moore 1903), but once more, a closer reading of seminal works of the period shows that this hypothesis overplays the metaphysical and semantic aspects of their arguments at the expense of their epistemological and normative aspects. This hypothesis is reflected in the prevalence of what may be called a "metaphysics first" approach, as opposed to an "epistemology first," to the interpretation of the moral philosophy of the period (see, e.g., Hurka 2014). The present chapter is an attempt to rethink this approach, which in my view suffers from a degree of anachronism. Third, were the philosophers in question right to turn away from these developments? It may be tempting to think that philosophers at the time turned away for reasons that more recent philosophy has either outgrown or transcended, but in fact their reasons for doing so turn out to display striking similarities with a way of responding to analogous arguments that remain widely accepted at the start of the twenty-first century (cf. Audi 1996; Shafer-Landau 2003; Peacocke 2004; Lillehammer 2016). They therefore remain directly relevant to contemporary discussions of morality and evolution.

In this chapter I critically discuss the dismissal of the philosophical significance of facts about human evolution and historical development in the work of one of the seminal figures in early "analytical" moral philosophy, namely W. D. Ross (1877–1971). I address Ross's views about the philosophical significance of the emerging human sciences of his time in two of his main works, *The Right and the Good* (1930) and *The Foundations of Ethics* (1939). Where the latter work contains a discussion and dismissal of Herbert Spencer's evolutionary ethics, the former work contains a discussion of French sociology that, although not explicitly directed as an example of evolutionary ethics as normally understood, is nevertheless directly relevant to (and to some extent foreshadows) contemporary responses to so-called evolutionary debunking arguments in ethics (see, e.g., Joyce 2005; Singer 2005; Berker 2009). For reasons that will become clear in what follows, it is largely irrelevant exactly what form the evolutionary theory to which Ross responds actually takes, e.g., whether it is a form of Darwinian natural selection or some other account of the emergence and development of our moral faculty.[3]

Ross on Spencer and Evolutionary Ethics

Ross explicitly confronts the philosophical challenges presented by evolutionary ethics in one of the less widely read sections of his Gifford lectures, published as *The Foundations of Ethics* in 1939. The main target of Ross's discussion in these sections is Herbert Spencer's evolutionary ethics, as read mainly through the spectacles of Moore's 1903 discussion of the same topic in *Principia Ethica*. In fact, there is substantial overlap between Moore and Ross with respect to which aspects of Spencer's work they address in their respective responses to Spencer's version of evolutionary ethics. For example, at one point Moore directly quotes a number of passages from Spencer's *Data of Ethics*, including the following:

No school can avoid taking for the ultimate moral aim a desirable state of feeling called by whatever name – gratification, enjoyment, happiness. Pleasure, somewhere, at some time, to some being or beings, is an inexpugnable element of the conception. (Spencer 1879, Sect. 16)

Moore's response to Spencer, in *Principia Ethica*, is as follows:

Mr. Spencer himself tells us his "proof" is that "reversing the application of the words" good and bad ... "creates absurdities" [Sect. 16] ... So ... he is ... a naturalistic Hedonist. (Moore 1903, 104–105)

What Moore appears to be saying is that Spencer is not, at bottom, an "evolutionary ethicist" (in the sense of someone who seeks to "derive" moral claims from naturalistic claims about human biology and the like) as much as a philosophical hedonist about the good, who also happens to think that evolution proceeds by producing more pleasure, and therefore more good on the whole. The fundamental case for Spencer's hedonism is not that it somehow "derives" from, or even accords with, the direction of evolution, but rather that its denial "creates absurdities." As Moore sees it, this is most plausibly read as an a priori claim about which ideas can be coherently affirmed together and is therefore one that can be made from the comfort of the armchair, hence without any prior appeal to, or knowledge of, evolutionary theory.

In his later discussion in *The Foundations of Ethics*, Ross makes exactly the same move. He claims that Spencer's "fundamental ethical theory" holds that "it is conduciveness to pleasure that is ... the real ground of rightness"; that "life always contains a surplus of pleasure over pain; and that conduciveness to life and conduciveness to pleasure always go together." These assertions, according to Ross, mean that Spencer's "fundamental theory turns out to be universalistic Hedonism, or Utilitarianism" and that, consequently, evolutionary ethics "need not be examined as a separate form of theory regarding the ground of rightness" (Ross 1939, 59). Once more, then, Spencer is said to be, at bottom, a philosophical hedonist, and not an "evolutionary ethicist" (as previously defined) at all. (The exact nature of Spencer's "hedonism" is a separate question I shall not pursue here. For present purposes, it matters less what Spencer actually thought about pleasure than what Ross and his philosophical contemporaries took him to think.[4])

Moore and Ross are by no means alone among moral philosophers of this period in responding to Spencer in this way. In *The Theory of Good and Evil*, published in 1907 (but substantially written some time earlier), Hastings Rashdall has a similar go at Spencer when he writes that "[m]orality essentially consists in the promotion of a good or ideal of life, the nature of which is discerned by our rational judgements of value," at least some of which are self-evident and a priori (Rashdall 1907, 401). Explicitly targeting Spencer, Rashdall notes that "[t]here are ... parts of Spencer's writings ... in which he would seem almost prepared to admit the simple, *a priori* unanalyzable character of the idea of Right" (Rashdall 1907, 368) and that "[h]is judgement that pleasure is the sole good is, in short, like all ultimate moral principles, an *a priori* judgement of value, true or false" (Rashdall 1907, 372). Rashdall, however, claims to find no evidence of this view in *The Data of Ethics*, which is the work cited by Moore and Ross on this topic (Rashdall 1907, 369) and which includes the very claim about pleasure being "an inexpugnable element of the conception" – a line that Rashdall himself quotes on page 379 of his book. On the whole, Rashdall's treatment of Spencer is characterized by a noticeably acidic tone, as when he claims that most of the empirical claims on which Spencer's argument really depend were "fairly well known before" the emergence of evolutionary theory (Rashdall 1907, 361). He adds that Spencer's data is based on "his experience of a Somersetshire village in 1834–6, and not upon any study of the habits either of the Amoeba or the 'peaceful Arafuras.'" He concludes that "[a]ll the biological and sociological apparatus [of Spencer's ethical theory] … was simply an afterthought, an attempt to invoke the supposed 'teaching of Science' in support of foregone conclusions" (Rashdall 1907, 397). Essentially the same response can be found in an essay (1902) by R. R. Marett, Rashdall's Oxford contemporary. Marett writes of "that other a priorist Mr Spencer": "insofar as it pretended to rest on history, 'rational utilitarianism' was a sham. Its appeal was never to veritable history, but to … the 'is really' of an *a priori* metaphysical naturalism" (Marett 1902, 246, 265).[5]

Either way, Marett, Moore, Rashdall, and Ross all claim that, insofar as Spencer's alleged philosophical hedonism has any legitimate claim to our philosophical attention, this is because its truth or falsity must be assumed to be knowable a priori, and quite independently of any theory of evolution, whether of a Darwinian, Spencerian, or any other kind. No development in evolutionary theory – past, present, or future – therefore has the potential to change our fundamental moral beliefs one way or the other. With respect to the basic truths of morality, evolutionary theory, like all empirical speculation about the nature and development of ethics, is "simply beside the mark" (Ross 1930, 15). This dismissive response to evolutionary ethics on the part of Ross and his philosophical contemporaries was a simple function of their a prioristic methodology.

Did Ross and his fellow a priorists get Spencer right, or did they fail to grasp the place of his claims about the foundations of moral knowledge in his wider science of human morality? The issue is not entirely straightforward, at least for two reasons. First, much of the inspiration of Spencer's ethics derives from a heterogeneous tradition of British moral philosophy that is neither always clear nor always consistent on this issue (see Schneewind 2003). Second, Spencer's views on morality changed substantially over time, usually without either explicit acknowledgment or revision of earlier views.[6] What is entirely straightforward is that, at some point in his career, Spencer did express a commitment to the sorts of claims that Ross and his fellow a priorists latched onto. Thus, in his 1853 paper "The Universal Postulate" (republished in part as "General Analysis" in *The Principles of Psychology* in 1855), Spencer writes that to "[m]ean what we mean by the word truth, we have no choice but to hold that a belief which is proved, by the inconceivableness of its negation, to invariably exist, is true" (Spencer 1853, 530; quoted in Francis 2007, 179) and that "the invariable existence of a belief [is] our sole warrant for every truth of immediate consciousness, and for every primary generalization of the truths of immediate consciousness … [and] our sole warrant for every demonstration" (Spencer 1853, 528–529; quoted in Francis 2007, 179).[7] Spencer's application of this "universal postulate" in these passages is meant to provide a reconciliation of his broadly empiricist philosophy with the "common sense" rejection of skepticism. The analogous application of the postulate to basic moral beliefs will have been tempting, if not irresistible. Hence, in *Data of Ethics*, Spencer claims that "there is no escape from the admission that in calling good the conduct which subserves life, and bad the conduct which hinders or destroys it ... we are inevitably asserting that conduct is good or bad according as its total effects are pleasurable or painful" (Spencer 1879, 28). He thinks that "[t]o prove this it needs but to observe how impossible it would be to think of them as we do, if their effects were reversed" (Spencer 1879, 31) – thus a "fundamental assumption" (Spencer 1879, 40) or "postulate" that is "universally accepted" (Spencer 1879, 45). Pleasure, concludes Spencer, "is as much a necessary form of moral intuition as space is a necessary form of intellectual intuition." It is, therefore, "an inexpugnable element of the conception" of morality (Spencer 1879, 46).

From the perspective of Ross and his fellow a priorists, whether the basic truths in question are described as "a priori self-evident principles," "first principles of reason," or "fundamental and indispensable assumptions," none of these designations are as important as the fact that no appeal to experience beyond the armchair, much less the theory of evolution, would seem to be necessary to arrive at them. Spencer's universal postulate is quite similar to the "ethical intuitionism" of Ross and his allies. They, therefore, quickly recognized that Spencer's evolutionary ethics were unanchored in any empirical proposition, evolutionary or otherwise.[8]

Ross on the French Sociologists

In the opening chapter of what may be his most widely read work, *The Right and the Good*, Ross briefly argues against the temptation to draw any significant moral or epistemological conclusions from the empirical study of the causes or functions of moral beliefs, whether those functions be biological, historical, or otherwise. His main target in the relevant passages (less than three pages in total, but strategically placed in the first chapter of the book) is the emerging science of sociology, here in the guise of Émile Durkheim and Lucien Lévy-Bruhl, both of whom Ross appears to have read (or at least have read about) in the original French.[9] Addressing their work and its significance for moral philosophy, Ross writes:

[I]t may be well to refer briefly to a theory which has enjoyed much popularity, ... the theory of the sociological school of Durkheim and Lévy-Bruhl, which seeks to replace moral philosophy by the "science de moeurs," the historical and comparative study of the moral beliefs and practices of mankind. It would be foolish to deny the value of such a study, or the interest of many of the facts it has brought to light with regard to the historical origin of many such beliefs and practices.... What must be denied is the capacity of any such inquiry to take the place of moral philosophy. (Ross 1930, 11–12)

According to Ross, the members of this "sociological school" are trapped in what he describes as an "inconsistency" (Ross 1930, 13). On the one hand, they tell us to accept "an existing code as something analogous to an existing law of nature ... and on this side the school is able ... to proclaim itself conservative of moral values" (Ross 1930, 13). On the other hand, by claiming that "any code is the product ... of bygone superstitions and ... out of date utilities, it is bound to create ... a skeptical attitude towards any and every given code" and thereby makes our moral beliefs vulnerable to a form of historical, sociological or developmental debunking (Ross 1930, 13). Some readers may see traces of "Hume's Law" or the so-called naturalistic fallacy as a diagnostic tool in the passages just quoted. Yet when Ross proceeds to deny the philosophical significance of the empirical facts that sociology has brought to light, what he offers is not an argument against inferring an "ought" from an "is" or identifying moral properties with natural properties, but instead an argument against a narrowly functional understanding of moral beliefs. Thus, he writes:

[T]he analogy which ... [sociology] draws between a moral code and a natural system like the human body ... is an entirely fallacious one. By analyzing the constituents of the human body you do nothing to diminish the reality of the human body as a given fact, and you learn much that will enable you to deal effectively with diseases. But beliefs have the characteristic which bodies have not, of being true or false, of resting on knowledge or being the products of wishes, hopes and fears. (Ross 1930, 13)

Ross's point is that a reductively functional understanding of moral (and other) beliefs fails to take account of the fact that beliefs have contents, the truth conditions of which cannot be assumed to be a simple function of the social or psychological role that these beliefs happen to play at any given time or place.

The comparison between moral codes and the human body can be traced to Durkheim's work, *The Rules of Sociological Method*, where Durkheim writes:

For societies, as for individuals, health is good and desirable; sickness on the other hand, is bad and must be avoided. If therefore we find an objective criterion, inherent in the facts themselves, to allow us to distinguish scientifically health from sickness in the various orders of social phenomena, science will be able to throw light on practical matters while remaining true to its own method.... The state known as health ... is a valuable reference point to guide our actions.... It establishes the norm which must serve as a basis for all our practical reasoning. (Durkheim 1982, 86–87)[10]

Yet to say that moral beliefs can be either healthy or unhealthy does not deny that they have evaluable contents or that the truth conditions of moral beliefs are distinctive in kind. Durkheim certainly recognizes that ethical thought is distinctively normative, as when he writes (in his review of Lévy-Bruhl's *Ethics and Moral Science*):

A science can reach conclusions which permit the establishment of norms, but is not normative by itself.... It is necessary, therefore, to renounce this contradictory conception of a normative science and to dissociate definitely science and practice. (Durkheim 1993, 30)[11]

Even so, perhaps Ross could be read as justly accusing his chosen interlocutors of describing the truth conditions of moral beliefs in implausibly relativistic terms, as for example, when Durkheim writes that "theoretical morality ... is quite simply a way to coordinate as rationally as possible the ideas and feelings which constitute the moral conscience of a definite period of time" (Durkheim 1993, 32). A comparable example of what Ross finds objectionable might be "[e]very system has its own rationality.... Roman morality has its rationale in Roman society just as our morality has its rationale in the nature of contemporary European society" (Durkheim 1993, 48).[12] Ross's response is telling. Instead of seriously considering the possibility that some ethical claims could nevertheless prove themselves to be justified (or perhaps true) even if they are in some sense "relative," Ross considers the alternative presented by "the sociological school" as the starting point of a journey into the epistemological abyss. He immediately goes on to say that the possible (and in some cases actual) mismatch between the real causes and function of moral beliefs and the truth conditions of those beliefs threaten to imply a universal skepticism about the claims expressed by those beliefs. He writes that "in so far as you can exhibit ... [moral beliefs] as being the product of purely psychological and non-logical causes of this sort, while you leave intact the fact that many people hold such opinions you remove their authority and their claim to be carried out in practice" (Ross 1930, 13). His discussion continues with the following oft-cited passage, with which many contemporary philosophers will be familiar (if only in part). Ross writes:

[I]f human consciousness is continuous, by descent, with a lower consciousness which had no notion of right at all, that need not make us doubt that the notion is an ultimate and irreducible one, or that the rightness (prima facie) of certain types of act is self-evident; for the nature of the self-evident is not to be evident to every mind however undeveloped, but to be apprehended directly by minds that have reached a certain level of maturity, and for minds to reach the necessary degree of maturity the development that takes place from generation to generation is as much needed as that which takes place from infancy to adult life. (Ross, 1930, 12)

What concerns Ross in this passage is that the views of "the sociological school" seem to imply that all our claims to moral knowledge are entirely without foundation. He is concerned to rebut the challenge that knowledge of the causes and functions of our moral beliefs will debunk those beliefs because their contents are unconnected, or at best, accidentally related, to the justificatory or veridical grounds that they are said to imply or presuppose.[13] Regardless of the truth-value of a given moral belief – or so the challenge goes – our justification for that belief would be undermined if it turns out that our possession of it has the wrong kind of relation to what would make it justified, or true. Ross writes:

[I]f anyone can show that A holds actions of type B to be wrong simply because (for instance) he knows such actions to be forbidden by the society he lives in, he shows that A has no real reason for believing that such actions have the specific quality of wrongness, since between being forbidden by the community and being wrong there is no necessary connexion.... He does not, indeed, show the belief to be untrue, but he shows that A has no sufficient reason for holding it true; and in this sense he undermines its validity. (Ross 1930, 14)[14]

Ross rejects the debunking challenge outright. His response is as follows: basic moral knowledge, in the form of self-evident universal principles, is a priori accessible to minds that have reached a certain stage of development (or civilization). It is therefore accessible to those minds regardless of any a posteriori knowledge they may or may not have of their own evolution or history, the social function of their moral sensibility, or any other contingent fact about particular moral systems, practices, or traditions. Furthermore, Ross appears to be confident that the human mind has, in fact, at the time and place of writing, reached a stage of development at which basic moral knowledge is accessible in this way and that some people (presumably including himself) actually demonstrate that they have access to such knowledge whenever they successfully exercise the relevant capacity (e.g., when they correctly rule out certain moral claims as absurd or implausible from the armchair). Ross writes:

[T]he human mind ... is competent to see that the moral code of one race and age is in certain respects inferior to that of another. It has, in fact, an a priori insight into certain broad principles of morality, as it can distinguish between a more or less adequate recognition of these principles.... [T]here is a system of moral truth, as objective as all truth must be, ... and from the point of view of this, the genuine ethical problem, the sociological inquiry is simply beside the mark. (Ross 1930, 14–15)[15]

Since, according to Ross, basic moral knowledge consists in the grasp of self-evident principles that are knowable a priori, any evolutionary or historical study of the moral beliefs of human groups is, consequently, beside the point. Not only do we not need a comparative study of different ethical systems, practices, traditions, or their evolution to recognize these truths; no such study would have the power to render such beliefs either justified or not justified.

Did Ross Argue against Straw Men?

In light of Ross's response to "the sociological school," it might be tempting to think that moral philosophy is faced with a dilemma. Either we are entitled to appeal to self-evident a priori principles to explain the objective credentials of our moral beliefs, or else we are stuck with an unpalatable form of moral relativism according to which our moral beliefs are at best valid strictly relative to a given society and, therefore, have no genuine objective credentials. It is quite clear that Ross thinks that his chosen interlocutors are stuck on the second horn of this dilemma, when he writes:

According to this school, or rather according to its principles if consistently carried out, no one moral code is any truer, any nearer to the apprehension of an objective moral truth, than any other; each is simply the code that is necessitated by the conditions of its time and place, and is that which most completely conduces to the preservation of the society that accepts it. (Ross 1930, 14)

To which Ross understandably responds that "the human mind is not content with such a view" (Ross 1930, 14).

From these remarks it is natural to infer that Ross's chosen interlocutors must have failed to grasp the epistemological implications of their explanatory project. This claim does not stand up to scrutiny. Far from having missed out on the epistemological need for an a priori foundation for moral knowledge in self-evident principles, Ross's interlocutors (rightly or wrongly): (a) explicitly reject the possibility of such a foundation, and (b) explicitly reject the claim that such a foundation is necessary to prevent the universal debunking of our moral beliefs. The idea that we must either be able to ground our moral knowledge on a priori self-evident principles or submit to some form of moral skepticism is (as they see it) based on a mistake.[16]

With respect to (a), Ross's chosen interlocutors explicitly argue that there is no prospect of providing our moral beliefs with an a priori foundation in a set of universal norms and principles because the norms and principles in question (even if we could eventually come to accept them as universally valid without exception) will inevitably be a posteriori projections of the norms and principles embodied in the actual system, practice, or tradition with which one begins. In *Ethics and Moral Science*, Lévy-Bruhl argues against the project of providing moral knowledge with an a priori foundation on exactly these grounds. According to Lévy-Bruhl, moral thought is essentially a "rational art" that consists in developing conceptual tools to address the practical challenges (of survival, mutual cooperation, etc.) presented by our contingent natural and historical circumstances and is therefore not a suitable object of a theoretical science in which answers to moral questions could be asymmetrically grounded in a set of self-evident principles knowable a priori. On the contrary, any rational ethical system, practice, or tradition "has to be entirely constructed" from the natural and historical materials with which we contingently find ourselves (Lévy-Bruhl 1905, 204). This process of construction will inevitably proceed by engaging critically with existing practice, which it will necessarily in some sense reflect, and therefore cannot be provided with an a priori and practice-independent foundation. In other words, if justice and ethics were understood as progressively developing, there would be "nothing a priori to authorize the affirmation that becoming is a progress" (Lévy-Bruhl 1905, 175).[17] Yet as Lévy-Bruhl sees it, none of this entails that our moral beliefs are incapable of improvement from one generation, or society, to another. What it does entail is that any such improvement is possible only in the light of a historically informed, and therefore a posteriori, engagement with some particular ethical system, practice, or tradition that instantiates it.

With respect to (b), Ross's chosen interlocutors explicitly deny that rational reflection on the historical causes and functions of our moral beliefs will necessarily debunk all, or even most, of those beliefs. On the contrary, they claim that such reflection has the potential to improve those beliefs, insofar as more informed beliefs can, when coherently integrated with existing beliefs, be more "rational," or sensible, (and therefore in some sense "better") than the beliefs they replace.[18] In *Ethics and Moral Science*, Lévy-Bruhl considers the possibility that if someone "learns that ... [morality] has no rational foundation, that the repulsion ... called forth by certain acts are explained by historical and psychological reasons," this will thereby "take from them both their prestige and their power" (Lévy-Bruhl 1905, 109–10).[19] His response to this challenge is as dismissive as Ross's own:

If ethical prescription does not hold its authority from a theoretical conviction or from a system of ideas, it could exist by its own strength.... The imperative character of the ethics now practiced not being derived from reflection, is scarcely enfeebled by it. (Lévy-Bruhl 1905, 111–112; cf. op. cit. 157)

In other words, the question whether our moral beliefs will survive critical reflection on facts about their causes and function is one to which the answer is, at least sometimes, affirmative. For example, we sometimes find ourselves in conditions where the reasons why we hold certain moral beliefs, and therefore those beliefs themselves, would critically withstand reflection on the nature of their own origins. (Just as we sometimes find ourselves in circumstances where they don't, e.g., cases where a moral belief can be shown to depend on a falsehood.) Durkheim is equally clear on this point, when he writes:

[O]ur moral beliefs are the products of a long evolution, ... they are the result of an endless succession of cautious steps, hard work, failures, and all sorts of experiences. We do not always perceive the causes that explain our moral beliefs because their origins are so distant and so complex. Therefore we ought to treat them with respect, since we know that humanity ... has not found anything better. We can be assured, for the same reasons, that we will find more wisdom accumulated in them than in the mind of the greatest genius. It would surely be childish to try to correct the results of human experience with our own limited judgement.... Morality would thus have quite enough authority in our minds since it would be presented to us as the summary and conclusion, however provisional, of human history. (Durkheim 1993, 135)

According to Durkheim, the extent of our justified confidence in our moral beliefs is partly a function of the extent to which they have already withstood reflection and challenge over time, including empirically informed reflection on their historical causes and function. Obviously, not every moral belief would pass this test, partly because some moral beliefs obviously conflict and partly because some moral beliefs may never have been suitably challenged. Yet the idea that knowledge of the causes and function of moral beliefs could somehow undermine our moral belief system as a whole is one that Durkheim is prepared to reject.

To conclude, not only did Ross take an unduly dismissive view of the philosophical significance of "the historical and comparative study of the moral beliefs and practices of mankind." He made no serious attempt to engage with those implications as they were actually conceived by his chosen interlocutors. In this respect, he is not alone – either among his, or among our own, philosophical contemporaries. What exactly to make of this fact is an important question, but one I cannot do justice to here. Instead, I will conclude this brief discussion of "the sociological school" by suggesting that their predominantly a posteriori approach to the study of moral thought may embody a philosophically respectable form of epistemological modesty. Once more in the words of Lévy-Bruhl:

To conclude, it seems to us that there is no answer to the demand: "Give us a system of Ethics!" because the demand has no object. Only the system they already have could be given to those who ask, because if another were suggested they would not accept it … [Yet] if ethical rational art does not offer us a "system of ethics" … it promises not the less to have important consequences; for, thanks to it, the ethical reality can be improved within limits impossible to fix beforehand. (Lévy-Bruhl 1905, 217)

Such an improvement, he claims, would involve "the custom of considering the ethics of a given society … in its necessary relation with the social reality of which it forms a part." On his view, this is a project that is "both modest and critical," rooted as it is in the continuous reassessment of our natural and historically conditioned moral beliefs (Lévy-Bruhl 1905, 230). In other words, even in the absence of an a priori self-evident foundation of our moral beliefs, we could still be entitled to hold onto (at least some of) them, and even (sometimes) improve on them. Yet any such improvement would always be the result of some empirically tractable engagement with some actual and historically embodied moral practice or tradition. Thus understood, it is compatible with the views of "the sociological school" that by thinking systematically about actual human moralities we can come to discover moral principles that apply at all times and in all places (at least with respect to beings like ourselves). What is not compatible with this view is that these moral principles would thereby have been shown to either have, or stand in need of, a purely a priori, or "fact independent," foundation (cf. Cohen 2003).[20]

Truth, Disagreement, and Confidence

As we have seen, therefore, both Ross and his chosen interlocutors were acutely aware of the moral and epistemological challenges arising from the fact of widespread difference and disagreement on ethical questions both "in different societies" and "within the same society" (Ross 1939, 17). In response, Ross claims that "on examination" such disagreement (at least among "sufficiently mature" minds) can be shown to depend "not on disagreement about fundamental moral principles, but partly on differences in the circumstances of different societies, and partly on different views which people hold, not on moral questions but on questions of fact," including (he might have added) facts about human evolution and historical development (Ross 1939, 18). Moreover, Ross claimed, "the very fact of difference of opinion is in itself evidence of the persisting confidence of all of us that there is an objective truth," even if we cannot be reasonably confident that we presently know what it is (Ross 1939, 19). Whether or not this claim is too optimistic, Ross's chosen interlocutors could in principle have agreed with him. Yet, even if they did not actually do so (I don't claim to have shown that they did), Ross arguably moved too quickly when he diagnosed the epistemologically subversive implications of their comparative, historical, and therefore essentially a posteriori approach to moral philosophy. He therefore also arguably underestimated the challenge of those who think that no philosophical account of moral knowledge is plausible that fails to seriously engage with the norms and principles of actual moral systems, practices, and traditions. Among these can be counted not only the French sociologists and their contemporary descendants in sociology, anthropology, and social psychology but also those who continue to argue for the normative significance of facts about human evolution and development, including contemporary proponents of some form of evolutionary ethics (see, e.g., Ruse 1986; Joyce 2005). Exposing the fault lines in Ross's response to these challenges not only helps make explicit what is at stake in these debates but also brings out the way in which contemporary discussions of ethics and evolution are themselves historically located in a particular intellectual tradition. How much further we have got in making sense of these issues in the century or so that has passed since Ross and his intuitionist contemporaries responded to Spencer and the French sociologists is an interesting question, worthy of further scrutiny.[21]

Notes

1 For two notable exceptions, see, e.g., Mackie 1977; Ruse 1986.

2 See, e.g., Cohen 2003; Parfit 2011.

3 For further elaboration on this claim, some of which relates directly to the discussion below, see Lillehammer 2016.

4 For further discussion, see Francis 2014.

5 Marett rejected this kind of naturalism, but also expressed a skeptical attitude toward the "unconditional dualism" of "is" and "ought" embodied in the a prioristic nonnaturalism with which his more influential contemporaries are associated (cf. Marett 1902, 246ff).

6 For further discussion of both points, see Francis 2007, 157–210.

7 Even if conceivability is an infallible guide to possibility (which is controversial), our ability to believe need not be an infallible guide to conceivability. Pursuing this issue further would take me too far afield here.

8 For a longer discussion of "the universal postulate" in Spencer's thought, see Francis 2007, 171–186.

9 The work to which Ross actually refers is Parodi 1910. The sociological work responded to by Ross was methodologically similar to that of Ross's own, to the extent that it was largely conducted from the armchair. (Indeed, the writers in question would have self-identified and also have been known as "philosophers" in a sense associated with that word at least in the continental parts of Europe.) Where it differs from Ross's work is in making no claim to be purely a priori.

10 Elsewhere, Durkheim writes: "Does it not happen constantly that the doctor becomes involved in problems for which physiology provides no solution? And what does he do then? He opts for the course of action which seems the most reasonable according to the present state of knowledge. Rational moral art will proceed likewise" (Durkheim 1993, 32). For more about this, see the section 'Did Ross Argue Against Straw Men?'

11 One possibility is that Durkheim's *The Rules of Sociological Method* is vulnerable to some version of Ross's argument but that, by the time he comes to review Lévy-Bruhl's book, there has been a significant development in his views. For discussion of this possibility, see R. T. Hall's introduction to Durkheim 1993.

12 Along similar lines, Lévy-Bruhl writes of ethical ideals as a "projection" of the "social reality of the epoch" that formulates them (Lévy-Bruhl 1905, 122) and that "the character of universality, attributed by the conscience of each individual to its dictates … is only the logical translation of imperious feeling" (Lévy-Bruhl 1905, 187). The latter claim might be interpreted as consistent with an "error theoretic" reading of moral beliefs (but see the following section).

13 On this point, Durkheim points out that "[h]istorical research can demonstrate that a certain moral practice is related to a belief which today is extinct and that the practice is thus entirely without foundation" (Durkheim 1993, 135). Along similar lines, Lévy-Bruhl asks: "[C]an it be doubted that a similar analysis of the Western conscience is possible? Do we regard the acts that we feel bound to do or not to do, obligatory or forbidden for reasons known to us and logically founded? No one would dare to affirm it in every case. We often explain them by motives that have nothing in common with their real origin. That observation has been made more than once with regard to those particular obligations which are the customs and conventions of society" (Lévy-Bruhl 1905, 70). In other words, some moral beliefs may be vulnerable to historical debunking. There is no suggestion by either Durkheim or Lévy-Bruhl that all of them are.

14 The appeal to a "necessary connexion" arguably overstates Ross's case. For a later development of such arguments on explicitly evolutionary terms, see Mackie 1977; Joyce 2005.

15 One question that Ross does not answer is, how, even if we do have access to a priori self-evident moral truths, we can be confident that we have actually accessed them, given the variable causes of our moral beliefs and the fact that self-evident truths need not be obvious? For discussion of the wider implications of this question, see Lillehammer 2011.

16 A comprehensive assessment of this claim falls outside the remit of this chapter. For a related discussion of the French (and other) sociologists on this point, see Abend 2010, 566–570. See also Lillehammer 2010.

17 Along similar lines, Durkheim writes: "Is there not really something strange in positing these high questions when we do not know yet ... what property rights are, or contract, or crime, or punishment, etc., etc. Perhaps the time for synthesis will come some day, but it hardly seems to have arrived yet. Therefore, the moral theorist can only respond by an admission of ignorance, to the often repeated question, 'What is, or better yet, what are, the ultimate principles of morality?'" (Durkheim 1993, 130).

18 Durkheim writes: "The more one becomes familiar with the laws of moral reality, the further along one will be in modifying it rationally, in saying what it ought to be" (Durkheim 1993, 31).

19 Lévy-Bruhl also writes (in the same passage and elsewhere) that ethics is "relative." What exactly he means by that, and exactly how it relates to various normative or meta-ethical views commonly known as "moral relativism," is a question worthy of more attention than I am able to give it here. For three relevant discussions, see Putnam 1981; Williams 1985; and Abend 2010.

20 On this reading, Lévy-Bruhl's position can be read as suggestive of a form of ethical constructivism. For a paradigm manifestation of constructivism in twentieth-century moral and political philosophy, see Rawls 1971, sections 1–9, 87.

21 Parts of this material have been presented at the Institute of Advanced Study at the University of Helsinki in 2013, at the Centre for Research in Arts Humanities and the Social Sciences (CRASSH) at Cambridge University in 2014, and at a workshop at Florida State University in 2015. I am grateful to the audience on those occasions and to numerous friends and colleagues for helpful comments and suggestions on previous versions.

2 Nietzsche's Rejection of Nineteenth-Century Evolutionary Ethics

Jeffrey O'Connell

Friedrich Nietzsche (1844–1900) was one of the first major philosophers to think seriously about the philosophical implications of Darwinism. This is particularly apparent in his writings on morality. He became interested in the project of applying an evolutionary frame to the study of morality early in his career, and remained interested in the project throughout his working life. But his position on the value of this project changed over time. In an early essay critical of the German theologian David Friedrich Strauss, Nietzsche rejected the possibility of grounding morality in the law of evolution but left open the possibility that the theory of evolution could account for the development of morality. The project of devising this sort of developmental account came to occupy Nietzsche for the rest of his career, though the account he ultimately settled on was formed in opposition to the kind of theorizing offered both by Darwin in *The Descent of Man* (1871) and by Herbert Spencer in *The Data of Ethics* (1879). This was due in part to the influence of his friend Paul Rée, himself an ardent Darwinian who nonetheless thought evolution insufficient as an explanation of morality. Rée's basic criticism, along with the influence of the German scientists Wilhelm Roux and William Rolph, led Nietzsche to develop his own distinctive theory of the development of morality, one that centers on the psychological doctrine of the will to power.

Nietzsche's Response to David Friedrich Strauss

In his book *The Old Faith and the New* (1872), David Friedrich Strauss argued that science, the new faith, had invalidated – and in some sense even replaced – religion, the old faith. Darwin assumed a starring role in this narrative, for his proof that humans evolved from a series of lesser forms meant, in the eyes of Strauss and others, that there could no longer be a serious question about whether or not we were the special, chosen children of God. Darwin had definitively debunked the biblical accounts of our genesis, and this fact cast considerable doubt on the entirety of the Bible, including the depictions of the Christian God portrayed therein. But these biblical accounts of God, along with his promises of heavenly reward and threats of eternal damnation, seemed necessary to ensure our adherence to the moral code vouchsafed for us in the Decalogue. Without God, we would lose the motivation to be moral. This was a key problem for Strauss, and indeed for many others in the nineteenth century. In the words of Dostoevsky, "if God does not exist, everything is permissible." But Strauss was not content to leave it at that. He wanted to show that the sanctity of morality could be re-derived using evolutionary theory itself. The trick, he thought, was to see that, despite our lowly beginnings and despite the mutability of species, all individual humans were members of a common "enduring kind" – that is, we all partook in the species "human being." Conscious recognition of this fact was enough to lift us out of the state of nature, conceived as a Hobbesian war of all against all, and into some higher participation in a biologically unified group. When we understood our inclusion in this group, we would understand that we were bound by a categorical imperative: act like a human. To act as a member of the group "human being" required treating other humans with respect simply in virtue of their participation in the same group (Strauss 1872, Vol. 2, 44–64).

Although it was in part Strauss's earlier book *The Life of Jesus* (1835) that led Nietzsche to his own atheism, Nietzsche emphatically rejected Strauss's categorical imperative, for he thought it was a disingenuous way to meet the challenge of what he called the death of God. There were two problems in particular that worried him. First was the issue of whether Strauss had really pulled off what he set out to do. His goal was to derive a moral code from his Darwinian view of nature, which he saw as the only option once the Christian view of the world had been deposed. On this score, Nietzsche made it clear that Strauss had failed:

With a certain rude contentment he covers himself in the hairy cloak of our ape-genealogists and praises Darwin as one of the greatest benefactors of mankind – but it confuses us to see that his ethics are constructed entirely independently of the question: "What is our conception of the world?" Here was an opportunity to exhibit native courage: for here he ought to have turned his back on his "we" and boldly derived a moral code for life out of the *bellum omnium contra omnes* and the privileges of the strong. (Nietzsche 1983, 29–30)

Instead of deriving a morality of competition from the competitive state of nature, Strauss did an about-face and told us that evolution taught us of our universal kinship with other human beings. Nietzsche was not suggesting that Strauss *should* have devised a morality of competition, nor, for that matter, was he even claiming that the theory of evolution was necessarily characterized by this Hobbesian quality of strife. Rather, he was pointing out that Strauss himself conceived of evolution in these terms, yet described morality in altogether different terms, thereby failing to accomplish his own goal of grounding morality in his view of nature.

But even if Strauss had succeeded in this task – that is, even if he had made the courageous move that Nietzsche chastised him for falling short of – he still would have been in the wrong. For Nietzsche thought that the very project of basing a prescription on a description was intellectually dishonest:

[A]n honest natural scientist believes that the world conforms unconditionally to laws, without however asserting anything as to the ethical or intellectual value of these laws: he would regard any such assertions as the extreme anthropomorphism of a reason that has overstepped the bounds of the permitted. (Nietzsche 1983, 31)

What Strauss *should* have done, he said, was "to take the phenomena of human goodness, compassion, love and self-abnegation, which do in fact exist, and derive and explain them from his Darwinist presuppositions" (Nietzsche 1983, 30). Instead of trying to secure a foundation for morality, Strauss should have stuck to the descriptive project of providing a naturalistic explanation of it. If not from a divine spark, whence human goodness, compassion, love, and self-abnegation? To ask where these things come from, why they exist, was to ask a very different sort of question from whether or not we ought to continue engaging in them. The latter question was certainly not one for the natural scientist, but the former might well be.

The Influence of Paul Rée on Nietzsche

The same year that he published his critique of Strauss, Nietzsche made the acquaintance of the German philosopher Paul Rée, who at that time was working on his own attempt to devise a naturalistic, descriptive account of morality. The two began a correspondence in 1875 and quickly became close friends – so close, in fact, that in 1876 they spent several months together at the home of a mutual friend in Sorrento. During this time, Rée was finishing up his second book, *The Origin of Moral Sensations*, while Nietzsche was working on his book *Human, All Too Human*.

Rée was an enthusiastic supporter of Darwinism, and he considered Darwin's ideas an essential part of the attempt to devise a naturalistic account of morality. As he said in the introduction to his work, "today, since Lamarck and Darwin have written, moral phenomena can be traced back to natural causes just as much as physical phenomena: moral man stands no closer to the intelligible world than physical man" (Rée 2003, 87). Particularly important was Darwin's explanation of the origin and development of the social instincts. According to Rée, these instincts became part of human nature in virtue of the advantage they provided to tribes in the struggle for existence: "When the members of a tribe of animals have a relatively strong social instinct, the stronger cohesion – caring and fighting for one another – that this provides gives this tribe an immense superiority in conflict with other tribes" (Rée 2003, 153). This was essentially a restatement of Darwin's own view of the origin of the social instincts, a view that Rée thought had serious philosophical implications. The mere existence of these instincts, he claimed, refuted both writers like Helvétius, who thought that all so-called altruistic actions were done for the sake of some ulterior egoistic end, and writers like Schopenhauer, who thought that altruism was only possible given the existence of some higher, metaphysical realm.

Neither Rée nor Darwin thought that an explanation of the social instincts was *sufficient* to explain morality, however. Morality was not just a matter of blind motivation; it required the capacity to reflect on past actions, to consider future actions, and to evaluate those actions as good or bad. This was only possible with a certain degree of intelligence. But while the capacity to make moral judgments required intelligence in addition to instinct, the *content* of our moral judgments, according to Darwin, was largely a product of the instincts themselves. On his view, when one reflected on a past situation in which one acted at variance with a social instinct, one would feel a sense of frustration at having blocked that instinct's satisfaction. This feeling of frustration, he said, "if weak we call regret, and if severe remorse" (Darwin 1871, 91). In other words, the moral emotion of remorse, which caused us to condemn as bad or blameworthy the action that brought it about, was constituted by the feeling of the frustration of the social instincts. Intelligence just aided our ability to feel this sensation before and after the act and to communicate our experience to others.

This analysis, of course, raised an immediate question about the moral authority of the social instincts. Why did the satisfaction and frustration of just these instincts inspire moral sensations and judgments and not the satisfaction and frustration of any other instinct? Why didn't we feel remorse after the frustration of our more egoistic impulses, for example, and call bad those things that frustrated them? According to Darwin, the difference was a matter of persistence. As he said, "The imperious word *ought* seems merely to imply the consciousness of the existence of a persistent instinct" (Darwin 1871, 92). The social instincts asserted themselves constantly, whereas other instincts were extinguished in the act. When we reflected on past situations in which we acted at variance with a social instinct, we would continue to feel its pull, and not the pull of the instinct that caused us to act. This constant presence, Darwin said, was the voice of conscience, telling us at all times that we ought to act for the benefit of others.

Rée rejected this aspect of Darwin's account. The instincts themselves did not inform either our moral sensations or our moral judgments. This operation was performed, instead, by intelligence. On Rée's view, the frustration of the social instincts would result merely in the feeling of frustration, which was no different in kind from the feelings that followed the frustration of any other instinct. The fact that these instincts were persistent meant only that we were more likely to experience the feeling of frustration several times after the fact. It did not mean that the feeling suddenly took on a moral nature. Rather, the frustration of a particular instinct would only acquire a moral quality if it already had been judged good or bad. And this prior evaluative act was carried out by intelligence, not instinct.

This insight led Rée to drive a wedge between instincts and moral judgments and to claim that fundamentally different processes gave rise to each:

One must reflect that the two things arise from different sources: non-egoism is innate, an inherited quality of our animal ancestors. The idea of its praiseworthiness, however, developed only at a certain stage of culture and then became, as it now is, a habit acquired by individuals in the course of their lives. (Rée 2003, 101)

Rée saw his own project as an attempt to describe the origin and development of our moral judgments, which he thought was fundamentally different from Darwin's project of explaining the origin and development of our instincts. To carry out this project, he used a blend of utilitarianism and psychological associationism. According to his view, moral judgments initially followed from the rational recognition of the utility of altruistic acts and motives. At some point in human history, due to the development of intelligence, we recognized that those who acted for the good of others were potentially useful, either to individuals or to the community as a whole. At first, such people were called good in virtue of the useful actions they performed, but eventually, "at a higher stage in the evolution of knowledge," we realized that altruistic motives were more useful and better at securing long-term peace than actions alone, and so came to call good those motives themselves. With this practice in place, we developed the habit of calling altruistic motives good and eventually forgot that they were originally called good in virtue of their utility. Thus, we started thinking that altruistic motives were simply good in themselves, regardless of the consequences of altruistic behavior (Rée 2003, 89–99).

In *Human, All Too Human*, Nietzsche presented his own first attempt at devising a naturalistic, descriptive account of morality. At this early stage, he was primarily interested in combatting the explanations of morality offered by metaphysical philosophers like Schopenhauer, who "assumes for the more highly valued things some miraculous origin, directly from out of the heart and essence of the 'thing in itself.'" These metaphysical speculations, Nietzsche thought, would soon give way to those offered by the "historical philosophizing" of people like Paul Rée and Nietzsche himself, who acknowledged that "everything has evolved; there are *no eternal facts*, nor are there any absolute truths" (Nietzsche 1986, 14-15). But in marked contrast to Rée, and despite the nod to the importance of evolution, Nietzsche's account contained not one reference to Darwin, and little in the way of biological speculation. Instead, Nietzsche focused on the sort of explanation that Rée himself offered for the development of moral judgments and went so far as to paraphrase Rée's account at several points:

At first we call particular acts good or evil without any consideration of their motives, but simply on the basis of their beneficial or harmful consequences. Soon, however, we forget the origin of these terms and imagine that the quality "good" or "evil" is inherent in the actions themselves, without consideration of their consequences....Then we assign the goodness or evil to the motives, and regard the acts themselves as morally ambiguous. (Nietzsche 1986, 43)

What mattered to Nietzsche were not biological facts, but cultural and psychological facts: facts about how utility was interpreted at various cultural stages, about habit formation, about the role of forgetting in the development of cultural practices, and about the association of ideas. That *Human, All Too Human* was so full of speculations of this kind was due in large part to Rée's influence. It was Rée, after all, who made the case that the instincts alone were insufficient to explain morality. We did not judge something to be right or wrong because we were impelled by a drive or set of drives to do so. Rather, we decided which drives were themselves right or wrong, good or bad. To explain such judgments, then, it was necessary to turn away from natural selection and toward some other explanatory mechanism. For Rée, this meant a turn toward a kind of utilitarianism married with associationism, and Nietzsche followed this to some degree in the late 1870s.

But while Nietzsche briefly experimented with Rée's explanatory schema, dependent on utility, habit, and forgetting, he did not merely adopt Rée's account wholesale. As he wrote to his friend Erwin Rohde, who had accused him of precisely that, "only look for *me* in my book and not for our friend Rée" (quoted in Donnellan 1982, 597). Even at this early stage Nietzsche was trying to construct his own account of the development of morality, one that would be quite distinct from Rée's. This was most apparent in Nietzsche's early adoption of psychological egoism, the view that egoistic desires such as the desire for pleasure and the avoidance of pain provided the sole source of motivation for our actions. Not only did Nietzsche think we were motivated to act in accordance with these egoistic impulses, but also he thought that we were motivated to make moral judgments in accordance with them. This led Nietzsche to reject Rée's distinction between altruism and egoism and to claim that all altruistic action is merely a "sublimated" form of egoism. Moreover, it led him to rethink the importance of utility for the formation of moral judgments and to emphasize instead the importance of pleasure as a criterion of value: we called good not what we perceived as useful, but rather what we perceived as pleasurable.

That Nietzsche adopted this doctrine might seem to take him back into territory that Rée had cautioned against; for if Nietzsche's view was that we were motivated to pursue pleasure, and that pleasure was the fundamental criterion of value, this would seem to result in the claim that our values were the products of a drive. Yet Nietzsche explicitly opposed his own view to the kind of view that Rée criticizes:

[T]he philosopher sees "instincts" in present-day man, and assumes that they belong to the unchangeable facts of human nature, that they can, to that extent, provide a key to the understanding of the world in general. This entire teleology is predicated on the ability to speak about man of the last four thousand years as if he were eternal, the natural direction of all things in the world from the beginning. But everything has evolved; there are *no eternal facts*, nor are there any absolute truths. (Nietzsche 1986, 14–15)

When Nietzsche said that "everything has evolved," he was thinking particularly of "the faculty of knowledge," and more specifically of the interpretations of pleasure and pain that this faculty gave rise to. Unlike Darwin's view that human nature comprised relatively stable instincts that picked out particular ends, some of which were determined in advance to be morally superior, Nietzsche's view was that we had one drive to pursue pleasure and to avoid pain, but that what we experienced as pleasure and pain was subject to change depending on how we interpreted the world. When he emphasized the importance of evolution, he meant the evolution of these interpretations.

As many have noted, Nietzsche's adoption of psychological egoism was influenced by the French moralists, including Helvétius and La Rochefoucauld, whose works Nietzsche and Rée read and discussed at length during their time together in Sorrento. But Nietzsche's view that pleasure was the fundamental criterion of value was likely inspired by Herbert Spencer. Nietzsche's letters and unpublished notes from the late 1870s and early 1880s reveal not only a familiarity with Spencer but also a surprising enthusiasm for his general project. And there are important parallels between Spencer's work and *Human, All Too Human.* Spencer held that moral judgments were tantamount to associations of pleasure and pain, associations that changed depending on one's external conditions of existence. Acts that were adaptive in a given set of conditions were pleasurable, and because we called good that which was pleasurable, acts that were adaptive were called good. Like Spencer, Nietzsche thought that moral judgments were essentially associations of pleasure and pain and that these associations were subject to change. But he rejected Spencer's claim that evolution, conceived as "the adaptation of inner relations to outer relations," was the driving force of this change. This idea was too close to the claim that nature itself gave rise to values, a claim that Nietzsche ardently opposed throughout his career. "Whatever has *value* in the present world has it not in itself, according to its nature – nature is always value-less," he said in the *Gay Science*, "but has rahter been given, granted value, and *we* were the givers and granters!" (Nietzsche 2001, 171).

This insistence on the valuelessness of nature was also the basis for Rée's initial critique of Darwin: on Rée's reading, Darwin had claimed that moral sensations and judgments emerged directly out of the social instincts, which seemed to suggest that a certain set of drives had an intrinsically moral character. This was unacceptable to Rée, and he endeavored to show that the drives themselves were morally neutral until they were imbued with value by us. This was done, at least initially, by our assessment of utility: some drives were seen as useful and therefore came to appear to us as valuable. In *Human, All Too Human*, Nietzsche simply swapped out utility for pleasure. Something became valuable if and only if it was interpreted as pleasurable. And just as the idea of something being useful, for Rée, "developed only at a certain stage of culture and then became, as it now is, a habit acquired by individuals in the course of their lives," so the idea of something being pleasurable, for Nietzsche, was a habit that developed at a cultural level: "displeasure is a habit that can be given up; many men do not feel it at all, even after the same actions that cause many other men to feel it. Tied to the development of custom and culture, it is a very changeable thing" (Nietzsche 1986, 44).

But while Nietzsche followed Rée in stressing the importance of cultural evolution for the development of morality, the two disagreed about the mechanism by which this development occurred. For Rée, the process was largely the work of intelligence: we rationally recognized utility as soon as our intelligence developed sufficiently for us to see it. For Nietzsche, in contrast, our assessments of pleasure and pain were often the work of potentially irrational forces and brute socialization:

The same drive evolves into the painful feeling of cowardice under the impress of the reproach custom has imposed upon this drive: or into the pleasant feeling of humility if it happens that a custom such as the Christian has taken it to its heart and called it good. (Nietzsche 1997, 26)

Throughout his early works, Nietzsche placed great emphasis on the cultural, political, and religious influences behind our interpretations of pleasure and pain, and thus behind our moral judgments. But while he stressed the prevalence of these forces over the operation of rationality, one of the most striking things about his early work was his faith in our ability to overcome the "errors" involved in these earlier interpretations, and to do so through reason itself. Whereas the Christian interpretation of existence established an association between the idea of selflessness and pleasure, and between the idea of selfishness and pain, Nietzsche claimed to have arrived at the truth that there is in reality no such thing as selflessness, a truth that he thought followed from the doctrine of psychological egoism: "No one has ever done anything that was solely for the sake of another and without a personal motive. How indeed could he do anything that was not related to himself, thus without an inner necessity" (Nietzsche 1986, 92). The recognition that all actions are impelled by an inner, personal necessity allowed us to see that selflessness was a false ideal, one that could never be attained. This in turn allowed us to break the associations of pleasure and pain established by Christianity and to drop the guilty conscience that they gave rise to: "If in the end man succeeds in convincing himself philosophically that all actions are unconditionally necessary and completely irresponsible, and if he takes this conviction into his flesh and blood, those vestiges of the pangs of conscience disappear, too" (Nietzsche 1986, 93).

In his critique of David Strauss, Nietzsche claimed that a proper explanation of morality would describe the origin of "human goodness, compassion, love and self-abnegation." By the time Nietzsche wrote *Human, All Too Human*, he came to see that this was not the right question to ask. For this project simply assumed that phenomena like compassion and self-abnegation were moral phenomena, an assumption that Nietzsche's early intellectual hero Schopenhauer insisted on, but that Nietzsche eventually came to reject. As his friend Rée pointed out, a proper explanation of morality would show not just how phenomena like compassion and altruism were possible but also why such things were valued in the first place. For Rée, they were valued because they were seen as useful. For Nietzsche, they were valued if and only if they were interpreted as pleasurable. A real explanation of morality, then, would explain our changing interpretations of pleasure and pain.

The Importance of Power in Nietzsche's Later Thought

As we have seen, Nietzsche shared with Rée the conviction that all of nature, which included our drives and inclinations, was valueless and was only imbued with value by us. Or as Nietzsche put it in *Beyond Good and Evil* in 1886, "there are no moral phenomena, only moral interpretations of the phenomena" (Nietzsche 1966, 108). This position was formed in opposition to both Darwin and Spencer. Against Darwin's claim that our non-egoistic impulses impelled us to value certain things over others, Rée claimed that we only came to value non-egoism on the basis of perceived utility. Against Spencer's claim that associations of pleasure and pain were established by the biological adaptation of inner relations to outer relations, Nietzsche held that associations of pleasure and pain were in fact established by cultural acts of interpretation. In his later work, however, Nietzsche dropped the doctrine of psychological egoism altogether and claimed instead that a multiplicity of drives were behind both our actions and our moral judgments and that power, rather than pleasure, was the fundamental criterion of value.

The seeds of Nietzsche's shift from pleasure to power as a criterion of value were already present in *Human, All Too Human*, where he said, in discussing the prehistory of the concepts good and evil, "The man who has the power to requite goodness with goodness, evil with evil, and really does practice requital by being grateful and vengeful, is called 'good.' The man who is unpowerful and cannot requite is taken for bad" (Nietzsche 1986, 47). But this idea, like his occasional appeals to utility, was at most an experimental sally of thought in comparison with the role of pleasure in the determination of values. By the time of *Daybreak* in 1881, however, power would come to dominate over pleasure and utility as the basic criterion of value. As he said quite directly, "When man possesses the feeling of power he feels and calls himself good: and it is precisely then that the others upon whom he has to discharge his power feel and call him evil!" (Nietzsche 1997, 189). For the Nietzsche of *Daybreak* and beyond, any naturalistic explanation of human behavior must take into account the feeling of power, which Nietzsche maintained "has become [man's] strongest propensity." This emphasis on power was likely influenced by Nietzsche's early philological training. His careful reading of Greek thinkers like Homer and Thucydides convinced him that the idea of utility as a criterion of value was unique to modern times and that, in earlier eras, the value of an act was judged according to the degree of power it afforded. As he said, "'Power which is attacked and defamed is worth more than impotence which is treated only with kindness' – that is how the Greeks felt. That is to say: they valued the feeling of power more highly than any kind of utility or good name" (Nietzsche 1997, 360).

In the *On the Genealogy of Morality*, Nietzsche contrasted power with utility as a direct challenge to Rée and went so far as to use Rée's explanatory schema as a paradigm example of inept moral theorizing. Nietzsche's own basic framework was set up in opposition to Rée's:

Now in the first place it is obvious to me that the actual genesis of the concept "good" is sought and fixed in the wrong place by this theory: the judgment "good" does not stem from those to whom "goodness" is rendered! Rather it was "the good" themselves, that is the noble, powerful, higher-ranking, and high-minded who felt and ranked themselves and their doings as good, which is to say, as of the first rank, in contrast to everything base, low-minded, common, and vulgar. Out of this pathos of distance they first took for themselves the right to create values, to coin names for values: what did they care about usefulness! The viewpoint of utility is as foreign and inappropriate as possible, especially in relation to so hot an outpouring of highest rank-ordering, rank-distinguishing value judgments: for here feeling has arrived at an opposite of that low degree of warmth presupposed by every calculating prudence, every assessment of utility – and not just for once, for an hour of exception, but rather for the long run. As was stated, the pathos of nobility and distance, this lasting and dominant collective and basic feeling of a higher ruling nature in relation to a lower nature, to a "below" – that is the origin of the opposition "good" and "bad." (Nietzsche 1998, 10–11)

What mattered in the creation of value judgments was not perceived utility, but rather what Nietzsche called "the pathos of distance," which was essentially the feeling of power by another name. Those who had power were originally compelled to call themselves good in virtue of this quality, and to call bad those who lacked it; alternatively, those who lacked power were motivated to invert this value standard by an act of "revaluation." This was the "slave revolt in morality," a discussion of which takes up much of the first section of *Genealogy*. But the key idea, that value judgments were essentially connected to the feeling of power, would inform the rest of Nietzsche's thinking about morality. As he said in *The Antichrist* (1895), "What is good? – All that heightens the feeling of power, the will to power, power itself in man. What is bad? – All that proceeds from weakness" (Nietzsche 1954, 2).

This move toward the feeling of power as a key explanatory device was coupled, in Nietzsche's mature period, with an increasing interest in the biological phenomena of life. From *Daybreak* onward, the concept of *drive* came to play a central role in the explanation of human behavior. Indeed, drives were said to explain not only our behavior but also our thoughts, feelings, and even our moral judgments. This might seem to put Nietzsche once again in danger of violating Rée's basic critique of Darwin, that the drives themselves are morally neutral until they are imbued with value by an independent process of evaluation. But Nietzsche never renounced this basic point. He was unwavering in his opposition to the idea that there was some subset of drives determined in advance by the process of biological evolution to inform the content of our moral judgments.

To get clear on Nietzsche's view, it is necessary to understand how Nietzsche conceived the concept drive. There is large agreement among scholars that this concept was for Nietzsche a disposition to act. To have a drive was just to be disposed to act in a particular kind of way, toward a particular kind of end. There is considerable disagreement, however, about the process by which Nietzsche thought we acquired drives. A plausible suggestion put forth by John Richardson is that Nietzsche conceived of drives as selected for by Darwinian evolution. We had a drive just insofar as it had proved useful in the struggle for existence (Richardson 2004, 35–45).[1] But for Nietzsche, an important implication of this fact was that all of our drives were just as good and useful as any other from the perspective of evolution. That is, they were all equally useful in the struggle for existence. If a drive was not useful, it would not be preserved; therefore, our drives had all, in a sense, proved their worth. It was with this point in mind that Nietzsche said, in *Gay Science*, "the evil drives are just as expedient, species-preserving, and indispensable as the good ones" (Nietzsche 2001, 32). This point, he thought, showed that any attempt to account for morality by appeal to Darwinian evolution would fail. The best that Darwin's theory could do was to show how we acquired our drives; it could not show how we came to regard certain drives as better than others. But morality was essentially a matter of evaluating ends and comparing relative worth; it was about sanctioning the activity of some drives above that of others: "Wherever we encounter a morality, we find an evaluation and ranking of human drives and actions" (Nietzsche 2001, 114).

Darwin, then, could not explain the moral evaluation of the drives. This was Rée's original critique, which led him toward a kind of utilitarian theory of development. But Nietzsche wanted to push back on the idea, central to Rée's analysis, that the process of evaluating drives could be undertaken by an autonomous, rational subject. There was no way to step back from our drives to evaluate them according to independent, objective standards like utility. Instead, he thought, the process of ranking must somehow be done at the level of the drives themselves: "While 'we' believe we are complaining about the vehemence of a drive, at bottom it is one drive which is complaining about another" (Nietzsche 1997, 109). If morality was essentially an "evaluation and ranking of human drives and actions," then this evaluation and ranking must be self-generated. Nietzsche's problem was to explain how this was possible: how could the drives arrange themselves in rank order?

Nietzsche's doctrine of the will to power was an attempt to answer this question. On Nietzsche's view, individual drives were engaged in a constant struggle for power over one another. The drive or set of drives that won this battle won the power to determine action and values. His thinking on this point was reinforced, and indeed guided, by his extensive reading in the natural sciences of his day. As Gregory Moore has shown, Nietzsche took over from nineteenth-century German cytology and embryology the idea that individual organisms comprise a multiplicity of parts. Particularly influential for Nietzsche was the embryologist Wilhelm Roux, who described this multiplicity of parts as engaged in a ceaseless internal struggle for existence during development. This struggle, in turn, explained the tendency to variation, a tendency Roux thought was insufficiently explained by the impact of the external environment (Moore 2002, 24–25, 37–38). Like Roux, Nietzsche thought that an internal struggle could explain a phenomenon that the relationship to the environment could not. But unlike Roux, he was not primarily interested in the biological tendency to variation. He was interested instead in the phenomenon of morality. The crucial issue was the tendency of the drives to arrange themselves in rank order.

The idea of an inner struggle for power among the parts comprising a biological organism gave Nietzsche a way to explain the emergence of values: moral judgments emerged out of the hierarchization of the drives. If each drive was selected for according to the end toward which it was directed, then a hierarchy of drives was essentially a hierarchy of ends, which started to look like a moral system. If an instinct of aggression won out in the struggle among the drives, aggressive acts would be valued more highly than acts of humility. Moral judgments would differ just insofar as the hierarchy of drives differed. But this still left open the question of how drives did in fact form a hierarchy. Why did certain drives attain the right to determine action and values, while others remained latent? The obvious answer, given Nietzsche's emphasis on the feeling of power, was that whichever drive resulted in the highest degree of the feeling of power would win out. And indeed, something like this seems to be Nietzsche's position. The will to power, while sometimes naming an inner-directed struggle for power among the drives, sometimes named an outer-directed struggle on the part of the organism for the feeling of power. For Nietzsche, the inner-directed will to power was a battle among the drives to service this outer-directed need.[2]

Nietzsche's theory of the inner-directed struggle was influenced, as we have seen, by the embryologist Roux. His theory of the outer-directed psychological motivation for power, on the other hand, was influenced by the zoologist William Rolph, who considered himself an anti-Darwinian but who was in reality more directly opposed to Herbert Spencer than to Darwin himself (Moore 2002, 47–48). As noted previously, Spencer held that human beings were fundamentally motivated to pursue pleasure and to avoid pain, and he assumed that evolutionary forces maintain a necessary connection between pleasure and life-preserving acts. Those acts that were adaptive were pleasurable; those that were maladaptive were painful. While human beings were directly motivated by pleasure and pain, the necessary connection between these states and the preservation of life ensured that our actions were in fact determined by the pursuit of the preservation of life. By pursuing pleasure and avoiding pain, we would naturally do whatever led to this ultimate end. We have seen that Nietzsche rejected the connection between adaptiveness and pleasure as early as *Human, All Too Human*. But Rolph gave Nietzsche the alternative he needed, an alternative that was rooted, as far as Nietzsche could tell, in sound scientific evidence. According to Rolph, the ultimate end of action was not the preservation of life, but rather the acquisition of power.

Nietzsche famously reiterated Rolph's critique of Spencer in *Beyond Good and Evil*:

Physiologists should think twice before positioning the drive for self-preservation as the cardinal drive of an organic being. Above all, a living thing wants to *discharge* its strength – life itself is will to power –: self-preservation is only one of the indirect and most frequent *consequences* of this. – In short, here as elsewhere, watch out for *superfluous* teleological principles! (Nietzsche 1966, 13)

Nietzsche's use of the phrase "drive for preservation," a phrase that Spencer did not use but which echoed Rolph's language, had the added benefit of grouping Spencer with figures like Spinoza and Schopenhauer, both of whom Nietzsche expressed an early allegiance to but later came to reject. As we have seen, crucial to Spencer's position was the claim that we would perform, and value, whatever action best secured the continuation of life, a relation that depended on the external conditions we found ourselves in. In a warring society, aggressive impulses would best secure continued life, and so aggression would be highly valued; in a peaceful, industrial society, cooperative impulses would do the trick, and so cooperation would be highly valued. In *Human, All Too Human*, Nietzsche replaced this reliance on passive adaptation with the active force of interpretation. In *On the Genealogy of Morality*, he again took up this theme and was no less determined to replace the passive nature of Spencer's view with his own theory of actively imposed values:

Under the pressure of [the democratic idiosyncrasy] one ... places "adaptation" in the foreground, that is to say an activity of the second rank, a mere reactivity; indeed life itself is defined as an ever more purposive inner adaptation to external circumstances (Herbert Spencer). In so doing, however, one overlooks the essential pre-eminence of the spontaneous, attacking, infringing, reinterpreting, reordering, and formative forces, upon whose effect the "adaptation" first follows; in so doing one denies the lordly role of the highest functionaries in the organism itself, in which the will of life appears active and form-giving. (Nietzsche 1998, 52)

The difference between this passage and Nietzsche's earlier thought is that, here, Nietzsche had in mind his theory of a multiplicity of drives vying for dominance over one another. A drive did not attain dominance over another by being better adapted to external conditions for the end of preserving life; the struggle was rather one that took place wholly internally according to the end of achieving a feeling of power. For Nietzsche, then, the feeling of power took over the role played by pleasure and pain on Spencer's view. Psychologically, we were motivated to gain a feeling of power. This was still a teleological explanation, insofar as it posited an end to be attained, but it was not *superfluously* teleological, in that it did not posit some further end like self-preservation that dictated the kind of thing felt to be an instance of power. The struggle among the drives, and thus the determination of moral judgments, was decided entirely by the feeling of power attained for the organism. Self-preservation was simply a frequent consequence of this more basic process.

Conclusion

Nietzsche's mature view, then, was that moral judgments emerged out of the hierarchization of the drives, a process that Nietzsche thought had little to do with either Darwinian or Spencerian evolution, but much to do with the feeling of power. But he only arrived at this view through a long process of reckoning with various nineteenth-century attempts to develop an evolutionary account of morality. Early on, he rejected David Friedrich Strauss's attempt to derive an ethical code from the law of evolution, but left open the possibility of describing the development of morality by starting from "Darwinian presuppositions." Through his friendship with Paul Rée, he learned of Darwin's own attempt to do just that, but agreed with Rée that the principle of natural selection was ultimately insufficient for this end. He also agreed with Rée that the attempt to explain "human goodness, compassion, love and self-abnegation" was not the most important issue. Far more important was an explanation of why we value such phenomena in the first place. To this end, Nietzsche initially turned to the doctrine of psychological egoism, but later rejected this doctrine in favor of his theory of the will to power. This theory was Nietzsche's mature attempt to provide a naturalistic, descriptive account of the development of morality.

Notes

1 Against Richardson, Janaway and others rightly point out that Nietzsche sometimes seems to suggest that drives can be the products of either a kind of Lamarckian evolution or a kind of cultural evolution. See Janaway 2007, 45–48.

2 Whether or not the drive for a feeling of power has been selected by natural selection is an issue of some debate. See Richardson 2004, ch.1 for an extended discussion of this issue.

3 American Pragmatism, Evolution, and Ethics

Trevor Pearce

Introduction

It is hard to miss the fact that the American pragmatists were influenced by evolutionary ideas – especially given Dewey's famous collection of essays, *The Influence of Darwin on Philosophy* (1910). It thus comes as a surprise that several recent works on Dewey's ethics contain almost no mention of evolution (Fesmire 2003; Lekan 2003; Pappas 2008; Stroud 2011). Jennifer Welchman, in her otherwise excellent book, effectively dismisses the biology connection, claiming that Dewey's use of terms such as "adaptation" does not indicate "a close acquaintance with, let alone comprehension of, either Darwinian theory or subsequent developments in the life sciences" (Welchman 1995, 121). This position contrasts with that of earlier scholars, who saw the link with biology as obvious and important (Gouinlock 1972, 237–266). Dewey's contemporaries agreed: Addison Webster Moore's *Pragmatism and Its Critics*, for example, emphasized "the central role of the conception of evolution in the development of pragmatism," and the last chapter of his book was a dialogue on ethics between an absolutist and an evolutionist (Moore 1910, vii–viii).

Turning to another pragmatist philosopher, Jane Addams, the story is similar. Although Mary Jo Deegan's *Jane Addams and the Men of the Chicago School* (1988), with its focus on the ecological approach of Chicago sociology, discussed Addams's evolutionary account of the urban environment, later treatments of her social ethics have tended to neglect this aspect of her views (Seigfried 1996, 1999; Anderson 2004; Hamington 2009; Cracraft 2012).

There have been promising exceptions to this trend. Some recent work on Dewey and Addams has emphasized their progressive evolutionary viewpoint (Teehan 2002; Eddy 2010; Green 2010; Fischer 2013). Likewise, in the most recent version of her *Stanford Encyclopedia of Philosophy* entry on "Dewey's Moral Philosophy," Elizabeth Anderson has added this nice summary passage:

Dewey situated reflective morality in a non-teleological Darwinian view of organisms' adaptation to environmental contingencies. Nature does not supply a telos or rule for human beings, but rather a constantly changing environment to which humans need to adjust by using their intelligence. (Anderson 2014a)

Nevertheless, most work on pragmatism and ethics has left these issues in the deep background.[1]

I suspect that many pragmatism scholars have been wary of evolutionary ethics because of its association, both today and in the past, with conservative political positions. Herbert Spencer is infamous for his "Social Darwinism," and E. O. Wilson equally so for his sociobiology (Spencer 1884; Hofstadter 1944; Wilson 1978; Kitcher 1985). However, this is just what is most interesting about the pragmatists: they were developing an evolutionary approach to morality that was explicitly opposed to the most famous evolutionary ethics of their day – that of Spencer. In particular, they were working with a richer and more dynamic notion of human evolution in complex social environments, one that has been unduly neglected.

This chapter seeks to remedy this neglect by providing an overview of pragmatist evolutionary ethics – specifically that of John Dewey and Jane Addams – and its intellectual context. To make things manageable, I will focus on pragmatist texts from 1890 to 1910, a period that includes Dewey's clearest statements of the relation between ethics and evolution as well as Addams's major works on social ethics. For much of this period, Dewey and Addams were both in Chicago: Hull House, Addams's settlement house at Halsted and Polk, was founded in 1889; a few years later, Dewey became chair of philosophy at the new University of Chicago, and he remained there until 1904. It was also a tumultuous period socially and economically, with what Alan Trachtenberg (1982) has called "the incorporation of America": urbanization, immigration, industrialization, depression, and all of the accompanying disputes between capital and labor (Feffer 1993). This social context was directly relevant to the evolutionary approach of Dewey and Addams.

The beginning of the chapter will provide some necessary background, discussing earlier works to which Dewey and Addams were reacting. First, I will give a brief overview of Spencer's evolutionary ethics, along with a few contemporary criticisms. I will then present Thomas Henry Huxley's account of the relationship between ethics and the struggle for existence. The biological basis of Huxley's position was attacked in different ways by Dewey and by the anarchist Pyotr Alekseyevich Kropotkin, whose views influenced Addams (Eddy 2010). In the third part, I will analyze Dewey's "evolutionary method" in ethics, a dynamic functionalist approach. Finally, in the last part of the chapter, I will discuss Addams's account of social ethics as an evolved response to the new environment of the industrial city. The evolutionary ethics of Dewey and Addams, developed around 1900, may seem a historical curiosity. Nevertheless, several philosophers inspired by Dewey – in particular Elizabeth Anderson and Philip Kitcher – have recently argued that a similar approach to ethics is our best option today.

Spencer's Evolutionary Ethics

Herbert Spencer introduced the idea of organism-environment interaction to the English-speaking world and popularized the term "environment" (Pearce 2010a, 2014b). His ethics was built on the idea of a correspondence between organism and environment. As he reminded readers in *Data of Ethics* (1879), he had defined life in *Principles of Biology* as "the continuous adjustment of internal relations to external relations" (Spencer 1864, 80; quoted in Spencer 1879, 19). For Spencer, the success and complexity of this organism-environment correspondence indicates a species' position on the evolutionary scale:

The life of the organism will be short or long, low or high, according to the extent to which changes in the environment are met by corresponding changes in the organism. Allowing a margin for perturbations, the life will continue only while the correspondence continues; the completeness of the life will be proportionate to the completeness of the correspondence; and the life will be perfect only when the correspondence is perfect. (Spencer 1864, 82)

More evolved species, said Spencer, are able to meet a wider and more complicated set of environmental challenges.

At the beginning of *Data of Ethics*, Spencer applied this lesson to conduct, or "the adjustment of acts to ends" (Spencer 1879, 5). On Spencer's view, conduct evolves as purposive acts lead to an improved correspondence with the environment. Spencer claimed that the highest form of conduct is not strictly individualistic, since the life of the species matters to evolution as well. Echoing the libertarian principle of his earlier work *Social Statics*, he praised "adjustments such that each creature may make them without preventing them from being made by other creatures" (Spencer 1879, 18; cf. Spencer 1851, 103). Even this was not the limit: "a still higher phase" in the evolution of conduct, for Spencer, is "mutual help in the achievement of ends ... either indirectly by industrial co-operation, or directly by volunteered aid." This "mutual aid," said Spencer, "increases the totality of the adjustments made, and serves to render the lives of all more complete" (Spencer 1879, 19–20). Thus, the highest species – humanity chief among them – have complicated cooperative societies.

But what does this evolutionary history have to do with ethics as traditionally conceived? On Dewey's later reading, Spencer's ethics offered three advances. First, it argued "that certain acts must be beneficial because furthering evolution, and others painful because retarding it," thus providing a "fixed objective standard" for happiness. Second, it reconciled intuitionalism and empiricism by showing that "certain moral ideas now innate or intuitive" are the result of human evolutionary history. Third, it reconciled egoism and altruism by demonstrating that evolution tends toward their coincidence:

The being which survives must be the being which has properly adapted himself to his environment, which is largely social, and there is assurance that the conduct will be adapted to the environment just in the degree in which pleasure is taken in acts which concern the welfare of others. (Dewey 1891, 67–71)

In other words, evolution has built us so as to take pleasure in unselfish acts.

Of course, Spencer was aware that people often behave in selfish and antisocial ways. But this will not be the case, he insisted, for "the completely adapted man in the completely evolved society." This "ideal social being" is one whose "spontaneous activities are congruous with the conditions imposed by the social environment formed by other such beings" (Spencer 1879, 275). As he put it several decades earlier in *Social Statics*, "the ultimate man should be one who can obtain perfect happiness without deducting from the happiness of others" (Spencer 1851, 413). The only reason we have not reached this point is that "the new conditions to which adaptation has been taking place have themselves grown up but slowly. Only when a revolution in circumstances is at once both marked and permanent, does a decisive alteration of character follow" (Spencer 1851, 414; on the notion of circumstances, see Pearce 2010a, 2010b). Right and wrong today, what Spencer called "relative ethics," is thus to be judged from the point of view of "absolute ethics" – that is, the "ideal code of conduct" that will guide the behavior of the fully evolved man in his "ideal social state" (Spencer 1879, 275, 280).

Spencer's contemporaries were unconvinced. Josiah Royce asked the obvious question:

Why is this coming state the highest? Does anyone say: Because it will come at the end of the physical process of evolution? Nay then, if every more advanced state is to be more acceptable, by such reasoning the sprouting potato or the incubating egg will always be more acceptable than the fresh potato or the fresh egg. Highest, as last, or as most complex, or even as most permanent, cannot be in meaning identical with the morally highest that we want defined for us. (Royce 1885, 75–76)

In other words, even if we accept that evolution is tending toward the state described by Spencer, why should the actions of people in that state be our benchmark for moral valuation?

Samuel Alexander, for his part, criticized Spencer's idea of "good conduct as an adaptation or adjustment to environment" (Alexander 1889, 267). Spencer's notion of adaptation was too static, according to Alexander:

Morality is rightly described as an adaptation of man to his social environment. But in using this conception we have to guard against the danger of slipping in an assumption that the environment is itself something fixed and permanent, according to which, as he gradually discovers its character, he must arrange his conduct – which is, to use a homely expression, the cloth according to which he must cut his coat. (Alexander 1889, 271)

On Alexander's more dynamic view, "the act of adaptation can only be understood as a joint action of the individual and his environment, in which both sides vary together"; moral progress occurs because "the act of adjustment implied in good conduct itself alters the sentiments of the agent, and creates new needs which demand a new satisfaction" (Alexander 1889, 271, 277). Spencer's "ideal social state" was not specifiable, according to Alexander, because these "new sentiments and new ideals of character … cannot be forecast in detail" (Alexander 1889, 268–269). Thus, although Spencer's evolutionary ethics was acknowledged as pioneering, it was also widely criticized – and Dewey was familiar with these criticisms (see Dewey 1891, 77–78).

Ethics and the Struggle for Existence

Spencer and Thomas Henry Huxley were friends and members of the same dining club, but they disagreed about the implications of evolution for politics and society. Spencer had infamously argued in *Social Statics* that state education and other positive government interventions undermined natural adaptive processes:

Let us never forget that the law is – adaptation to circumstances, be what they may. And if, rather than allow men to come in contact with the real circumstances of their position, we place them in artificial – in false circumstances, they will adapt themselves to these instead; and will, in the end, have to undergo the miseries of a re-adaptation to the real ones. (Spencer 1851, 353–354)

For Spencer, the struggle for existence is a good thing: it pushes us toward the ideal social state previously mentioned.

Spencer and Huxley debated the issue of government regulation in the early 1870s, with Huxley accusing Spencer of misconstruing the analogy between organisms and societies (Huxley 1871; Spencer 1871). Huxley returned to the topic in 1888, arguing that the struggle for existence was a central feature of primitive rather than civilized society (for political context, see Helfand 1977, 161–170). In primitive times, said Huxley, "life was a continual free fight, and beyond the limited and temporary relations of the family, the Hobbesian war of each against all was the normal state." "The ethical man," in contrast, "devotes his best energies to the object of setting limits to the struggle." On Huxley's view, "the chief end of social organization" is to mitigate or abolish the struggle for existence (Huxley 1888, 165–166). He pointed out, however, that the misery of the urban poor was leading people to question the success of that organization and even the end itself:

The animal man, finding that the ethical man has landed him in such a slough, resumes his ancient sovereignty, and preaches anarchy; which is, substantially, a proposal to reduce the social cosmos to chaos, and begin the brute struggle for existence once again. (Huxley 1888, 171)

In their earlier debate, Spencer had insisted he was no anarchist; after all, he was in favor of negative governmental regulation (Spencer 1871, 638). But from Huxley's point of view, opposition to institutions such as state education was a turning away from civilization – and although Spencer and Huxley were not all that far apart when it came to biology and ethics, this essay effectively ended their friendship (Richards 1987, 314–316).

Huxley's comment about preaching anarchy was likely also a veiled reference to the Russian anarchist Pyotr Alekseyevich Kropotkin, then living in London, who a year earlier in the same journal had claimed that "anarchy proves to be in accordance with the conclusions arrived at by the philosophy of evolution" (Kropotkin 1887, 243). Kropotkin had also pointed out that despite its author's denials, Spencer's evolutionary philosophy tended toward anarchism, which Kropotkin glossed as "the no-government system of socialism" (Kropotkin 1887, 238, 244).

After reading Huxley's essay, Kropotkin went on the offensive, attempting to undermine the biological basis of Huxley's position in a series of papers much discussed by historians (Kinna 1992; Girón 2003; Borrello 2004, 16–22; Eddy 2010; Harman 2010, 9–37; Hale 2014, 206–251). Huxley had claimed that "from the point of view of the moralist the animal world is on about the same level as a gladiator's show" and that "primitive men" had been engaged in a "war of each against all" (Huxley 1888, 163, 165; quoted in Kropotkin 1890, 339). Recall that on Spencer's view, "mutual aid" was the highest phase of the evolution of conduct (Spencer 1879, 20). Kropotkin went further, claiming that mutual aid was important not just among civilized peoples but across the animal kingdom:

Sociability is as much a law of nature as mutual struggle. Of course it would be extremely difficult to estimate, however roughly, the relative numerical importance of both these series of facts. But if we resort to an indirect test, and ask Nature "Who are the fittest: those who are continually at war with each other, or those who support one another?" we at once see that those animals which acquire habits of mutual aid are undoubtedly the fittest. (Kropotkin 1890, 339–340)

In the most obvious example, "the ants and termites have renounced the 'Hobbesian war,' and they are the better for it" (Kropotkin 1890, 344). Kropotkin found similar lessons in human evolutionary history. Huxley had claimed, as we have seen, that the family was the only respite for "primitive men" from the "Hobbesian war" (Huxley 1888, 165); Kropotkin countered that "societies, bands, or tribes – not families – were … the primitive form of organization of mankind and its earliest ancestors" (Kropotkin 1891, 540). Marshaling evidence from both biology and ethnology, Kropotkin denied Huxley's claim that an all-consuming struggle for existence characterized pre-civilized life. On Kropotkin's reading of the natural world and human history, cooperation was just as important, if not more important, than competition (for the broader Russian context of this claim, see Todes 1989).

Several years before Kropotkin's "mutual aid" series, Huxley had declined to engage him in the same journal on another issue: "I have neither brains nor nerves, and the very thought of controversy puts me in a blue funk!" (Huxley to James Knowles, June 1, 1888, in Huxley 1900, Vol. 2, 213). Huxley did not debate Kropotkin even after the Russian's explicit attack, but he did return to the topic a few years before his death. He argued in a famous lecture on "Evolution and Ethics" that

the practice of that which is ethically best – what we call goodness or virtue – involves a course of conduct which, in all respects, is opposed to that which leads to success in the cosmic struggle for existence. In place of ruthless self-assertion it demands self-restraint; in place of thrusting aside, or treading down, all competitors, it requires that the individual shall not merely respect, but shall help his fellows; its influence is directed, not so much to the survival of the fittest, as to the fitting of as many as possible to survive. (Huxley 1893, 33)

That is, ethics is directly opposed to the cosmic evolutionary process.

But how could society and ethics be truly independent of evolution, as Huxley seemed to imply? In a "Prolegomena" to his lecture, published a year later, Huxley tried to clarify his position with an extended analogy. Huxley asked his readers to imagine the weeds and gorse of the downs, which "by surviving, have proved that they are the fittest to survive." This is the "state of nature," the result of the cosmic process. But if someone walls off an area of the downs and plants a garden, its existence depends on human intervention:

That the "state of Art," thus created in the state of nature by man, is sustained by and dependent on him, would at once become apparent, if the watchful supervision of the gardener were withdrawn, and the antagonistic influences of the general cosmic process were no longer sedulously warded off, or counteracted. (Huxley 1894, 9–10)

Just as in this case the "horticultural process" is antithetic to the cosmic process, so in society the "ethical process" combats the cosmic process (Huxley 1894, 13; cf. Huxley 1893, 34). Huxley even argued that in "the most highly civilized societies," where "the ethical process has advanced so far as to secure every member of the society in the possession of the means of existence, the struggle for existence, as between man and man, ... is, ipso facto, at an end" (Huxley 1894, 35–36).

When Kropotkin claimed that mutual aid played a key role not only in modern societies but also throughout the biological world, he was attempting to undermine Huxley and Spencer's restriction of this feature to "civilized societies" or the "higher phase" of ethics: where the two English thinkers saw a break, Kropotkin saw continuity. Dewey published a response to Huxley in 1898 that made a different argument in favor of continuity. He summarized Huxley's view as follows: "The rule of the cosmic process is struggle and strife. The rule of the ethical process is sympathy and co-operation.... The two processes are not only incompatible but even opposed to each other" (Dewey 1898, 323). But whereas Kropotkin had undermined Huxley by arguing that cooperation was also a major factor in the cosmic process, Dewey focused instead on the "man against nature" image that Huxley seemed to be promoting.

Exploring the garden analogy further, Dewey suggested that we should not see the garden as somehow opposed to nature:

We do not have here in reality a conflict of man as man with his entire natural environment. We have rather the modification by man of one part of the environment with reference to another part. Man does not set himself against the state of nature. He utilizes one part of this state in order to control another part. (Dewey 1898, 325)

Social progress, said Dewey, does not involve "building up an artificial world within the cosmos" (Huxley 1893, 35); it "consists essentially in making over a part of the environment by relating it more intimately to the environment as a whole; not, once more, in man setting himself against that environment" (Dewey 1898, 326). Thus we should not think of the struggle for existence as having ended – it has merely changed as the environment has changed. As social and ethical beings, we live in an environment that is largely our own creation, and thus Dewey could use Huxley's own words against him:

The conditions with respect to which the term "fit" must now be used include the existing social structure with all the habits, demands, and ideals which are found in it. If so, we have reason to conclude that the "fittest with respect to the whole of the conditions" is the best; that, indeed, the only standard we have of the best is the discovery of that which maintains these conditions in their integrity. The unfit is practically the anti-social. (Dewey 1898, 326; internal quotation from Huxley 1893, 33)

According to Dewey, then, whether in modern society or anywhere else in the biological world, fitness is relative to the environment (cf. Huxley 1893, 32).

Instead of interpreting "the term 'fit' … with reference to an environment which long ago ceased to be," we need to acknowledge that "the environment is now distinctly a social one, and the content of the term 'fit' has to be made with reference to social adaptation" (Dewey 1898, 328; for more on the Dewey-Huxley debate, see Teehan 2002).

Both Kropotkin and Dewey, therefore, fought against any radical separation between earlier and later phases of human evolution. Kropotkin used empirical data in an attempt to show that cooperation has always featured in the success of humans and other animals: we should not see cooperation as having replaced competition with the advent of ethics. Dewey argued that the struggle for existence – the cosmic evolutionary process – continually changes its form as organisms interact with and modify their environment: we should not see ethics as having stopped evolution, or society as separate from nature.

Dewey's Evolutionary Method

Dewey was profoundly influenced by Spencer's account of the organism-environment relationship, even though he ended up with a quite different picture (James 1904, 2; Godfrey-Smith 1996, 66–130; Pearce 2014b, 23–27). He also followed Spencer in thinking of good conduct as adaptation to environment: the moral situation according to Dewey is often one where "an act which was once adapted to given conditions must now be adapted to other conditions. The effort, the struggle, is a name for the necessity of this re-adaptation" (Dewey 1898, 333). But Dewey's "readaptation" and "readjustment" were a deliberate counter to Spencer's "adaptation" and "adjustment" – like Alexander, Dewey had a more dynamic and dialectical vision of the organism-environment relation (Pearce 2014a; see also Sullivan 2001, 12–40).

Hence, despite the criticism detailed previously, Dewey thought that Huxley was inspired by a great truth about morality: it involves conflict and tension. Dewey sided with Huxley against Spencer on this point, echoing Royce's ridicule of Spencer's ideal social state as one in which there is "nothing but a tedious cooing of bliss from everybody" (Royce 1885, 74):

There are many signs that Mr. Huxley had Mr. Spencer in mind in many of his contentions; that what he is really aiming at is the supposition on the part of Mr. Spencer that the goal of evolution is a complete state of final adaptation in which all is peace and bliss and in which the pains of effort and of reconstruction are known no more. (Dewey 1898, 334)

For Dewey, however, the conflict was not between the ethical process and the cosmic process, as Huxley would have it, but instead between past and present:

This, I take it, is the truth, and the whole truth, contained in Mr. Huxley's opposition of the moral and the natural order. The tension is between an organ adjusted to a past state and the functioning required by present conditions. And this tension demands reconstruction. (Dewey 1898, 333)

This idea of reconstruction – and the related notions of readaptation and readjustment – was at the heart of Dewey's "evolutionary method" in ethics.

Dewey's most thorough review of this method was his two-part article, "The Evolutionary Method as Applied to Morality" (1902). He began the essay with a long detour into the philosophy of science, linking the experimental method in science and the evolutionary method in ethics. Experiments, said Dewey, are designed to isolate "the exact conditions, and the only conditions, which are involved in [a phenomenon's] coming into being" and are thus applications of a genetic method. Knowledge of these conditions fulfills the promise of science – "intellectual and practical control" (Dewey 1902, 108–109). That is, if we know how to generate a given phenomenon, we can intervene to create or maintain it. But what of the distinction between the natural and historical sciences, with which Dewey would have been familiar (Tufts 1895)? The conditions that make possible the formation of water from its constituents seem quite different from the conditions that led Julius Caesar to cross the Rubicon. Dewey insisted, however, that this difference has more to do with our interests than with reality. After all, each molecule of water that we experimentally generate is strictly speaking unique. It is just that "we do not care scientifically for the historical genesis of this portion of water: while we care greatly for the insight secured through the particular case into the process of making any and every portion of water." In contrast, at least in some of our moods, we care less about the causes of civil war in general than we do about Caesar's particular case: "There is a peculiar flavor of human meaning and accomplishment about him which has no substitute or equivalent" (Dewey 1902, 111). Thus according to Dewey, the experimental method and the evolutionary/historical method are both versions of the genetic method.

How can this method be used in ethics? For Dewey,

history, as viewed from the evolutionary standpoint, ... is a process that reveals to us the conditions under which moral practices and ideas have originated. This enables us to place, to relate them. In seeing where they came from, in what situations they arose, we see their significance. (Dewey 1902, 113)

Dewey thought, for example, that the early stages of the history of ethics "provide us with a simplification which is the counterpart of isolation in physical experiment" (Dewey 1902, 124). The evolutionary method assumes, said Dewey, "that norms and ideals, as well as unreflective customs, arose out of certain situations, in response to the demands of those situations" (Dewey 1902, 356). For example, in an earlier essay Dewey had claimed that the opposing schools of Hellenistic philosophy were a response to social changes, nicely illustrating the evolutionary approach:

With the growth of the Macedonian and Roman supremacies, the welfare and customs of the local community came to mean less and less to the individual. He was thrown back upon himself for moral strength and consolation.... Both [the Stoic and Epicurean schools] are concerned with the question of how the individual, in an environment which is becoming more and more indifferent to him, can realize satisfaction. (Dewey 1894, 881–882)

In short, social norms and ethical theories are responses to the environment.

This historical analysis was of more than merely antiquarian interest: "We are still engaged in forming norms, in setting up ends, in conceiving obligations. If moral science has any constructive value, it must provide standpoints and working instrumentalities for the more adequate performance of these tasks" (Dewey 1902, 356). If we understood the function and adequacy of historical norms and theories, wrote Dewey, this could help us "guide and control the formation of our further moral judgments"; "whatever ... can be learned from a study of the past, is at once available in the analysis of the present" (Dewey 1902, 357, 370). In a discussion of moral intuitions, for example, Dewey argued that

if we can find that the intuition is a legitimate response to enduring and deep-seated conditions, we have some reason to attribute worth to it. If we find that historically the belief has played a part in maintaining the integrity of social life, and in bringing new values to it, our belief in its worth is additionally guaranteed. But if we cannot find such historic origin and functioning, the intuition remains a mere state of consciousness, a hallucination, an illusion, which is not made more worthy by simply multiplying the number of people who have participated in it. (Dewey 1902, 358)

That is, moral intuitions are empty unless they can be explained as successful responses to concrete environmental problems, either today or in the past.

Dewey also used his evolutionary method to understand moral progress:

It is the lack of adequate functioning in the given adjustments that supplies the conditions which call out a different mode of action; and it is in so far as this is new and different that it gets its standing by transforming or reconstructing the previously existing elements. (Dewey 1902, 368)

Moral progress occurs when we demand "that a way of conceiving or interpreting the situation cease to be mere idea, and become a practical construction" (Dewey 1902, 368). It is only through "failure from the standpoint of adjustment," and subsequent readjustment, "that history, change in quality or values, is made" (Dewey 1902, 367). The winners in this process, according to Dewey, are those values that actually help us resolve our current social problems; the losers are "surds, mere survivals, emotional reactions" (Dewey 1902, 370).

Dewey's picture might be styled dynamic functionalism. He thought that moral norms had specific functions at their origin and that sometimes those functions persisted. But just as Darwin took the traditional notion of adaptation in biology and made it dynamic, Dewey argued that the function of moral norms is rarely static: not only do we inevitably discover that they do not meet all of our needs, but those needs themselves also change as we build new and more complicated social institutions and environments. Dewey often used technological metaphors: "The logic of the moral idea is like the logic of an invention, say a telephone" (Dewey 1902, 366). As Dewey wrote in an earlier work, "the invention of the telephone does not simply satisfy an old want – it creates new. It brings about the possibility of closer social relations, extends the distribution of intelligence, facilitates commerce" (Dewey 1891, 208). The new device solves certain problems and meets certain needs, but it also changes the environment and thus creates new problems and new needs. Adjustment leads to readjustment, as technology and environment "vary together," in Alexander's phrase.

Specific ethical decisions do make use of moral theories, said Dewey, but only as tools for addressing a particular concrete problem (or type of problem): "theory is used, not as a set of fixed rules to lay down certain things to be done, but as a tool of analysis to help determine what the nature of the special case is" (Dewey 1892, 595). Thus, in another of Dewey's earlier works, he claimed that in deciding what to do in a difficult new ethical case, we should draw on moral theory in the same way that an engineer building a tricky new sort of tunnel employs the rules of mechanics: these rules and theories do not dictate our action, they inform it, along with all the specific facts of the case, our individual and social goals, and so on (Dewey 1892, 594–595). In Dewey's dynamic functionalism, then, we can think of morality as an evolving technology, with innovations that stand or fall on the basis of how well they help us adapt to the ever-changing social environment.

Addams's Social Ethics

Hull House, the settlement house that Jane Addams co-founded in a poor immigrant neighborhood of Chicago in 1889, was a response to a new social environment – "an experimental effort to aid in the solution of the social and industrial problems which are engendered by the modern conditions of life in a great city" (Addams 1893, 22). The task of the settlement residents was to arouse "the social energies which too largely lie dormant in every neighborhood given over to industrialism," and thus she stressed "its flexibility, its power of quick adaptation, its readiness to change its methods as its environment may demand" (Addams 1893, 23). Like Dewey, Addams was influenced by evolutionary ideas. In *Democracy and Social Ethics*, which Marilyn Fischer (2013) has described as "an evolutionary idealist text," Addams used the Spencerian language of adaptation and adjustment, although (like Dewey's) her picture was more dynamic: "to attain individual morality in an age demanding social morality, to pride one's self on the results of personal effort when the time demands social adjustment, is utterly to fail to apprehend the situation" (Addams 1902, 2–3). As this quotation indicates, Addams argued that a shift from individual ethics to social ethics was needed as a response to the new environment of the modern industrial city.

In a chapter on the relationship between parents and daughters, for instance, Addams wrote of the tension between familial and social obligations. She praised the historical institution of the family but declared that, in "periods of reconstruction," we must "enlarge the function and carry forward the ideal of a long-established institution." More specifically, she argued that

the family in its entirety must be carried out into the larger life. Its various members together must recognize and acknowledge the validity of the social obligation. When this does not occur we have a most flagrant example of the ill-adjustment and misery arising when an ethical code is applied too rigorously and too conscientiously to conditions which are no longer the same as when the code was instituted, and for which it was never designed. (Addams 1902, 78–79)

Since more and more daughters were receiving a full education, said Addams, it no longer made sense to deny them their role as citizens of the world. What was needed, according to Addams, was "an adaptation of our code of family ethics to modern conditions" (Addams 1902, 82–84).

In another chapter, echoing a letter quoted by Kropotkin in his "mutual aid" series, Addams noted that "a very little familiarity with the poor districts of any city is sufficient to show how primitive and genuine are the neighborly relations" (Addams 1902, 19; Eddy 2010, 31; cf. Kropotkin 1896, 928). In the neighborhood where she worked, poorer residents were often outraged by the calculating approach of charities. Charity methods, said Addams, were thus condemned as being too scientific, but for her they were "not scientific enough." Before they "had become evolutionary and scientific," botany and geology had consisted of dry classification; charity work, according to Addams, was in a parallel "pseudo-scientific" stage, one that said, "Don't give" unless certain boxes were checked. The solution was "to apply this evolutionary principle to human affairs," that is, to understand the new social conditions of urban poverty and jointly reconstruct both our ideals and those conditions:

The young woman who has succeeded in expressing her social compunction through charitable effort finds that the wider social activity, and the contact with the larger experience, not only increases her sense of social obligation but at the same time recasts her social ideals. (Addams 1902, 64–69)

Our virtues, said Addams, needed to be socialized.

In her next book, *Newer Ideals of Peace*, Addams argued more explicitly for "the ideals of genuine evolutionary democracy" (Addams 1907, 60). She was in favor of more "local self-government" and, like Dewey, saw political institutions as adaptive techniques:

As the machinery, groaning under the pressure of new social demands put upon it, has broken down, ... we have mended it by giving more power to administrative officers, because we still distrusted the will of the people. We are willing to cut off the dislocated part or to tighten the gearing, but are afraid to substitute a machine of newer invention and greater capacity. (Addams 1907, 34–35)

She also declared that the demand by the well-off for protection from "the many unsuccessful among us" betrayed ignorance of "the historic method" – it was "not to have read the first lesson of self-government in light of evolutionary science." Politics, she said, needed to adapt "to new and strenuous conditions" (Addams 1907, 63). Drawing from her experience at Hull House, for example, she attacked the exclusion of immigrant communities from local government. Public health problems such as tuberculosis could not be solved, Addams wrote, without "the intelligent cooperation of the immigrants themselves." The immigrant, she continued, represents the "type which is making the most genuine contribution to the present growth in governmental functions, with its constant demand for increasing adaptations" (Addams 1907, 74–75). Thus, while Dewey was teaching Addams's work and inserting the evolutionary method into ethics textbooks, Addams was putting the method into practice in Chicago (Dewey 2010; Dewey and Tufts 1908, 321–322).

Conclusion

In Spencer's evolutionary ethics, good conduct was successful adjustment to the environment. Huxley attempted to argue that ethics was opposed to the evolutionary process, but Kropotkin and Dewey replied that he was working with an overly simplistic notion of evolution – one that neglected the widespread importance of cooperation and ignored "the evolution of environments" (Dewey 1898, 339). Dewey's dynamic functionalism saw ethics as an adaptive technology, modifying the environment even as it changed in response to it. Addams put a similar view to work in her social activism, arguing that ethics and politics had to change in response to the new conditions of industrial life.

Although an update and defense of Dewey's and Addams's views would require a chapter of its own, it is worth noting that several philosophers have recently adopted similar theories. Elizabeth Anderson, for many years, has argued in favor of a pragmatist approach to ethical inquiry in which our morals are continually modified in light of "our experiences in living out the lives our ethical principles prescribe for us" (Anderson 1998, 16). In a more explicitly evolutionary vein, Philip Kitcher's book, *The Ethical Project* (2011), pursues a neo-Deweyan functionalist approach to morality. Both Anderson and Kitcher emphasize moral progress and "experiments in living" (Anderson 1991; Kitcher 2011, 104–137, 209–252; Anderson 2014b). Thus the experimental evolutionary approach of Addams and Dewey is still a live option today.

Note

1 Eddy's *Evolutionary Pragmatism and Ethics* (2016), a book-length treatment of the evolutionary ethics of Dewey and Addams, appeared as this chapter was going to press. For a review, see Pearce (Forthcoming).

4   Social Darwinism and Market Morality: A Modern-Day View

Naomi Beck

Social Darwinism has almost as bad a reputation as evolutionary ethics. It smacks of evolutionary fatalism and the kind of biological determinism we usually associate with outdated, nineteenth-century theories, such as Herbert Spencer's. Elsewhere (Beck 2013), I argued that the identification of Social Darwinism with ideologies that support inequality distorts the views of early proponents of evolution, including Darwin himself. To him, as to the co-discoverer of natural selection Alfred Russel Wallace, the application of evolutionary principles to the social domain did not entail denial of aid to the poor and the weak. It is true that both Darwin and Wallace agreed that the multiplication of the unfit would be undesirable because it would lead to a deterioration of the human race. But neither advocated active weeding out of the unhealthy and slow members of society, even though both felt quite comfortable referring to the studies on eugenics conducted by Darwin's cousin, Francis F. Galton.

The rise of Nazism with its host of horrors performed in the name of racial supremacy discredited appeals to evolution and to the principle of "survival of the fittest" outside of biology. The term "Social Darwinism," which came into fashion after 1940 (Hodgson 2004), acquired a distinctly pejorative connotation. This explains the vehement attack on Edward O. Wilson's call (1975) to revive Darwin's original proposal (1859, 488) and throw light on the origin of humans, their history and social behavior. In the second half of the twentieth century, most political thinkers stayed clear of evolutionary arguments in their theories – most, but not all. A notable exception is Nobel laureate in economics Friedrich August von Hayek. Writing around the time Wilson made the claim that ethics be "removed temporarily from the hands of the philosophers and biologicized" (Wilson 1975, 27), Hayek offered an alternative. He challenged Wilson's position and argued that the leading role in the evolution of morality belongs to culture, not genetics (Hayek 1979, 153–157). In the final years of his life, Hayek attempted to work out the details of his theory of cultural evolution, dedicating his last, unfinished, book to the "basic question … *how does our morality emerge, and what implications may its mode of coming into being have for our economic and political life?*" (1988, 8; italics in the original).

In what follows, I examine Hayek's claims as an example of modern-day Social Darwinism. Hayek would have no doubt objected to this designation. He specifically criticized nineteenth-century Social Darwinists for their simplistic theories and overreliance on biology in their explanations. But if by Social Darwinism we adopt the general, and more common, definition of a position that advances supposedly scientific arguments in the name of partisan views that are anything but neutral and disinterested, then Hayek's theory fits the bill. And it offers a vantage point in the recent past for exploring some of the problems that have given the Social Darwinist approach to evolutionary ethics its bad reputation.

Capitalism Is Not the Answer

Asking questions is the beginning point of all research ventures. Asking the right questions is key to their success or failure to come up with defendable, convincing answers. There is a telling difference between Darwin and Hayek in their manner of approaching the question of moral evolution. Darwin was eager to find out whether his theory could account for an observable trait: the human moral sense. He did not endeavor to define morality, preferring to refer cursorily to notions of duty and sacrifice and to defer to the authority of Kant in the matter. Darwin then declared: "This question has been discussed by many great writers of consummate ability; and my sole excuse for touching on it is the impossibility of here passing it over, and because, as far as I know, no one has approached it exclusively from the side of natural history" (Darwin 1871, 71). Whether Darwin was successful or not is debatable to this day. But one thing is clear: he was careful not to confuse the descriptive and the normative elements of this question, or to attempt a reinterpretation of ethics under the dictates of the struggle for existence and the survival of the fittest. His goal was to develop a natural historical account of morality that respects the principles of the theory first exposed in the *Origin of Species*.

The task was not an easy one. Explaining morality and altruistic behavior with the aid of natural selection seemed a long shot. How could prosocial behavior, which decreases the survival chances of the benefactor, be selected? To resolve this problem, Darwin came up with a complementary mechanism: community selection. He claimed that sometimes selection acts on the group as a whole rather than on individuals. When selective forces are sufficiently strong – for instance in tribal wars – the conditions are met for favoring traits that do not contribute directly to individual survival (such as sacrificing one's life for the sake of others) but help group survival, and in so doing eventually contribute to the well-being of its members. Community selection, along with other factors such as a desire to be praised by one's fellows and avoid blame, a natural inclination toward sympathy, and the acquired habit of offering help in the hope of getting help in return, accounted for the paradox of human morality. Darwin was also cautious not to choose the seemingly easy path of equating natural selection with free competition, and adaptation with economic viability, notwithstanding his affinity with British individualism (see Schweber 1980; Hodge 2009) and the fact that he adopted Spencer's expression "the survival of the fittest" in later editions of the *Origin of Species.*

Spencer, and also Hayek, assumed a different attitude. Both had the clear objective of turning the explanation of the evolution of morality into a justification of their own system of values and political position. Here, I will focus on the latter, as his evolutionary views are less well known, albeit Hayek's considerable influence in recent times and his explicit attempt to link evolution and capitalism. This is apparent in the seemingly neutral query previously quoted concerning the origins of morality and the implications its mode of coming into being have for our economic and political life. In reply Hayek wrote: "The contention that we are constrained to preserve capitalism because of its superior capacity to utilize dispersed knowledge raises the question of how we came to acquire such an irreplaceable economic order" (Hayek 1988, 10). This bias in favor of capitalism – Hayek in effect provided the answer before posing the question – led him to develop a theory of cultural evolution, which meant to serve a preexistent definition of modern civilization. This definition described the advantages of modern civilization in a strikingly similar way to those of the free market.

In *The Constitution of Liberty* Hayek described civilization in the following manner:

Most of the advantages of social life, especially in its more advanced forms which we call "civilization," rest on the fact that the individual benefits from more knowledge than he is aware of. It might be said that civilization begins when the individual in the pursuit of his ends can make use of more knowledge than he himself acquired and when he can transcend the boundaries of his ignorance by profiting from knowledge he does not himself possess. (Hayek [1960] 1971, 21)

Hayek portrayed the free market in the exact same way, namely as a decentralized structure that allows us to use knowledge shared by all (and available to none in its entirety) in the most effective manner (Hayek 1988, 15). It naturally followed that modern civilization equaled the free market, while the advantages of the social life Hayek spoke of were first and foremost the fruits of economic growth. In a revealing footnote Hayek declared: "What is true of economics is also true of culture generally: it cannot remain stationary and when it stagnates it soon declines" (Hayek 1979, 206). The "Great Society," as Hayek referred to modern civilization, was great in the literal sense; it relied on economic growth.

To substantiate this exclusive definition of modern civilization, Hayek turned to biology. "It is quite possible," he wrote, "that one kind of system … is so much more effective than all others in producing a comprehensive order for a Great Society that, as a result of the advantages derived from all changes in the direction towards it, there may occur in systems with very different beginnings a process corresponding to what biologists call 'convergent evolution'" (Hayek 1976, 40). Elsewhere Hayek specified: "There may exist just one way to satisfy certain requirements of forming an extended order – just as the development of wings is apparently the only way in which organisms can become able to fly (the wings of insects, birds and bats have quite different genetic origins)" (Hayek 1988, 17). Although Hayek did not deny the existence of a plurality of cultures, he insisted that only the free market is compatible with advanced civilization, and he believed that cultural evolution spontaneously progressed toward this goal.

Others have noted that this conception of evolution, which vehicles an implicit assumption of the perfectibility of society, reveals a strong resemblance to the philosophy of Herbert Spencer (see Gray 1984, 103–109; Hodgson 1991, 77; 1993, 179–181). Spencer conflated evolution and progress in a famous article from 1857, which provided the inspiration for his multivolume *System of Synthetic Philosophy*. But Hayek almost never referred to Spencer. The reason is no doubt connected to his desire to dissociate himself from Social Darwinism, as evidenced by a particularly depreciatory footnote that identifies Spencer's philosophy with Social Darwinism and accuses him of "having spoiled a good argument by the crude and insensitive way in which he applied it" (Hayek 1958, 243–244, note 21). If Hayek meant to decry Spencer's teleological understanding of evolution, he was badly placed. His theory, as much as that of his nineteenth-century forerunner, assigns a direction and a goal to cultural evolution. As we shall soon see, Hayek considered any digression from the perceived trajectory of evolution as contrary to the natural course of things. This was the essence of his attack on socialism, deemed at best anachronistic and more often detrimental to progress.

Hayek hoped to dodge accusations of teleology by claiming that a teleological interpretation of evolution such as his is "entirely in order so long as it does not imply design by a maker but merely the recognition that the kind of structure would not have perpetuated itself if it did not act in a manner likely to produce certain effects, and that it has evolved through those prevailing at each stage who did" (Hayek 1967, 77). This type of argument hides a fundamental error, which paleontologist Stephen J. Gould termed "adaptationism." Gould pointed out that contingent events can play a crucial role in both biological and cultural evolution and cautioned against the human tendency to seek patterns and find cause and meaning in all events. Evolution does not lead to optimality, and there is no good reason to presume "that everything must fit, must have a purpose, and in the strongest version, must be for the best" (Gould 1987, 68). Hayek appeared to have presumed exactly this when he spoke of "the twin ideas of evolution and the spontaneous order" (Hayek 1984, 177). To him, evolution was the creation of a beneficial order. Otherwise, he would not have considered Bernard Mandeville a precursor of evolutionary thinking (Hayek 1984, 180). And though Hayek stated that he "carefully avoided saying that evolution is identical with progress" (Hayek 1979, p. 168), he also asserted that while the extended market order is not "good in some absolute sense ... *it does enable us to survive and there is something perhaps to be said for that*" (Hayek 1988, 70; italics in the original).

In fact, in Hayek's eyes, there was a lot to be said for that. Hoping to escape teleology and the unseemly equation of evolution with progress, Hayek stumbled right into the reductionist trap. In his effort to explain why we are constrained to preserve capitalism, he reduced both biological and cultural evolution to a process of population growth that depends on free competition for its material foundation.

The Reductionist Trap

On various occasions, Hayek emphasized the differences between biological and cultural evolution. He decried the assumption "that any investigator into the evolution of human culture has to go to school with Darwin" (Hayek 1988, 23) and proclaimed: "A nineteenth-century social Darwinist who needed Darwin to teach him the idea of evolution was not worth his salt. Unfortunately some did, and produced views which under the name of 'social Darwinism' have since been responsible for the distrust with which the concept of evolution has been regarded by social scientists" (Hayek 1973, 23). It is a pity Hayek did not have better knowledge of Darwin's theory. Like many of his fellow economists, he ignored Darwin's views on cultural evolution and apparently was not aware of the pains Darwin took not to confound the latter with biological evolution (see Marciano 2007). Equally problematic was Hayek's superficial view of biological evolution itself. "Variation, adaptation and competition," Hayek declared, "are essentially the same kind of process," and "all evolution rest[s] on competition" (Hayek 1988, 26). This interpretation conflated biological evolution with the economic idea of free competition, turning it into a theory of growth rather than change. Hayek indeed erroneously claimed that Darwin got the basic idea of evolution from economics, most particularly from reading Adam Smith in 1838. Rather curiously, he also rejected the contribution of Malthus's theory on population to evolutionary thought.

Darwin, let us recall, was forthright with regard to the prominent role Malthus's claims occupied in his theory. He commented about natural selection: "It is the doctrine of Malthus applied with manifold force to the whole animal and vegetable kingdoms" (Darwin 1859, 63). According to Malthus, population grows at an infinitely greater rate than the natural resources needed for its subsistence. As a result of this uneven ratio, any increase in the productivity of the land is outpaced by increase in population. Malthus believed that equilibrium is restored when population growth is checked through periodic catastrophes such as war, famine, and disease. Hayek argued that this theory was perhaps applicable in Malthus's time, when human labor could be regarded as a more or less homogeneous factor of production, i.e., when wage labor was all of the same kind, employed in agriculture, with the same tools and the same opportunities. But in modern conditions, the Malthusian threat of worldwide pauperization is irrelevant. The key to understanding Malthus's error is differentiation: "man's greatest achievement, leading to his other distinct characteristics" (Hayek 1988, 126). In a situation where labor is diversified, an increase in population makes further increases possible because differentiation allows for the division of labor, which in turn heightens productivity. Thanks to the increase in output, more mouths can be fed and population can grow. In a denser population, even greater specialization is possible, which is then followed by a more elaborate division of labor, resulting in further increases in productivity that facilitate further population expansion and so on in a "self-accelerating" cycle (Hayek 1988, 122).

Contrary to Malthus, Hayek concluded that over much of human history, population growth was a "self-stimulating" rather than a "self-limiting" cause (Hayek 1988, 126). Thus, even though Hayek presumed that "the close interconnections among evolution, biology and ethics" were related to the problem of scarcity of resources, he did not accord Malthus's theory the same weight as Darwin or his followers did. Nor did he lament the ills engendered by the struggle for existence (Hayek 1988, 15), which both advocates and detractors of evolutionary ethics deplored. Darwin confessed: "It is impossible not to regret bitterly, whether wisely is another question, the rate at which man tends to increase; for this leads in barbarous tribes to infanticide and many other evils, and in civilised nations to abject poverty, celibacy and to the late marriages of the prudent" (Darwin 1871, 180). These late marriages hinder the most virtuous members of society from procreating at the fast rate of the reckless, degraded, and vicious ones, who are not as prudent in delaying procreation until they can support themselves and their children in comfort (Darwin 1871, 174). T. H. Huxley took a firmer stand. In an article on the struggle for existence in human society he wrote: "So long as unlimited multiplication goes on, no social organisation which has ever been devised or is likely to be devised, no fiddle-faddling with the distribution of wealth, will deliver the society from the tendency to be destroyed by the reproduction within itself, in its intensest form, of that struggle for existence, the limitation of which is the object of society" (Huxley [1888] 2011, 212–213). A few years later Huxley delivered a famous lecture in which he denounced evolutionary ethics on similar grounds (as discussed later in this chapter).

Unlike Hayek, many in the twentieth century agreed with Malthus's analysis. Ecologist Garrett Hardin, whose work on ethics Hayek cited (1988, 15), authored an influential paper in 1968 in which he claimed that population growth in a finite world is a problem that has "no technical solution." Huxley's grandson Julian, with whose views Hayek was clearly familiar (Hayek 1988, 121), cautioned: "Early in man's history the injunction to increase and multiply was right. Today, it is wrong, and to obey it would be disastrous.... The spectacle of explosive population-increase is prompting us to ask the simple but basic question, *what are people for?* And we see that the answer has to do with their quality as human beings, and the quality of their lives and their achievements" (Huxley 1961, 24; italics in the original). This affirmation is almost the exact antithesis of Hayek's claim: "Whatever men live *for*, today most live only *because* of the market order. We have become civilized by the increase in our numbers just as civilization made that increase possible: we can be few and savage, or many and civilized" (Hayek 1988, 133; italics in the original). Hayek indeed contended that all of the knowledge contained in books would be of little use to humanity without numbers sufficient to fill the jobs demanded for extensive specialization and division of labor.

This view dictated a very specific perception of morality. Hayek argued that there is no point in asking whether the actions that contribute to demographic growth are good in some absolute sense, "particularly if thus it is intended to inquire whether we like the results" (Hayek 1988, 120). The biological necessity to exist prior to the existence of human wishes and values rendered these last of secondary concern. Life's main purpose, Hayek stated, is itself: "life exists only so long as it provides for its own continuance" (Hayek 1988, 133). This statement held for biological as well as cultural evolution. Their ultimate goal is to "maximize ... the prospective stream of future lives" (Hayek 1988, 132). But if this were the case, then Hayek's attack on sociobiology's gene-centered approach is unwarranted. When life's sole purpose is to continue itself, all that matters is reproduction. In the final analysis, we are nothing but vehicles for propagating our genes. Unlike Wilson, who recognized that morality poses a dilemma and challenges evolutionary thinking, Hayek inadvertently became the spokesman for a profoundly biological conception of cultural evolution. He not only admitted the material substratum of morality but also transformed it into a value in itself. The definitive value really, because it was the condition for all others. Quantity prevailed over quality, and Hayek put a positive spin on Marx's famous dictum, declaring without any hint of irony: "capitalism created the proletariat: it gave and gives them life" (Hayek 1988, 124). From this perspective it was immoral to do anything that might interfere with free competition and economic growth as this would reflect directly on the quantity of life.

Like it or not, the current world population already exists. Destroying its material foundation in order to attain the "ethical" or instinctually gratifying improvements advocated by socialists would be tantamount to condoning the death of billions, and the impoverishment of the rest. (Hayek 1988, 120)

It all fits nicely if one accepts the definition of progress as growth, evolution as competition, and cultural development as a process of competition that leads to growth. But the explanandum got lost on the way. Hayek completely sidestepped the thorny issue of whether evolution by natural selection could account for pro-social behavior, and redefined morality in a way that made it a derivative of biology. This explained away the problem rather than elucidate it. Hayek's evolutionary narrative did not help resolve this failing. It redefined morality as the rules of the free market while negating the validity of what we usually regard as moral values.

From Love Thy Neighbor to Profit-Seeking

Hayek postulated that cultural evolution involved a shift from a primitive morality based on solidarity and altruism to an advanced morality comprising the rules of the free market. The first evolved as an adaptation to life in small groups of hunters-gatherers, whose members knew each other and had shared goals that coordinated their activities. This type of morality prevailed during the longer part of our species' history. But since it was limited to the members of one's group, it could not accommodate the rise of modern, big societies in which cooperation takes place in a largely anonymous manner. The process of cultural evolution required therefore repression of our primordial altruistic drives. Under this light, modern civilization could be viewed as unnatural: it did not conform to humans' biological endowments (Hayek, 1988, 19). Yet Hayek argued that to designate as moral the instincts that welded together the small group would be a mistake because innate reflexes have no moral quality: "And 'sociobiologists' who apply terms like altruism to them (and who should, to be consistent, regard copulation as the most altruistic) are plainly wrong. Only if we mean to say that we *ought* to follow 'altruistic' emotions does altruism become a moral concept" (Hayek 1988, 12; italics in the original).

It was important for Hayek to maintain that a mismatch exists between humans' natural inclinations and the rules of the extended order of modern civilization. Otherwise he would not have been able to denounce opponents of the free market on the basis of his evolutionary arguments. Hayek blamed them for not understanding that they attack the very foundations of modern civilization. In seeking greater solidarity and equality, these social reformers disregard the efficacy of the rules that made growth and progress possible in the first place. Hayek in effect described the various steps in the transition to market economy and modern civilization – e.g., the recognition of private property through land enclosures, allowance of competition with fellow craftsmen in the same trade, and the lending of money for interest – as "breaches of that 'solidarity' which governed the small group" (Hayek 1979, 162). The powers that be "generally resisted rather than assisted changes conflicting with traditional views about what was right or just." This explains the feelings of guilt or bad conscience, which often accompany the acquisition of material success in the free market. Such sentiments are vestigial impulses that disserve progress now and might have hindered it in the past, when "continued obedience to the command to treat all men as neighbors would have prevented the growth of an extended order" (Hayek, 1988, 13). The striving for financial gain, though it makes us withhold from known, needy neighbors what they might require, helps us "serve the unknown needs of thousands of unknown others" (Hayek 1979, 162–165). Profitability, Hayek clarified, works as a signal that guides us toward the most fruitful activities, and "only what is more profitable will, as a rule, nourish more people, for it sacrifices less than it adds" (Hayek 1979, 146).

To explain how the transition from a morality of "love thy neighbor" to the free market order took place, Hayek evoked the idea of group selection. His inspiration came from Oxford biologist-turned-sociologist Alexander M. Carr-Saunders, though the latter used it to claim that selection favors restrictions on reproduction rather than growth (see Carr-Saunders 1922). Hayek did not refer at all to Darwin's views on group selection. He was also little concerned by the fact that most biologists rejected group selection as a result of G. C. Williams's attack on V. C. Wynn-Edwards's adaptation of this idea to the animal domain (see Wynn-Edwards 1962; Williams 1966). What appealed to Hayek in the idea of group selection was the possibility of depicting cultural evolution as a process "guided not by reason but by success" (Hayek 1979, 166). According to Hayek's narrative, the free market emerged because groups of individuals that have adopted certain rules of behavior displaced those who did not. In line with Edmund Burke's reasoning, which Hayek hoped to buttress with evolutionary arguments, he defended the wisdom of the ages against the private stock of individual reason and claimed:

To understand our civilization, one must appreciate that the extended order ... arose from unintentionally conforming to certain traditional and largely moral practices, many of which men tend to dislike, whose significance they usually fail to understand, whose validity they cannot prove, and which have nonetheless fairly rapidly spread by means of an evolutionary selection – the comparative increase of population and wealth – of those groups that happened to follow them. (Hayek 1988, 6)

Using this specific interpretation of group selection allowed Hayek to pretend that the rules of behavior of the free market were not a matter of choice. "Man has been civilized very much against his wishes," he wrote. "It was the price we had to pay for being able to raise a larger number of children" (Hayek 1979, 168). Hayek's evolutionary account failed, however, to explain why the selective pressures pushing toward growth and market morality were stronger than those favoring entrenched human inclinations such as solidarity. After all, the weight of small-group traditional moral practices was much heavier than that of the rules of capitalism. Here again Hayek let his ideological bias trump evolutionary logic. He attempted to describe cultural evolution as an "accelerating development of civilization ... that took place during the last 1 percent of the time during which *Homo sapiens* existed" (Hayek 1979, 156). In that short period, cultural evolution "swamp[ed] genetic evolution" because it relied on the faster mechanism of transmission of acquired rules of behavior through imitation and learning (Hayek 1988, 16). Hayek's account thus put the emphasis on a very short chronological period – the rise of free-market society – which he treated as a quantum leap that completely changed the rules of the game. In so doing, he discounted the views and beliefs of preceding millennia, during which time human morality evolved along different lines and affirmed that "we may not like the fact that our morals have been shaped mainly by their suitability to increase our numbers but we have little choice in the matter" (Hayek 1988, 134).

Apart from being unpersuasive, this reasoning plunged Hayek into the murky waters of the naturalistic fallacy, namely, the confusion of "is," or what selection favors, with "ought," or what selection should favor, or rather, what we should favor. Hayek was aware of this potential criticism. "I do not claim," he wrote, "that the results of group selection ... are necessarily 'good' – any more than I claim that other things that have long survived in the course of evolution, such as cockroaches, have moral value" (Hayek 1988, 27). But given Hayek's equation of modern civilization with the extended market order, and his conflation of growth with progress, it is hard to take this declaration at face value. Moreover, Hayek clearly believed that survival was indicative of moral value for he subjected human reason and ideals to inherited customs and traditions.

It is not only in his knowledge, also in his aims and values, that man is a creature of civilization; in the last resort it is the relevance of these individual wishes to the perpetuation of the group or the species that will determine whether they will persist or change. It is, of course, a mistake to believe that we can draw conclusions simply because we realize that they are the product of evolution. But we cannot reasonably doubt that these values are created and altered by the same evolutionary forces that have produced our intelligence. All that we can know is that the ultimate decision about what is good or bad will be made not by individual human wisdom but by the decline of the groups that have adhered to the "wrong" beliefs. (Hayek [1960] 1971, 32)

With respect to evolutionary ethics, Hayek adopted exactly the kind of approach T. H. Huxley condemned. The latter argued that nature was no moral guide and could not provide justification for good or bad behavior: "The thief and the murderer follow nature just as much as the philanthropist. Cosmic evolution may teach us how the good and the evil tendencies of man may have come about: but in itself, it is incompetent to furnish any better reason why what we call good is preferable to what we call evil than we had before" (Huxley [1893] 2011, 80). Huxley contrasted evolution with the "ethical process" that, according to him, comprised the essence of social progress. Unlike Hayek, for whom competition was a procedure of discovery that promised beneficial social outcomes, Huxley rejected the struggle for existence as the basis for a modern, progressive society (see Hayek 1978). To Huxley, society had a "definite moral object:" the achievement of "mutual peace" by setting limits to the struggle for existence and reining in the "unceasing competition" that is the rule of nature (Huxley [1888] 2011, 203–204; [1893] 2011, 13). In practical terms, this meant the establishment of a welfare state that would take care of the poor and maintain the conditions necessary for social stability. As an alternative to the "gladiator's show" that is the reality of evolution, Huxley advocated the creation of an "artificial world" guided by the ideals of equality and solidarity (Huxley [1888] 2011, 168; [1893] 2011, 83).

Hayek shared Huxley's claim that evolution cannot be held accountable to human norms of justice. But instead of calling for a clear separation of the two, he advised discarding our idealistic view of the latter:

The fruitless attempt *to render a situation just* whose outcome, by its nature cannot be determined by what anyone does or can know, only damages the functioning of the process itself. Such demands for justice are simply inappropriate to a naturalistic evolutionary process.... *Evolution cannot be just.* To insist that all future change be just is to demand that evolution come to a halt. Evolution leads us ahead precisely by bringing about much that we could not intend or foresee, let alone prejudge its moral properties. (Hayek 1988, 74; italics in the original).

In the name of this perception of evolution Hayek embraced a view of justice that equated it with the operation of impersonal free market forces and resulting inequalities. He also repudiated human agency from cultural development, thus negating the validity of our moral inclinations.

The Human Factor

We, humans, have an ingrained sense of justice and deep-seated moral dispositions as well as cooperative tendencies. Recent findings from research in neurobiology and from studies of nonhuman primates confirm this claim (see Bowles and Gintis, 2006, 2011). Darwin recognized it as well and believed that, in part, these leanings were the product of an evolutionary process. This is why he argued that had this process taken a different turn, human morality would have looked very differently:

I do not wish to maintain that any strictly social animal, if its intellectual faculties were to become as active and as highly developed as in man, would acquire exactly the same moral sense as ours.... If for instance, to take an extreme case, men were reared under precisely the same conditions as hive-bees, there can hardly be a doubt that our unmarried females would, like the worker-bees, think it a sacred duty to kill their brothers, and mothers would strive to kill their fertile daughters; and no one would think of interfering. (Darwin 1871, 73)

The fact is that we are not bees. Our moral sense is one that upholds prosocial behavior and forbids killing family members. In addition, one of the specificities of our species lies in its highly developed cognitive abilities and the capacity to influence our existence and development to a much greater extent than other living beings. Like Darwin, Hayek acknowledged that we have an ingrained *Rechtsgefühl* (sense of justice), which he compared to our inborn *Sprachgefühl* (sense of language) (Hayek 1967, 45). But he reserved a very uneven treatment for the two. The sense of language was his choice example for our innate capacity to follow unformulated rules, such as those of grammar and syntax, without being conscious of the underlying elements organizing these patterns. Social learning assumed the same form, Hayek argued. It is predominantly based on imitation: "Much of what we can do rests on skills or aptitudes or propensities acquired by following examples, and selected because they proved successful, but not deliberately chosen for a purpose" (Hayek 1978, 292). The sense of justice was not awarded the same status. Hayek denounced our so-called atavistic conceptions of justice and the desire for a fair distribution of wealth as obstacles to progress. Instead, he advocated a procedural conception of justice, which concerned only the rules of the game. Whether or not these rules bring about a desirable result from a human point of view was beside the point. What mattered was that these rules follow the precepts of free competition, even if this means the strong inevitably get more than the weak (Hayek 1973, 141).

Hayek attacked specifically political philosopher J. Rawls's influential *Theory of Justice* (1971), which offered a modern interpretation of Rousseau's social contract. According to Rawls, the principles of justice could be retrieved with the help of an artificial device, a "veil of ignorance," behind which all members of society are placed. Under this condition, Rawls believed, people would agree that each should have an equal right to the most extensive basic liberty compatible with a similar liberty for others and that inequalities in the distribution of wealth should benefit the worst-off members of society. Hayek opposed this view, declaring: "If, at some earlier date, some magic force had been granted the power to enforce, say, some egalitarian or meritocratic creed ... the evolution of civilization [would have been] impossible" (Hayek 1988, 74–75). Proponents of social justice must accept that in a process governed by impersonal forces, the expectations of some will be ruined, and they might suffer unmerited failure due to a multitude of factors that no one controls. In short, in Hayek's eyes, there was a clear trade-off between progress and the welfare state. What he did not explicitly assert, but others noted, was that by recommending the free market as a purely procedural coordination mechanism, one that is capable of reconciling people's conflicting individual ends while avoiding potential strife, Hayek overlooked that this mechanism follows the maxim: "to each according to his/her market value" (see Kley 1994, 24). He tried to present this maxim as an unbiased, instrumental view of justice, but in reality it is no less contentious than the conception of social justice. Its defense, like that of any alternative principle, must rest on a moral philosophical argument.

Hayek hoped to substantiate his position by exposing the unfeasibility of social justice due to evolutionary constraints resulting in the limitedness of human rationality. Reason, he claimed, was created by cultural selection. And the development of civilization "has largely been made possible by subjugating the innate animal instincts to the non-rational customs which made possible the formation of larger orderly groups of gradually increasing size" (Hayek 1979, 155, 166). It was therefore nonsensical as well as contrary to the principles of cultural evolution to think that humans are able to positively influence this process. We, humans, cannot shape cultural development according to our wishes, as proponents of social reforms would have us believe. This is so because the rules of the free market on which our modern civilization relies emerged in a process of selection independent of human intentions and ideals and "simply because they enabled those groups practising them to procreate more successfully and to include outsiders" (Hayek 1988, 16). To think otherwise was a "fatal conceit," as the title of Hayek's last, unfinished book had it. He decreed that a proper understanding of cultural development necessitated "completely discard[ing] the conception that man was able to develop culture because he was endowed with reason" (Hayek

1979, 156). But in effect this was Hayek's working hypothesis rather than an inference from his theory.

Wishing to uncover the "errors of socialism" (the subtitle of his book), Hayek worked his way from politics back to evolutionary theorizing. The main reason he was attracted to the latter resided in what he considered to be an "astonishing fact, revealed by economics and biology, that order generated without design can far outstrip plans men consciously contrive" (Hayek 1988, 8). Accordingly, his account of cultural group selection excluded human choice from this process (see Beck 2012). But his interpretation of evolution has more in common with the modern economic concept of the invisible hand than with Darwin's claims, or the claims of those who further developed the idea of group selection (see Wilson 2004). And Hayek's narrative left wide gaps concerning the specific modalities by means of which the shift from a small group morality to market ethics occurred. Why the latter prevailed over the former? How it emerged? What motivated in the first place individual players to adopt customs whose benefits they did not understand? Hayek thought that by pushing to the sidelines the human factor he did valuable service to the project of evolutionary ethics:

Social Darwinism is wrong in many respects but the intense dislike of it shown today is also partly due to its conflicting with the fatal conceit that man is able to shape the world around him according to his wishes. Although this ... has nothing to do with evolutionary theory properly understood, constructivist students of human affairs often use the inappropriateness (and such plain mistakes) of Social Darwinism as a pretext for rejecting any evolutionary approach at all. (Hayek 1988, 27)

In reality, Hayek's own mistakes – the use of evolutionary theory to defend a political cause and the confusion of explanation with justification – did more to protract the bad reputation of evolutionary ethics that is associated with Social Darwinism than to rehabilitate it. His theory best serves as a modern-day cautionary tale of the obstacles to avoid when applying evolutionary logic to the sociocultural domain.

5 The Path to the Present

Abraham H. Gibson

Historians may well cringe at the title of my chapter. Ever since Cambridge historian Herbert Butterfield coined the phrase "Whiggish history" back in the 1930s, the guild has considered it professionally uncouth to study the past with reference to the present, to approach the historical record with an agenda in search of antecedents. One is on far safer ground studying the past on its own terms, without reference to the present (Butterfield 1931; Mayr 1990; Jardine 2003; Besaude-Vincent 2014). Yet the Whiggish interpretation of history generally presumes that one side has prevailed, that winners have set about cherry-picking their favorite historical precursors, whereas more than a century of inquiry into the complex relationship between evolution and ethics has failed to produce any such consensus. On the contrary, the world's foremost experts on the evolutionary origin of altruism endorse a variety of radically different explanations, and the field of contemporary sociobiology is positively rife with controversy (Abbot et al. 2011; Nowak et al. 2010; Okasha 2010).

Furthermore, this turmoil is nothing new. Biologists have clashed over the purported evolutionary origins of ethical behavior for more than 150 years. They have offered a variety of explanations, some of which have even enjoyed broad support from time to time, but these explanations have just as often inspired heated, even heartfelt, protests. In an effort to contextualize the current controversy, this chapter will review the past 150 years of sociobiological thought. It will focus on one particular institution, Harvard University, where many of these battles have played out (Ruse 2011), but it will also place the research conducted at Harvard within a larger national and international context. Philosophers and sociologists feature in this story, but most of the clamor has arisen from the biological sciences. Generations of biologists have studied the evolutionary origins of cooperation, and their investigations have spanned the entire tree of life. Despite this broad interest, however, no other branch of biology has devoted more energy to the question of altruism than the entomological sciences. This is not an accident. No other organisms on Earth are more compulsively cooperative than the social insects. No other organisms lay bare the naked tension between the individual and the group, between the part and the whole (Ratniecks et al. 2011).

Before our narrative shifts to Cambridge, Massachusetts, it is important to acknowledge that scientists had already sought to explain the origin of altruism for thousands of years prior to the nineteenth century. The debate reaches at least as far back as Plato and Aristotle, but it was not until British naturalist Charles Darwin published *On the Origin of Species* that scientists began to seriously study altruism in an evolutionary context. As historians and biologists are both well aware, Darwin acknowledged that social insects represented the most serious "special difficulty" for his proposed theory of natural selection (Darwin 1859, 236–242). He was especially perplexed by the so-called neuter insects, which sacrificed reproduction in service to the colony. Since these animals did not reproduce, it was not clear how they inherited their altruistic tendencies, or how they passed these traits on to future generations. Despite these challenges, Darwin ultimately insisted that natural selection could, in theory, target groups just as easily as individuals.[1] Darwin may have acknowledged that cooperation could prove evolutionarily advantageous, but his successors were more likely to emphasize competition. When Darwin's celebrated bulldog Thomas Henry Huxley waxed philosophical about the relationship between evolution and ethics in the late nineteenth century, for example, he portrayed life as a constant struggle, and he portrayed nature as an amoral arbiter (Huxley 1894; Ruse 2009b).

When it came to questions about social evolution, however, Darwin generally deferred to one of his other countrymen, philosopher Herbert Spencer (Darwin 1872, 428; Haines 1991; Allhoff 2003), who wrote extensively on the purported evolutionary origins of sociality (Osborn 1894; Hofstadter 1944). He published one of his most influential essays, "The Social Organism," in 1860, just a few months after Darwin published the *Origin of Species* (Spencer 1860). In the essay, Spencer explained that human societies resemble biological organisms in three conspicuous ways: they generally advance from a state of simplicity to a state of greater complexity, they generally become more heterogeneous over time, and their respective constituent parts generally grow ever more interdependent over time. Spencer believed that these similarities were more than incidental, insisting that societies and organisms were both governed by the exact same natural laws. He believed that this integrative tendency explained apparent acts of altruism in nature and that it also provided the evolutionary process with a nested, hierarchical structure. He also studied the advanced sociality in certain insect colonies, though he felt that comparisons to human societies were necessarily limited. "Each of them is in reality a large family," he wrote about insect colonies. "It is not a union among like individuals substantially independent of one another in parentage ... but it is a union among the offspring of one mother, carried on, in some cases, for a single generation and in some cases for more." Spencer thus distinguished insect sociality, which was based on kinship, from human sociality, which transcended kinship (Spencer 1881, 6–7).

As is well known, biologists at Harvard University reacted to Darwin's theory of evolution by natural selection in radically different ways. For example, Harvard botanist Asa Gray, who had corresponded with Darwin prior to the publication of the *Origin* and had orchestrated the book's publication in the United States, became an outspoken supporter of evolution. He served as Darwin's foremost champion in the United States, and his support culminated with the publication of an aptly titled book, *Darwiniana* (Gray 1877; Browne 2010). In marked contrast, Harvard zoologist Louis Agassiz recoiled from the theory of evolution, which he regarded as completely and dangerously false (Dupree 1959; Wolfe 1975; Morris 1997). Subsequent generations have obviously sided with Gray, essentially casting anti-evolutionary dogmatism out of the biological sciences (if not the general public) in less than a generation. It should be noted, however, that accepting evolution did not, as a rule, mean that one also accepted Spencer. Harvard's great psychologist William James routinely castigated Spencer as unscientific and far removed from academic debates, but he could hardly silence the man (Crippen 2010; McGranahan 2011). As a result, Spencer remained the most widely read philosopher in the United States throughout the second half of the nineteenth century, and his voluminous writings helped convert innumerable Americans to an evolutionary worldview (Francis 2014; Francis and Taylor 2015).

For an entire generation of biologists who came of age in the late nineteenth-century United States, Spencer was at least as influential as Darwin. This was certainly true of Harvard's first great entomologist William Morton Wheeler, who cited both Darwin and Spencer with great frequency. Wheeler was already the world's foremost authority on ants when he arrived at Harvard to serve as Professor of Economic Entomology at the Bussey Institute in 1907 (Parker 1938). Soon after his arrival, Wheeler began promoting a new theory of insect sociality. He first dared suggest that ant colonies were true organisms when he published *Ants* in 1910, and he later expanded on the idea during a now famous lecture at the Marine Biological Laboratory in Woods Hole (Wheeler 1910, 1911). Much as Spencer had once insisted that human societies were true organisms, Wheeler offered several reasons for believing that ant societies were also true organisms. He observed, for example, that foraging workers served the colony's nutritive demands. He also noted that ant colonies were reproductively differentiated, just like traditional metazoans. In this case, the queen ant represented the germ line, while the nonreproducing members of the colony (i.e., the rest of the colony) represented somatic cells. What is more, he continued, the colony experiences both ontogenetic and phylogenetic growth, just like other organisms. Finally, he noted that the defensive caste, the soldier ants, helped maintain the colony's "individuality" (Wheeler 1911).[2]

Although Wheeler was an entomologist by trade, he was interested in more than just ants. In fact, he insisted that the same social impulse that stirred ants to cooperate with one another suffused all living things at every level of the biological hierarchy. "Every organism manifests a strong predilection for seeking out other organisms and either assimilating them or cooperating with them to form a more comprehensive and efficient individual," he wrote, adding that "one of the fundamental tendencies of life is sociogenic." Furthermore, he insisted, the same cooperative impulse that rendered insect colonies true organisms had also produced other true organisms, including cells, metazoa, and multispecies communities, which he called "biocoenoses." Alas, while his vision was certainly compelling, he provided no mechanism to explain the origin of sociality, a fact that he himself recognized. "We are still confronted with the formidable question as to what regulates the anticipatory cooperation, or synergy of the colonial personnel and determines its unitary and individualized course," he acknowledged (Wheeler 1911, 139–142).

Wheeler thought that he had finally solved the riddle in 1918 when he observed a peculiar phenomenon taking place among the ants he was studying. He noticed that ants shared foodstuffs and liquids with one another in reciprocal fashion, and he concluded that this process, which he dubbed "trophallaxis," explained the "source of the social habit" in ants and other insects (Wheeler 1918). There are at least two reasons why he felt so certain. First, he was eager to distinguish himself from people like Henri Bergson and Hans Driesch, both of whom were chastised for resorting to vitalism (Driesch 1908; Bergson 1911). Trophallaxis provided Wheeler with a strictly materialist explanation for the origin of cooperation that made no reference to vitalism. Second, he was struck by the idea that the circulation of liquids and other foodstuffs throughout the colony was perfectly analogous to the circulation of blood throughout metazoan organisms (Wheeler 1923, 260). That being said, there were several factors working against his hypothesis. For example, bees are social insects, yet they do not practice trophallaxis. Wheeler explained this away by saying that bees might have abandoned the practice after achieving eusociality, but it was a rather weak explanation. Furthermore, his insistence on a strictly materialist explanation actually undermined his claim that multispecies communities, which shared no physical bonds, also represented true "organisms."

Wheeler continued to insist that ant colonies were true organisms throughout the 1920s. What is more, he placed this revelation within a much larger framework, one that purportedly explained all life on Earth. In short, he felt that biologists had placed far too much emphasis on competition and that cooperation had played an equally important role in the evolutionary process. Previous generations had been preoccupied with struggle, he explained, but "to us it is clear that an equally pervasive and fundamental innate peculiarity of organisms is their tendency to cooperation, or 'mutual aid'" (Wheeler 1923, 3). Elsewhere, he repeated the claim that there was something "fundamentally social in living things," (Wheeler 1926, 437–438), and he insisted that this sociogenic impulse suffused all living things (Wheeler 1926, 437–438). Throughout the tree of life, he wrote, this cohesive tendency has resulted in individuals integrating with one another to produce new individuals at a higher level, and those new individuals have, in turn, continued to integrate with still more individuals at still higher levels. All life on Earth therefore comprised innumerable individuals emerging and integrating at several different nested levels. Wheeler labeled this grand process "emergent evolution," a phrase he borrowed from English evolutionary psychologist C. Lloyd Morgan (1923). According to Wheeler, "emergence ... signifies neither the manifestation or unveiling of something hidden and already existing ... nor some miraculous change ... but a novelty of behavior arising from the specific interaction or organization of a number of elements ... which thereby constitute a whole, as distinguished from their mere sum" (Wheeler 1928, 14). Like trophallaxis, the theory of emergent evolution thus provided Wheeler with a strictly materialist explanation that had no use for vitalism.

Wheeler's thoughts on the significance of kinship are surprisingly difficult to pin down. Like contemporary supporters of kin selection, he promoted trophallaxis on explicitly individualistic grounds. "A decidedly egoistic appetite, and not a purely altruistic maternal anxiety for the welfare of the young, constitutes the potent 'drive' that initiates and sustains the intimate relations of the adult ants to the larvae," he once remarked (Wheeler 1923, 172). In similar fashion, he once wrote that only members of the same species were capable of "true societies" (Wheeler 1927, 31). This statement suggests that he believed genetic proximity played some role in sociality. Wheeler sometimes contradicted himself, though. For example, he labeled the multispecies ecological community a "true organism" throughout his entire career (Wheeler 1911). What is more, he regarded the social bonds between members of the same species as indistinguishable from those between members of dissimilar species and suggested that the term "symbiosis" applies equally well to both homogenous and heterogeneous partnerships (Wheeler 1923, 195). Moreover, he noted that it was not uncommon for "alien organisms" of different species to fall into the insect colony's social vortex. In Wheeler's estimation, this "tendency to consociation with strange organisms" made the mixed colony a "super-superorganism" (Wheeler 1926, 437). It is difficult to reconcile these statements with kin selection, which credits familial bonds with providing the initial impetus for sociality.

Wheeler was not the only person at Harvard with an interest in the evolution of ethics, and he was not the only one to promote organicist philosophies. On the contrary, many of his Harvard colleagues from across the academic spectrum proffered very similar ideas. In 1924, Wheeler's friend and colleague, zoologist George Howard Parker, published an essay in the *Philosophical Review* describing a biological principle he termed "organicism." As he explained it, an individual's constituent parts were always less important than the manner in which those parts interact. "The processes of nature are dependent upon the way in which natural elements are assembled and interrelated," he wrote (Parker 1924a, 601). Integration was essential. It is worth noting that when Parker published a similar essay in *Science* later that same year, he provided an addendum acknowledging that he had recently happened across Lloyd Morgan's phrase, emergent evolution, and that he viewed the concept as nearly identical to organicism (Parker 1924b). Others articulated similar theories. Walter Bradford Cannon, a physician in the Harvard medical school, developed the concept of "homeostasis" during this time, publishing his ideas in *The Wisdom of the Body*. Not unlike Spencer's concept of "dynamic equilibria," Cannon's concept of homeostasis rested on efficient integration (Cannon 1932; Cross and Albury 1987). Meanwhile, English mathematician Alfred North Whitehead, who arrived at Harvard in 1924, developed a sweeping vision of the universe that emphasized dynamism, development, and process, and he referred to this elaborate metaphysical system as a "philosophy of the organism" (Whitehead 1925; Peterson 2011). In similar fashion, physiologist Lawrence J. Henderson reached explicitly "organismic" conclusions when he studied the "fitness of the environment." As Henderson once explained, the organism is "an autonomous unit in which every part is functionally related to every other and exists as the servant of the whole" (Henderson 1917, 21).

Harvard's scientists and scholars were on the cutting edge when it came to studying the relationship between evolution and ethics, but many different researchers in many different places scattered across the United States asked similar questions and drew similar conclusions. At the University of Chicago, ecologist Warder Clyde Allee published several books on the spatial optimization of social and ecological relations and its implications for the origin of sociality. "The first step toward social life in lower animals is the appearance of tolerance for other animals in a limited space," he wrote (Allee 1927, 387). Once organisms were in the presence of other organisms, however, ecological interactions were more or less inevitable. Elsewhere, he was even more explicit, insisting that "there is a general principle of automatic cooperation which is one of the fundamental biological principles" and that "the tendency toward a struggle for existence is balanced and opposed by the strong influence of the cooperative urge" (Allee 1938, 50, 210). Several of Allee's colleagues at Chicago shared his views, but none of them was more enthusiastic than termite expert Alfred Emerson, who trumpeted the organicist perspective with aplomb (Henson 2008). Emerson's research on termites had convinced him that integrated individuals could form still larger individuals and that "these ascending hierarchies of integrated units with their special character-istics form the basis of the concept of emergent evolution." He was obviously not the only one who endorsed emergent evolution, but his understanding of the theory differed from others in one subtle but important way. He glimpsed that natural selection could demarcate biological individuals at various levels, boldly insisting that "natural selection acts upon the integrated organism, superorganism or population" (Emerson 1939, 196).

While organicism proved most popular in the entomological sciences, other fields of study showcased similar trends. At the University of Wyoming, for example, plant ecologist Frederic E. Clements wrote that the tightly governed ecological relationships among different species of plants (and, indeed, ani-mals) were strikingly similar to the physiological bonds that demarcated indi-vidual organisms (Clements 1905; Clements 1916). Like so many others, he emphasized the importance of coordination, writing that "the organization of the family group from the lowest to the highest organisms suggests the extent to which cooperation can be made to overrule competition" (Clements 1949, 36). Not too far away, University of Colorado physician Ivan Emmanuel Wallin hypothesized that the mitochondria found in all nucleated cells were once free-living organisms that had long since integrated with their eukary-otic hosts (Sapp 1994). Noting that all organisms contained mitochondria within their cells, he wrote that "micro-symbiosis appears to be a universal biological phenomenon" (Wallin 1927). Meanwhile, American philosophers Edward G. Spaulding and Roy W. Sellars both endorsed explicitly organicist interpretations of the evolutionary process (Spaulding 1918; Sellars 1922). Finally, citing research on the psychological lives of unicellular organisms, the appropriately named biologist Herbert Spencer Jennings, who trained at Harvard, famously declared that the theory of emergent evolution was the "declaration of independence" for the biological sciences and "a hopeful uplifter" for all humanity (Jennings 1906; 1927, 22, 19; Russell 2015).

Meanwhile, scientists and scholars beyond Harvard and beyond the United States were also interested in the evolution of ethical behavior during the interwar period. In fact, many different people all around the world promoted very similar theories. I have already mentioned Lloyd Morgan (1923), but there were others in Great Britain who articulated much the same (Haldane 1932). For example, Haldane's countryman, zoologist Julian Huxley (grand-son of the aforementioned Thomas Henry Huxley) likewise promoted emergent evolution, insisting that cooperation was evolutionarily significant. Huxley's interest in the subject derived from his experience studying sponges. These organisms were phylogenetically far removed from social insects, yet Huxley observed strikingly similar phenomena. Repeating Henry Van Peters Wilson's famous reaggregation experiment (Wilson 1907), he concluded that sponges are descended from "cells which existed as free-living and independent individuals." Over the course of countless generations, these cells ceded more and more of their individuality to the colony, thereby enabling multicellularity. "Each [cell] preserves a considerable measure of independence," Huxley observed, "and is yet subordinated to the good of the whole. This resulted in the metazoan type of structure, where the individual is built up out of a number of cells instead of one." Significantly, Huxley recognized the exact same process at other levels of the biological hierarchy. "So it comes to pass," he wrote, "that the continuous change which is passing through the organic world appears as a succession of phases of equilibrium, each one on a higher average plane of independence than the one before, and each inevitably calling up and giving place to one still higher" (Huxley 1912, 92, 116). Emergence was not part of the scientific lexicon when he conducted his sponge experiments, but he was among the first to promote the idea during the interwar period (Huxley [1923] 1970). "Intelligence seems to have played as important a part in evolution as brute force, and cooperation has contributed as much as competition," he wrote (Huxley 1942, 87; Phillips 2007; Smocovitis 2008).

Suggestively, scientific interest in the evolutionary origins of cooperation was most pronounced in all those places recently scarred by war. Anne Harrington and Cheryl Logan have each described the surprising popularity of organicist theories in Germany and Austria, respectively, but organicism was equally popular in the Allied nations during the interwar period (Harrington 1996). In 1919, for example, the hot topic in French universities was not the forthcoming Paris Peace Conference that would broker an end to five years of bloodshed, but rather the recent publication of Paul Portier's scandalous biological treatise on microscopic organisms, *Les Symbiotes* (Portier 1918). Portier dared suggest that nucleated cells owed their origin to subcellular organisms that had integrated within one another in the ancient past, though the idea proved highly controversial. In 1924, Russian botanist Konstantin Kozo-Polyansky offered similar hypotheses, suggesting that the chloroplasts in plant cells were once free-living organisms. Accordingly, the cell was not an individual, but rather a consortium. "As we study bacterial consortia, we witness the evolution of the cell," he wrote. In similar fashion, he insisted that each multicellular organism was in actuality a "federation of organisms" (Kozo-Polyansky 1924 [2010], 19, 110). Finally, but perhaps most significantly, the onetime prime minister of South Africa, General Jan C. Smuts, wrote that an innate tendency toward integration suffused the evolutionary process, and he even coined a new word, "holism," to describe this tendency. "Holism … underlies the synthetic tendency in the universe, and is the principle which makes for the origin and progress of wholes in the universe," he wrote. Organisms were not merely individuals, he added, but also well-integrated wholes. "An organism is fundamentally a society in which innumerable members cooperate in mutual help in a spirit of the most effect-ive disinterested service and loyalty to each other," he wrote. This was as true of unicellular organisms as multicellular ones. "In the cell there is implicit an ideal of harmonious cooperation, of unselfish mutual service, of loyalty of each to all" (Smuts 1926, ix, 85).

Alas, just when it seemed that this transcontinental, transdisciplinary philosophy might finally reveal some hidden truth about the evolutionary significance of cooperation, the entire world descended back into the hell-ish depths of war. World War II changed just about everything everywhere. Allied mastery of the atom secured victory over the Axis forces, and wholesale faith in reductionism permeated post-WWII biology no less than post-WWII physics. Scientists and scholars showed little patience for organicist philoso-phies that previous generations had eagerly endorsed. "Although the term superorganism has a venerable pedigree," Peter Corning writes, "it became a pariah among biologists during the middle years of the twentieth century and was widely criticized as an inappropriate, even mystical metaphor" (Corning 2005, 187). Consequently, emergent evolution was deemed too vitalistic for the new world order (even though its practitioners explicitly disavowed vital-ism), and so the theory was grouped with other discredited ideas and ejected from mainstream evolutionary biology. Other events yielded similar results. For example, the evolutionary synthesis that congealed between 1936 and 1947 artfully united Darwinian evolution with Mendelian genetics and fore-told a far more reductionist age in the biological sciences (Mayr and Provine 1980; Smocovitis 1992). It is perhaps worth mentioning that the person who labeled this period of evolutionary biology the "modern synthesis," Julian Huxley, never fully surrendered his organicist inclinations. As late as 1947, by which time most of his colleagues had long since rejected anything that reeked of organicism, Huxley reaffirmed his belief that the evolutionary pro-cess was emergent in nature. "Now and again there is a sudden rapid passage to a totally new and more comprehensive type of order or organization, with quite new emergent properties, and involving quite new methods of further evolution," he wrote (Huxley and Huxley 1947, 120).

By mid-century, however, Huxley and his organicist ilk were in the minor-ity. Science had entered an exceedingly reductionist period, a turn that was especially conspicuous at Harvard, which had once served as the epicenter of holistic theorizing in the biological sciences. The first significant change occurred in 1956, when the Department of Biology added American molecular biologist James Watson to its faculty. Alongside English molecular biologist Francis Crick, Watson had recently discerned the structure of deoxyribo-nucleic acid (DNA), and the dramatic discovery accelerated the biological sciences' wholesale shift toward reductionism (Watson and Crick 1953). In short order, molecular biology began to monopolize most of the department's funding and institutional appointments. Leading the charge was Watson, who was convinced "that biology must be transformed into a science directed at molecules and cells and rewritten in the language of physics and chemistry" (Wilson 1994, 219).

Meanwhile, the very same year that Harvard hired Watson, the Department of Biology also hired American entomologist Edward O. Wilson at the same rank and position. Wilson viewed biology from a radically different perspective. He had first taken an interest in social insects in 1945, when he purchased a copy of Wheeler's *Ants*, which included the author's famous colony-as-organism theory (Wilson 1994, 94). When Wilson enrolled at Harvard to pursue his doctorate in biology, he chose as his advisor paleoentomologist Frank Carpenter, whose own advisor was none other than William Morton Wheeler. For this reason, philosopher Michael Ruse has referred to Wheeler as Wilson's "intellectual grandfather" (Ruse 2011). Indeed, it is difficult to overstate the effect that Wheeler's research had on the aspiring naturalist. "Although Wheeler died in 1937, when I was still a little boy, I have studied his research so closely and heard so much about his life since that I feel as though I also personally knew him," Wilson recently remarked (Wilson 2012, 139). Indeed, Wilson wasted little time endorsing the "superorganism" metaphor in his research. In one essay on the evolution of polymorphism, he observed that "the ant colony behaves as a superorganism, the basic unit upon which natural selection operates" (Wilson 1953, 153–154). His organicist tendencies did not go unnoticed. When Julian Huxley visited Harvard in 1954, he asked to visit Wilson, who was still just a graduate student at the time (Wilson 1994, 313). Aside from Huxley, however, the professional response to Wilson's proclamation was generally tepid.

At the same time that Wilson was in the market for reductionist explanations, another grad student, British biologist William Hamilton, provided the answer. In 1964, Hamilton introduced his theory of "inclusive fitness," which claimed that an organism could improve its chances of genetic success by cooperating with close family members. As evidence he cited ants, whose haplodiploid method of reproduction meant that colony members shared approximately three-quarters of their genes with their siblings (Hamilton 1964a, 1964b). Though Wilson initially dismissed the theory as flawed, he quickly fell under its enchantment (Wilson 1994, 320). He came to the conclusion that inclusive fitness (later relabeled "kin selection") explained insect eusociality with satisfying precision. Ants do not assist one another for any selfless reason, he concluded. Instead, they help the colony because doing so increases the likelihood that some of their genes will survive. In other words, altruism is a gene-centered strategy, and kinship is an essential prerequisite of cooperation. When Wilson published his first major monograph, *The Insect Societies*, he insisted that kinship was "of the greatest importance" (Wilson 1971, 6).

In the last chapter of *The Insect Societies*, Wilson hinted that comparing the evolution of social insects to the evolution of other social organisms might prove useful. Four years later, in 1975, he published *Sociobiology*, a massive book that summarized decades of research on the evolution of social behavior among a variety of radically different life forms, from microbes to mammals, including, significantly, humans. He promoted an explicitly gene-centric model. "The organism does not live for itself," he wrote, "the organism is only DNA's way of making more DNA." Significantly, he credited kinship with providing the initial impetus for sociality. "How can altruism, which by definition reduces personal fitness, possibly evolve by natural selection?" he asked. "The answer is kinship; if the genes causing the altruism are shared by two organisms because of common descent, and if the altruistic act by one organism increases the joint contribution of these genes to the next generation, the propensity to altruism will spread throughout the gene pool" (Wilson 1975, 3).

Wilson's decision to include humans in *Sociobiology* all but guaranteed that the book would generate interest. Sure enough, the sociobiology controversy that erupted in the 1970s was among the most famous academic battles of the twentieth century. Critics complained that Wilson assigned far too much credit to genes, and they accused him of politicizing his science to defend the status quo. In a now famous letter to the *New York Review of Books*, an influential group of scholars wrote that Wilson's work on sociobiology merely sought to exonerate the powers that be from accepting responsibility for social problems. They implied that his science was designed to protect "existing privileges for certain groups according to class, race or sex" (Allen et al. 1975, 43–44). Fanning the controversy, Wilson's most passionate critics were located within his own department at Harvard. Perhaps no one was more vocal in his opposition than Wilson's former collaborator, biologist Richard Lewontin, who accused Wilson of deep "ethnocentric bias." What is more, he dismissed sociobiology as a "caricature of Darwinism" (Lewontin 1976). In response, Wilson was forced to defend his honor, reiterating his published belief that the environment influences human culture more than genes (Wilson 1976). Despite the controversy, biologists largely accepted Wilson's central assertion, that social dynamics (human or otherwise) were best understood within an evolutionary context, and most even accepted his proposed mechanism, kin selection (Segerstale 2000; Jumonville 2002; Alcock 2003; Losco 2011).

Around the same time that Wilson was highlighting the importance of kin selection, the man who first introduced Wilson to the theory, William D. Hamilton, was having second thoughts. His doubts had actually begun in 1968, when an American geneticist named George R. Price contacted him out of the blue to ask him several questions about the purported origin of cooperation. By the time the two men finally spoke on the telephone a year later, Price explained that he had reworked Hamilton's equations to show that genetic relatedness was not an essential prerequisite of sociality (Harman 2010, 223). Applying covariance mathematics to evolutionary dynamics demonstrates that altruism can evolve anytime that between-group selection overrides within-group selection. This insight reframed the entire evolutionary process as a nested hierarchy, each level governed by natural selection. During that initial phone conversation in 1969, Price confessed that his derivation was something of a "miracle" and that he was especially excited about the equation's implications for group selection (Harman 2010, 221). Hamilton initially received this claim with skepticism, but within a few months he acknowledged conversion. "I am enchanted with your formula," Hamilton wrote to Price. "I really have a clearer picture of the selection process as a result. In its general form I can see how we might use your formula to investigate 'group selection'" (quoted in Harman 2010, 227).

Hamilton expounded on the Price equation and its implications for group selection in a very interesting chapter of an edited volume titled *Biosocial Anthropology*. He remarked that he had always resented how Maynard Smith's label, kin selection, had become synonymous with his own label, inclusive fitness. Hamilton insisted that his theory was broader than mere kinship. Citing Price's work on the nature of evolutionary dynamics, he showed that unrelated groups could also conceivably achieve sociality. "It obviously makes no difference if altruists settle together because they are related (perhaps never having parted from them) or because they recognize fellow altruists as such," he wrote (Hamilton 1975, 141). So long as all of the relevant actors possess the requisite genes for socialization, he reasoned, it would not really matter whether or not those genes were obtained via recent shared ancestry. Hamilton insisted that this revelation did not obliterate his notion of inclusive fitness, but rather broadened it, emboldened it, and, most importantly, distinguished it from kin selection.

Wilson estimates that he first began to question the legitimacy of kin selection in the early 1990s. It had always bothered him that kin selection failed to explain eusociality among termites, who reproduced via diplodiploidy, and the discovery of other eusocial species that reproduce via diplodiploidy weakened his faith further still. It was also around this time that he became enchanted by Lynn Margulis's theory of endosymbiosis, which states that the first nucleated cells owed their origin to an ancient symbiosis between prokaryotic partners. Wilson understood that if Margulis's theory were true, it would mean that dissimilar, unrelated individuals can sometimes cohere in evolutionarily stable relationships (Wilson 1992, 245). By 2005, he had begun to openly question kin selection's explanatory power. "It is theoretically possible, and may well occur in nature, that colonies evolve by selective favoring of genes that prescribe group formation with altruistic workers in a manner that has little or nothing to do with kinship," he wrote (Wilson 2005, 163). By 2008, he had grown more confident, coauthoring multiple papers with fellow biologist David Sloan Wilson (no relation) and throwing his full support and celebrity behind a new explanation, "multilevel selection theory" (Wilson and Wilson 2007, 2008).

While these papers generated murmurs among biologists, the controversy did not fully erupt until the summer of 2010, when Wilson and fellow Harvard biologists Martin Nowak and Corina Tarnita published an article in *Nature* that rejected kin selection in dramatic fashion. "Relatedness does not drive the evolution of eusociality," they baldly asserted. Citing more than forty pages of mathematical equations as evidence, the trio hypothesized that "[n]atural selection targets the emergent traits created by the interactions of colony members" (Nowak et al. 2010, 1059–1061). In response, more than 100 biologists have signed letters reaffirming the importance of kin selection, effectively rejecting Wilson's rejection (Abbot et al. 2011). Significantly, though perhaps not surprisingly, several of the signatories worked at Harvard. When Wilson expanded his views in *The Social Conquest of Earth*, Oxford biologist Richard Dawkins wrote a disparaging review entitled "The Descent of Edward Wilson" (Dawkins 2012; Wilson 2012). In response, Wilson has started referring to Dawkins as a "science journalist" both in public and in print (Wilson 2014). The controversy has long since spilled out of academia and into the public sphere, where it has been described as a "scientific gang fight" (Jensen 2010).

Multilevel selection resembles emergent evolution in at least one important way. Both theories examine the evolution of organisms at different levels of the biological hierarchy, though it should be noted that Wheeler expressed ambivalent feelings about using the word "levels" to describe different grades of emergence. "These sections have been called levels," he wrote. "The word is not very apt since it conveys a spatial and static metaphor, whereas emergents must be regarded as intensively manifold spaciotemporal events" (Wheeler 1928, 22). Structural similarities notwithstanding, multilevel selection differs from the organicist theories that preceded it in important ways. Spencer believed that the evolutionary process was hierarchical in nature, and he therefore intuited the most basic aspect of multilevel selection theory. Nevertheless, his insistence that organisms inherit their ancestors' acquired characters meant that he never fully grasped natural selection's pervasive agency. Wheeler likewise intuited evolution's hierarchical process, and his grasp of ecological physiology was second to none. Even so, he always regarded natural selection as a purely destructive force, one that could cull, but not create, variation. He credited trophallaxis with initiating sociality among ants, a decision that helped distance him from neovitalists, but he never provided a convincing mechanism to explain the social impetus in other organisms at other levels. Thanks to Hamilton's theory of inclusive fitness, Wilson long assumed that kinship was the answer and that genetic relatedness explained the origin of sociality. According to this view, natural selection targets the gene exclusively, and apparent selection at other levels is merely illusory. While many remain intransigent, Wilson is now convinced that natural selection, not kin selection, provides the initial impetus for sociality and that this is true at every level of the biological hierarchy.

It is perhaps ironic that the scientific debate over the evolutionary origins of ethical behavior remains in a state of near-constant turmoil, but it should no longer come as a surprise. Ever since Darwin first articulated natural selection more than 150 years ago, biologists and philosophers have struggled to explain the precise mechanisms that facilitate cooperation among disparate parts, and, as a result, the field of sociobiology has remained in a state of near-constant controversy. Many biologists agree with Wilson that multilevel selection provides the most likely explanation for the origin of eusociality, though at least as many continue to endorse kin selection. It remains to be seen whether multilevel selection can actually explain the evolution of ethical behavior and, thus, whether it can succeed where so many theories have failed. For the time being, the field remains, as ever, in turmoil.

Notes

1 My use of the world "group" here is deliberate, though many scientists and scholars would no doubt prefer that I use the word "family." There is sharp disagreement over whether or not Darwin considered kinship a prerequisite for sociality. For examples, see Dugatkin 1997; Borrello 2005; Eldakar and Wilson 2011; Gardner 2011.

2 Appropriately, Wheeler's essay appeared in a special issue of the journal dedicated to Wheeler's longtime mentor Charles Otis Whitman, who had once inverted Spencer's famous metaphor about societies being organisms, asking if organisms should also be considered societies (Whitman 1891, 1894; Wheeler 1911).

Part II

For Evolutionary Ethics

6 Darwinian Evolutionary Ethics

Michael Ruse

What Is Evolutionary Ethics?

Some ideas are not simply wrong; they are morally and aesthetically rather grubby. You know that people who push them almost certainly have issues of one sort or another – generally with authority or, more specifically, with certain racial groups or some such thing. For the first 100 years after Darwin published his *Origin of Species* in 1859, most Anglophone philosophers felt very much that way about evolutionary ethics, the attempt to explain and justify moral feelings and behaviors on the basis of our simian pasts (Ruse 2009a, b). Thus in the first year of the journal *Mind*, we find the noted utilitarian philosopher Henry Sidgwick (1876) arguing that when it comes to ethics, evolution is just not that relevant. His student G. E. Moore (1903) famously made evolutionary ethics the paradigmatic example of wrong-headed arguments, the worst of all possible ways of committing the so-called naturalistic fallacy. And his student C. D. Broad (1944) was still at it thirty or forty years later.

Prima facie this is all a little bit odd, because everyone recognizes that the world after Darwin is very different from the world before Darwin (Ruse 2017). Even those who went on being religious saw that much, and those who had already jettisoned conventional religious belief saw only confirmation that the path they had taken was correct. Surely, therefore, evolutionary theory in general and Darwin's theory of evolution through natural selection in particular were going to have some major implications, and nowhere more than on the ways in which we think and behave about right and wrong? And yet, not so. There are various reasons why evolutionary ethics found little favor among the philosophers (Cunningham 1996). Whether these reasons are good or bad, they are matters of history and dealt with in detail in other contributions to this volume, so I will not dwell on them here. The question here is with the revival of evolutionary ethics and the positive case that can be made.

I should say that I see the main issue is that of getting over the Humean is/ ought distinction, of which I take the naturalistic fallacy to be a variant – the claim that you cannot derive matters of value from matters of fact – and I see two different approaches to the problem. One favored by my coeditor Robert

J. Richards (this volume) that I think can be found in Herbert Spencer (1892) and then in a chain including Julian Huxley (1943) – the biologist grandson of Thomas Henry Huxley and older brother of the novelist Aldous Huxley – down to Edward O. Wilson (1975, 1978). Here, the is/ought distinction is simply denied or downgraded. It is argued that the world itself has value – not just the forests and the fish in the ocean, but the very mountains and lakes and seas – and so almost expectedly human values emerge through the evolutionary process. It is a view going back through the German Romantics to Plato, particularly the *Timaeus*, and it is a view coming forward to Rachel Carson and *Silent Spring* and more recently James Lovelock and the Gaia hypothesis (Ruse 2013). I suspect that nine-tenths of my fellow members of the Sierra Club subscribe to it. And although I do not myself accept it, no longer do I think it silly or necessarily philosophically crude. A case can be made, is made very ably by Richards, and I shall leave things at that. Or rather, I shall leave it to other contributors to this volume to make the contrary case and leave things at that.

The other approach is to accept fully the is/ought distinction but (in the language of sports) to do an end run around it. If I say that this is the position found in Darwin that is really only a half-truth because I don't think that Darwin is interested in the problems of philosophers, particularly those of justification, then it is at least a half truth because the position does depend heavily on Darwin's thinking, especially that of the *Descent* (1871). Hints can be found in later writers, particularly those of the paleontologist George Gaylord Simpson (1964), and then later the philosopher (and my fellow student at the University of Rochester) Jeffrey Murphy (1982). I think J. L. Mackie (1978, 1979) was going in the same direction but he died before he could articulate his thinking fully. I myself set out as a fervent disciple of Moore, a position I articulated in *Sociobiology: Sense or Nonsense* (1979b). However, a detailed discussion review of that book by Mackie (1980) – gratifyingly favorable for a totally unknown philosopher that yet took me to task on biology and morality – set me thinking in new ways. Taking advantage of the fact that I was unknown and hence had no reputation to lose, I announced my new position in an article in the science-religion journal *Zygon* (1986) and then in a book *Taking Darwin Seriously: A Naturalistic Approach to Philosophy* (1986). I should say that the responses were so gratifyingly awful that I collected them together in a collage, making it the backdrop for invitations to my fiftieth birthday party – "Come and celebrate fifty years of unbroken success" – and sent copies to all of my critics. I should say that most of them responded in the good-humored way that the offer was extended. Already people were moving from ignoring me to writing refutations in journals that would never accept anything by me.

The case I make for evolutionary ethics is simple – too simple say the critics. From Darwin on it has been virtually a truism that evolution by natural selection promotes "altruism." By this it is understood that the key to evolutionary success is adaptation – features that help their possessors to survive and reproduce – that behavioral features are as important as physical features, and while at times strife and combat may be good adaptive strategies, often cooperating pays major dividends. Half a cake is less than the whole cake but better than no cake at all. It is worth noting that the 1960s saw a quantum leap in interest by evolutionary biologists (all Darwinians) in social behavior and a number of powerful models to explain "altruism" were devised (Ruse 1979b). These included "reciprocal altruism" – you scratch my back and I will scratch yours – an idea with roots in the *Descent*, and "kin selection" – help to relatives rebounds vicariously with the success of your own shared genes – an idea not found in Darwin because it requires understanding of modern genetics.

Note that I put "altruism" in quotes because this is not necessarily literal altruism – Mother Teresa altruism, where people consciously try to do the right thing. It extends to all social behavior of a reciprocal kind and indeed the paradigm examples are the hymenoptera – the ants, the bees, and the wasps – and no one thinks these creatures to be reflective. However, the claim is made by evolutionary biologists – starting with Darwin in the *Descent* – that genuine altruism is something promoted by natural selection to make us humans good "altruists." "It must not be forgotten that, although a high standard of morality gives but a slight or no advantage to each individual man and his children over the other men of the same tribe, yet that an advancement in the standard of morality and an increase in the number of well-endowed men will certainly give an immense advantage to one tribe over another" (1, 166). Hence, "There can be no doubt that a tribe including many members who, from possessing in a high degree the spirit of patriotism, fidelity, obedience, courage, and sympathy, were always ready to give aid to each other and to sacrifice themselves for the common good, would be victorious over most other tribes; and this would be natural selection" (1, 166). And so it follows that, "At all times throughout the world tribes have supplanted other tribes; and as morality is one element in their success, the standard of morality and the number of well-endowed men will thus everywhere tend to rise and increase." Although it does not affect the discussion here, I should note that Richards and I differ over the interpretation of "tribe" (Richards and Ruse 2016). I take tribes to be groups of interrelated humans and, hence, Darwin is promoting a kind of proto-kin selection – only proto because he didn't have genetics – whereas Richards thinks that tribal members need not be related and Darwin is invoking selection at the level of the group (without specification of relatedness).

So much for the basic science. Since I started arguing for evolutionary ethics thirty years ago, so-called evolutionary psychologists have done much work on the evolution of the moral sentiment and its behavioral manifestations (Ruse 2009a). Much of this work is referred to elsewhere in this volume and so I will leave things at that. I think it very important and very interesting, but it does not affect the basic case I am making. I want now to turn to the philosophical questions that – in the spirit of Philosophy 100 – I take to be two. First, there is the question of substantive or normative ethics: what should I do? Second, there is the question of metaethics: why should I do what I should do? To take a clear example of an ethical system, Christianity, at the substantive level, we have dictates like "love your neighbor as yourself," and at the meta level (the *Euthyphro* problem notwithstanding), we have appeals to God's Will – "you ought to behave morally because this is what God wants."

Darwinian Substantive Ethics

In the case of the Darwinian approach to evolutionary ethics, at the level of substantive ethics, I don't think there is much cause for concern or special thinking. It is often believed that traditional evolutionary ethics – so-called Social Darwinism – promotes attitudes favorable to warfare and extreme laissez faire economics – "widows and children to the wall and let the robber barons take all." But although there are certainly instances of such writing, by and large the prescriptions are much more inclined to cooperation and helpfulness (Russett 1976; Richards 1987; Ruse 2009a, 2009b). Among the robber barons, John D. Rockefeller gave huge amounts of money to the fledgling University of Chicago, and Andrew Carnegie supported public libraries, where the poor but gifted child could go and learn. In any case, traditional evolutionary ethicists tend to fall more into the Spencerian camp, where what evolves is bound to have value, and so if there is a harshness to the prescriptions, it is thought that this is simply being realistic about what is possible, given human nature.

The more Darwinian approach is basically going to be one of common sense morality – help others, try to avoid cheating by yourself and others, and so forth. Widows and children deserve more attention and help than prosperous businessmen. Interesting, the most influential American ethicist of the second half of the twentieth century, John Rawls (1971), picked up on evolutionary biology and suggested that his position is one that emerges from the evolutionary process. It will be remembered that he argued for "justice as fairness" and, to achieve fairness, invited us to put ourselves behind a "veil of ignorance," where we would not know where we stood in society. If we knew we were going to be healthy, white females, then we would promote the well-being of healthy, white females, but it might be that we are sickly, black males, and then where would we be. The fair society is the one that maximizes the benefits for all, or at least minimizes the lack of benefits for all. This does not necessarily lead to equality – a fairer society might well be one where we pay our football coaches ten times what we pay our university presidents – but it is one where all benefit in some sense as much as possible. Rawls agreed that it was most unlikely that a group of elders set up such a social contract in the past, but he thought it likely that natural selection working on the genes might have done such a job.

In arguing for the greater stability of the principles of justice I have assumed that certain psychological laws are true, or approximately so. I shall not pursue the question of stability beyond this point. We may note however that one might ask how it is that human beings have acquired a nature described by these psychological principles. The theory of evolution would suggest that it is the outcome of natural selection; the capacity for a sense of justice and the moral feelings is an adaptation of mankind to its place in nature. As ethologists maintain, the behavior patterns of a species, and the psychological mechanisms of their acquisition, are just as much its characteristics as are the distinctive features of its bodily structures; and these patterns of behavior have an evolution exactly as organs and bones do. It seems clear that for members of a species which lives in stable social groups, the ability to comply with fair cooperative arrangements and to develop the sentiments necessary to support them is highly advantageous, especially when individuals have a long life and are dependent on one another. These conditions guarantee innumerable occasions when mutual justice consistently adhered to is beneficial to all parties. (Rawls 1971, 502–503)

Rawls only went so far with Darwin. Here he was writing about the origin of morality. He was still writing about substantive ethics. When it came to the justification of morality – metaethics – he pulled back. "These remarks are not intended as justifying reasons for the contract view" (Rawls 1971, 504).

Does this kind of naturalistic approach to substantive ethics make everything somewhat relative? The answer is yes and no. In an important sense, substantive ethics is very much a function of human nature, which means that if human nature were other than it is, ethics would be other than it is. Darwin put his finger on precisely this point. "If, for instance, to take an extreme case, men were reared under precisely the same conditions as hive-bees, there can hardly be a doubt that our unmarried females would, like the worker-bees, think it a sacred duty to kill their brothers, and mothers would strive to kill their fertile daughters; and no one would think of interfering." He continued: "The one course ought to have been followed, and the other ought not; the one would have been right and the other wrong" (Darwin 1871, 73–74). I don't see this as much cause for concern or regret because this is what you get with a naturalistic ethics, including, let us note, Thomistic natural law ethics. If human sexual organs had been other than they are, then all of the prohibitions about masturbation and anal intercourse would be otiose. Imagine if men could only ejaculate if first the penis was stimulated by saliva from the female's mouth or some such thing? And before you think that these are just the dirty-minded fantasies of an aged philosopher, I would point out that there are all sorts of even weirder cases in nature, for instance, necessitating group sex with other species. So much for the immorality of bestiality. (Check out the mating behavior of the Amazon molly for this one.)

All of this being said, note that no one is claiming that you can just do what you want to do. If it feels okay, then it is okay. The whole point about cooperation is that everyone has to be in it together, otherwise it does not work. It is the same as with language. Because I come from the English middle classes, I speak English better than anyone I know – with the possible exception of the Royal Family. But living down here in the American South, it frequently doesn't do me much good because people cannot understand a word of what I am saying. So I have to slow down and start again. There has to be some basic equivalence whether in language or in morality. It may all be subjective, but it cannot be completely relative.

Does that mean taking (what I am calling) a Darwinian approach to substantive ethics makes no big difference to what we already believe or claim? Here are three points to think about. First, some ethicists – Peter Singer (1972) springs to mind – argue that we have equal obligations to all and some would even extend this to nonhumans. Darwinians would certainly never say we have no obligations to all and any humans, and perhaps even some nonhumans, but there would be a differential. You are surely going to feel stronger obligations to family and then to friends and acquaintances and only finally to strangers. Good Samaritans are to be praised, but the feelings of the father toward the prodigal son are primary. This is not a new insight of Darwinism. Although I do not think him an evolutionist, it is right there in David Hume.

In like manner we always consider the natural and usual force of the passions, when we determine concerning vice and virtue; and if the passions depart very much from the common measures on either side, they are always disapproved as vicious. A man naturally loves his children better than his nephews, his nephews better than his cousins, his cousins better than strangers, where everything else is equal. Hence arise our common measures of duty, in preferring the one to the other. Our sense of duty always follows the common and natural course of our passions. (Hume 1739, 483–484)

Nor is it a sentiment confined only to nonbelievers. In *Bleak House* (1853), the good Christian Charles Dickens makes a similar point. He is totally scornful of the philanthropist Mrs. Jellyby, who devotes all of her attention to the well-being of an African tribe – the "Borrioboola-Gha venture." In so doing, she ignores the needs of her own children and husband, not to mention the needy members of her own society, notably Jo the crossing-sweeper.

Second, I do not expect to find that substantive morality always works in a totally rational fashion. The name of the Darwinian game is being better than the competitors, not being perfect in some absolute sense. I fully expect to find tensions, especially with the kinds of artificial cases so beloved of philosophers. Suppose you are a prisoner of war in Germany faced with the dilemma of bribing a guard to be able to escape. On the one hand, you have vital information needed by the allies. Utilitarianism suggests bribe. On the other hand, Kantianism deplores the way in which you are using the guard as a means to your ends and therefore suggests no bribe. Generations of philosophy students have sweated over how best to answer this paradox. My sense is that biology approves of maximizing happiness and it also approves of treating people as worthwhile in their own right. Generally these do not come into conflict. By and large, most of us do not spend our days in prison camps worrying about the morality of corrupting guards. Normally, these things bring on no big conflict. I help a student – I maximize her happiness and I treat her as a person. But sometimes at the edges they do conflict and we just have to live with that. Generally, expectedly, the problem cases are not common. Remember that when we do have complex moral problems, often it is because the science or circumstances are complex, not the morality. I should save life when I can; the problem is about the right course of treatment.

I suspect other notorious paradoxes are open to a similar robust dismissal. The trolley paradox, for instance, where we might happily pull a switch to save six people over one but would not throw the one on the line to save the six, seems to me to be such a case. Formally the two situations are the same. It is just that our biology makes us easier about pulling switches because they are not part of our evolutionary past. Not being mean to neighbors, even when the end result might be better if we are, is part of our biology. Too often, the ends turn out anyway not to be quite what we thought they might be. So there just isn't a definitive rational solution. It is more a matter of muddling through, which has obviously on average worked pretty well in the past.

One third point: if we really are the end point of a long, slow, law-bound process of evolution through natural selection, rather than the favored creation of a Good God at the end of the Sixth Day, then – given its crucial importance in everyday life – it really ought to make some difference to our moral understanding. Keen to stress the respectability of Darwinian ethics, for many years I downplayed such possibilities. Now, having written a book that focuses on the literary response to Darwinism, I think I underplayed (actually, totally overlooked) the extent to which, in a Darwinian world, vigor and success are valued (Ruse 2017). I don't mean now a reversion to crude Social Darwinism, but that while it is obviously not enough on its own, being prepared to "have a go" is thought morally admirable. The Darwinian has very mixed feelings about Jesus's claim: "Blessed are the meek." No one wants to deny that the meek can have great moral worth, but seeking out meekness, as it were, is not that admirable. A novel like George Gissing's *New Grub Street* makes this point very clearly. One writer (Edwin) is very talented but basically isn't prepared to make the effort to succeed. He ends up dead. The other chap (Jasper) is far less talented but with a zest for life.

"You hear?"

Marian had just caught the far-off sound of the train. She looked eagerly, and in a few moments saw it approaching. The front of the engine blackened nearer and nearer, coming on with dread force and speed. A blinding rush, and there burst against the bridge a great volley of sunlit steam. Milvain and his companion ran to the opposite parapet, but already the whole train had emerged, and in a few seconds it had disappeared round a sharp curve. The leafy branches that grew out over the line swayed violently backwards and forwards in the perturbed air.

"If I were ten years younger," said Jasper, laughing, "I should say that was jolly! It enspirits me. It makes me feel eager to go back and plunge into the fight again."

"Upon me it has just the opposite effect," fell from Marian, in very low tones. (Gissing 1891, 63)

In the end, Marian loses out, and Jasper ends up with Edwin's wife and the editorship of a prestigious journal and is able to move forward and contribute to society. It isn't just a matter of success, but a feeling that, if you are going to do good, you had better have the energy and drive to do it properly. In *The House of Mirth* (1905), the American novelist Edith Wharton chronicles the decline of her heroine, not a bad woman at all, and in respects genuinely kind, but unable to make decisions and in the end the victim of a (possibly suicidal) overdose. Wharton doesn't exactly despise the heroine, but rather looks down on her. Wharton herself exemplified her own philosophy, in the Great War doing huge amounts for Belgian and French refugees, but driven by her own energies and refusals to take no for an answer. Of course the Bible exemplified this point fully. Think King David, for example. Or Judith, if you want a warrior or Ruth if you want someone faithful but with real guts and devotion. She was not about to slink back to Moab.

Darwinian Metaethics

What of justification? What of metaethics? Here's the rub. If the is/ought barrier is impenetrable, then you simply cannot get justification out of the evolutionary process or its products. And if you refuse to look elsewhere – Rawls for instance is a Kantian thinking that morality emerges as a necessary condition for rational interaction – then there is no justification. The point is that the Darwinian does not take this as a mark of failure, but the starting point. There is no objective justification of substantive ethics. One is an ethical skeptic, meaning not skeptical about substantive ethics but skeptical about justifications – skeptical to the point of nonbelief.

In a way, of course, this is not so very radical a conclusion. The emotivists, analyzing moral sentiments in terms of emotion – "I don't like rape. Boo hoo, don't you like rape either" – were also moral skeptics in this sense. In a moment, we shall see a bit of a difference with emotivism, but leave this for a moment. First, let us turn to the obvious objection (Nozick 1981). "You have derived substantive ethics from the evolutionary process," says the critic.

These are basically scientific arguments and let us for the sake of argument grant you these. You still have not shown us why ethics has no foundation. Take the comparable case of epistemology. You want to argue that the claim that rape is wrong has no objective foundation because our beliefs that rape is wrong emerge from the process of natural selection. Why should we not equally argue that the speeding train bearing down on us has no objective foundation because our beliefs in the train emerge from the process of natural selection? Those that believed in trains survived and reproduced and those that did not, did not.

There is a difference. All roads lead to trains. If evolution does not make you believe in trains, you are going to die before reproducing. You might not know of the trains in the way that we do. Perhaps we could use a kind of radar like the bats. But in the end, trains win out. For moral claims, like love your neighbor, it is not the same. There is no absolute direction to evolution through selection. The equivalence to trains is not the love commandment but some kind of cooperation. If you could get the cooperation in a different way, then so be it. But suppose that, instead of "love your neighbor," you had "hate your neighbor but recognize that your neighbor thinks it a sacred duty to hate you, so you had better get along." This might well work, and indeed I call it the Dulles system of morality, after Eisenhower's secretary of state (John Foster Dulles) who hated the communists but knew they felt the same way about him and so cooperated right through the 1950s at the height of the Cold War. Perhaps you don't have any morality at all. You do it all on game-theoretic reasoning. Although whether this is practically viable is another matter. Morality may not always work but it gives you a quick and dirty solution to the cooperation problem and that is enough to go on.

There is nothing sacred about the way we have evolved any more than there is anything sacred about having five digits rather than eight (as was the case with some early vertebrates). So there is nothing sacred about fixing on to "love your neighbor" rather than "hate your neighbor." Hence, it seems now that you are faced with two options. Either you go straight to arguing that substantive morality has no backing. It is, as I have called it, an illusion of the genes put in place by natural selection to make us good cooperators (Ruse and Wilson 1986). To make us "altruists," nature has made us altruists. Or you say that there exists an objective morality but as like as not we don't know about it! Perhaps morality demands that we hate our neighbors but we deluded fools think that we should love our neighbors. In my opinion, if it is possible for humans to live full and satisfying lives doing the very opposite of what is demanded by objective morality, this gets pretty close to being a reductio ad absurdum of objective morality.

What about the objection of mathematics? It too seems nonnatural like substantive morality, yet we not only believe in it but also think that it could be the true mathematics. It is not an illusion of the genes, for if we don't obey it, we will be in big trouble. In response, I would reject mathematical Platonism, thinking mathematics exists out there independent of our thought or very existence. I just don't see how you could link up human brains with such a world of pure rationality. So at most (and at least), mathematics has to be something like relations between objects or some such thing. In other words, like morality, mathematics can be used but it doesn't refer to anything – meaning any (objective) thing. The difference with mathematics is that I am not sure that alternative mathematics (akin to Dulles morality) works as well as the mathematics we have now. If there are such alternatives, mathematics is no objection, and if there are not, then the parallel between mathematics and morality breaks down.

So stressing what I think the critics often miss, the nondirectionality of Darwinian evolution, I argue that there are no reasons to think that we have homed in on an objective morality and good reasons to think that there is no such morality to home in on. Of course, I have taken a somewhat Platonic stance on the objectivity of morality, namely that it exists outside our perceptions – the falling tree in the forest when no one is around – in the sense of rational entities as suggested by Plato or the will of an objective God or some such thing. What if you argue that the objectivity is more Kantian, not existing independently but as the necessary conditions of cooperative behavior and living? I have less of a quarrel here although I still don't think it works. I would concede that perhaps all morality ultimately must share the same formal structure of cooperation. If there is no reciprocation, it will not work. But as Kant (1959) himself pointed out, this is not enough for morality. We need something filling it out, something referring to human nature. And it is here that I argue that both Christianity and the Dulles form of morality would do the job, and there is no reason to think that the evolutionary process will necessarily lead us to one option rather than the other. So my conclusion holds.

Why, finally, is such a strong and sensible argument of mine so difficult for so many intelligent people to grasp? Simply because their biology is working flat out to make my argument seem unconvincing. Go back to emotivism and ask why so many people – and I was very much one here – found it not just false but somehow rather immoral. Rape really is wrong, and not just a matter of emotions and directives. If it isn't objectively wrong, then it wasn't wrong for those without those emotions. Russian soldiers raping German women when they invaded East Prussia. So what is going on here? Morality has no foundation and yet I criticize philosophies that argue that morality has no foundation. The crucial point is that, although morality has no foundation, our biology makes us think that it has. To use an ugly term of Mackie, we "objectify" morality. The meaning of morality incorporates objectivity. "Rape is wrong" means it really is truly, objectively wrong to rape. It is not a matter up for grabs – we may not always obey morality but that is another matter. Its meaning is that transgression is wrong.

There are obvious reasons why biology would make us think this way. If we did not, if we simply could puzzle out that morality has no foundation, we would start to ignore it and soon others would too and before long it would break right down. So selection has tweaked the meaning of morality. I refer to it as the Raskolnikov problem, for it shows that what I have told you will not at once make you free to rape and pillage. In *Crime and Punishment* the young student murders for gain. The detective knows that he has done it but waits until he confesses himself. The truth does not set you free, or at least it does in one sense (confessing) but not in another (recognizing that morality has no foundation). We are biologically disposed to think morality objective, and so, even if philosophically we can puzzle out otherwise, we cannot live by denying our human nature. As always, Hume is ahead of us here. You worry that morality has no ultimate basis?

Most fortunately it happens, that since reason is incapable of dispelling these clouds, nature herself suffices to that purpose, and cures me of this philosophical melancholy and delirium, either by relaxing this bent of mind, or by some avocation, and lively impression of my senses, which obliterate all these chimeras. I dine, I play a game of backgammon, I converse, and am merry with my friends; and when after three or four hours' amusement, I would return to these speculations, they appear so cold, and strained, and ridiculous, that I cannot find in my heart to enter into them any farther. (Hume 1739, 175)

Psychology trumps philosophy!

Conclusion

I once read a book in which the author stated in the preface with the sincere hope that nothing he was about to say was original. I am not quite that modest, but I know what he meant. Ultimately, it is all footnotes to Plato. More immediately, I see Darwinian evolutionary ethics as very much in the British empiricist tradition. It is, as I like to say, David Hume brought up to date by Charles Darwin. I am proud to be their spokesman.

7 Human Morality: From an Empirical Puzzle to a Metaethical Puzzle

Richard Joyce

Human Moral Thinking: An Empirical Puzzle

It is a striking feature of human psychology that we assess the world in moral terms. We do not merely like and dislike things – we judge certain actions to be obligatory, certain people to be wicked, certain states of affairs to be deserved, and so on. This is a phenomenon that cries out for explanation, for it seems no great feat of imagination to envisage an intelligent social species that does nothing of the sort. Picture a community of beings who are inclined to cooperate with each other (albeit conditionally) because they really like each other, who positively enjoy laboring in service of their fellows' welfare, in whom natural selection has simply expunged any temptation to defect. These beings may not even have the cognitive capacity to grasp our basic moral concepts; maybe we couldn't explain *just deserts* to them if we tried. Such creatures would be quite different from humans, obviously, though it is difficult to say with confidence how different. Our sense of fellow feeling is fainter than theirs, our pro-social resolve more susceptible to undermining temptations, and, of course, we have the capacity that they lack: to wield the peculiar suite of conceptual tools necessary for moral judgment.

What processes have brought it about that we humans are as we are and not like these imaginary creatures? What explains our unusual tendency to evaluate our world, each other, and ourselves in moral terms? The answer "natural selection" is almost certainly correct, but it's a coarse-grained answer that obscures much of significance. Natural selection is the process responsible for giving us a brain capable of moral thinking (compared, say, to a wombat's brain), but that is not to say that the capacity for moral thinking is an adaptation. It is possible that, rather, human moral thinking emerged as a by-product of psychological capacities that evolved for other purposes. By analogy, if one were wondering why humans can drive cars but wombats cannot, then "natural selection" is in some sense the right answer, but it doesn't follow that the capacity to drive cars was selected for somewhere in the hominin lineage. Rather, our ability to drive cars supervenes on various traits (largely lacked by wombats) that evolved for other fitness-pertaining reasons – such as visual perception, hand-eye coordination, cognitive mapping, etc.

Distinguishing a trait that is an adaptation from a trait that is a by-product of adaptations is a difficult business, for no amount of careful examination of the intrinsic features of an organism can settle the matter. Consider Stephen Jay Gould's architectural analogy of spandrels: triangular regions of a church's wall that are the inevitable outcome of mounting a dome on a ring of arches (Gould and Lewontin 1979); the spandrels are not a design feature but a by-product. However, we can imagine an eccentric architect inspired by a vision of beautiful spandrels; in the resultant building it is the arches and the dome that are the by-products. Picture the two churches side by side: one for which the spandrels are the by-product, one for which they are the primary design feature. The two churches could be brick-for-brick identical; determining which features are by-products requires discerning something of the architect's intentions. In the case of Darwinian evolution, determining whether a trait is an adaptation or by-product requires discerning something of the selective forces that were operational during the period of the trait's emergence. And the problem is, of course, that in the latter case it is usually more reasonable to speak of "educated guessing" than "discerning."

Things would stand differently if we had evidence that the trait emerged relatively recently – too recently for natural selection to operate – for then we could rule out the possibility of its being a genetic adaptation. But we have no such evidence in the case of human moral thinking. As far back as the historical record stretches we find moral judgments. The Egyptian *Book of the Dead* includes the words spoken by the deceased to Osiris: "I have not sinned against men.... I have not wrought evil" (Wallis Budge 1901, 360). The sixth tablet of the *Epic of Gilgamesh* recounts Ishtar describing actions in terms that are translated variously as "evil," "despicable," "foul," or "abominable." There is no evidence to suggest that we should not project human moral thinking back far earlier into prehistory than these first-recorded instances.

Thus we are really in no position to explain moral thinking with any confidence. Note that I am not referring to the explanation of particular moral judgments or moral systems. One might, for example, be able to explain why the ancient Egyptians thought sibling incest morally permissible (perhaps even in some circumstances obligatory) whereas modern Western cultures tend to find it morally repugnant. As to the question "Why does Ptolemy judge incest to be morally permissible?" a reasonable answer may well be "Because he was taught this." I, however, am scrutinizing the matter through a wider lens than this, one focused on explaining the trait that these different cultures and individuals share: both we and the ancient Egyptians (and every other known culture) employ a conceptual framework that categorizes actions in terms of (inter alia) permission, obligation, and repugnance. Particular moral systems are no doubt shaped by various historical and environmental factors and learned by individuals; the question is whether there is an evolved psychological capacity dedicated to this kind of learning. (After all, wombats can't learn moral systems.) To the extent that we do not know whether this more general trait is an adaptation or a by-product, our ability to explain human moral thinking remains fundamentally incomplete.

The difficulties inherent in distinguishing adaptations from by-products have not prevented plenty of plausible speculation on the genealogy of human moral thinking. Some, including myself, have advocated the nativist hypothesis that mechanisms dedicated to moral evaluation emerged as psychological adaptations in the human lineage (see Ruse and Wilson 1986; Dwyer 2006; Joyce 2006a; Mikhail 2011). Others have argued that moral thinking is a by-product of psychological mechanisms that evolved for other purposes (see Nichols 2004; Prinz 2007, 2008, 2009; Ayala 2010; Machery and Mallon 2010). It is not my intention on this occasion to attempt to adjudicate this complex and tricky debate; I'm inclined to think that the evidence necessary to do so remains unavailable. Rather, what I shall do is make some very broad comments about what both sides of the debate seem to have in common and then, on this basis, move the discussion into the province of metaethics.

Genealogical Hypotheses Sketched

What advantage did moral thinking provide our ancestors? If moral thinking is a genetic adaptation, then it is the product either of individual selection or group selection (assuming the two are distinct). If the former, then moral thinking provided its bearers with some fitness-enhancing benefit (relative to the competition). If the latter, then moral thinking provided its bearers' relevant group with some fitness-enhancing benefit (relative to the competition).

One can apparently find the latter view expressed by Darwin:

Although a high standard of morality gives but a slight or no advantage to each individual man and his children over other men in the same tribe, ... an advancement in the standard of morality will certainly give an immense advantage to one tribe over another. (Darwin [1879] 2004, 157)

I say "apparently" because Darwin's mention of "a high state of morality" is vague (for our purposes); it is unclear at this point in his discussion whether he is targeting the trait of moral *judgment* in particular.[1] This is often what one finds in those who advocate a group-selectionist account of the evolution of "human morality" (e.g., Gintis et al. 2008; Krebs 2011; Wilson 2012): they turn out to be attempting to explain human *cooperation* or human *altruism* – neither of which necessitates the capacity to evaluate the world in distinctively moral terms such as obligation, wickedness, desert, and so on. (Remember that the amoral creatures described earlier were cooperative and altruistic.) Nevertheless, advocates of such views may well assent to the natural thought that moral thinking operates largely in the service of cooperation. In other words, if certain forms of human cooperative behavior emerged via a process of group selection, then plausibly moral thinking emerged as a proximate mechanism for governing that behavior. This provides an immediate rough answer to the question posed at the start of this section: the advantage that moral thinking provided our ancestors was that it encouraged certain group-adaptive forms of cooperation.

A similar answer is also widely endorsed by those who eschew group selection and seek to explain the evolution of morality in terms of individual selection (see Ruse 1986; Frank 1988; Irons 1996; Joyce 2006a, 2006b; Churchland 2011): here again it is widely accepted that the adaptive function of moral thinking was its contribution to social cohesion. Individual-selectionists face the additional challenge of explaining how cooperation produces benefits that accrue *to the individual* – in particular, they face the problem of why a disposition to defect would not better serve the individual – but evolutionary biology has made great strides in resolving this challenge in recent decades. Kin selection, reciprocity, and mutualism (inter alia) are processes by which individual-level selection can favor forms of cooperation, and therefore to the extent that moral thinking can contribute to these forms of cooperative behavior, we have a broad answer to how moral thinking might have been advantageous to our ancestors on an individualistic basis.

There remains an important puzzle as to *how* moral thinking contributed to cooperative behavior. The puzzle can be framed by asking why humans didn't evolve simply to *want* to cooperate (albeit conditionally). One can answer by saying that humans did evolve to want to cooperate (albeit conditionally), and moral thinking is one special psychological mechanism dedicated to regulating this desire. Conceiving of an action as one that "must be done whether I like it or not" – as opposed to categorizing it simply as furthering one's desires – is likely to strengthen one's resolve to perform it; the former is less prone to self-sabotaging temptations. Moral thinking has a distinctive emotional profile: failure to perform an action that one simply wants to do leads to *regret*, failure to perform an action that one judges to be morally obligatory leads to *guilt*, and guilt encourages motivation to restore social equilibrium (Tangney and Fischer 1995). Moreover, observing another person perform an action that one simply wishes she or he hadn't done (e.g., having won a race that you wanted to win) prompts different emotions than observing someone perform an action that one judges to be morally wrong (e.g., having won the race by cheating). The latter licenses not only a different emotional reaction but also a different social response – a social response aimed at reestablishing and maintaining social order. Because of this essential social aspect of moral thinking, even an action that one might be inclined to think of as entirely self-regarding can contribute to social coordination. If a community were to judge wearing a type of footwear, say, to be not merely foolish or unfashionable but *morally wrong*, then members of the community would feel warranted in taking an interest in any footwear transgressors. Abiding by the moral footwear norm could thus be used as a marker of solidarity and social inclusion – as a form of cooperation. For empirical evidence that moral judgment strengthens cooperative tendencies, see Bandura 1999, Tenbrunsel and Smith-Crowe 2008, van der Linden 2011, Simpson et al. 2013, and Bartels et al. 2016.

Even those who take moral thinking to be a by-product rather than a dedicated adaptive mechanism tend to agree with this general picture of its operating in the service of cooperation. Francisco Ayala (2010), for example, argues that morality emerged not as an adaptation but as an exaptation from traits like the human abilities to anticipate consequences, to make value judgments, and to choose between alternative courses of action. Yet, when speculating about why moral thinking would thus emerge, his explanation is in terms of "patterns of action beneficial to the tribe or social group" (Ayala 2010, 9019) and "promoting social stability and success" (Ayala 2010, 9021). Jesse Prinz (2007, 2008, 2009) also argues for the non-nativist view of moral thinking as a by-product of psychological traits like the capacities for higher-order emotions (i.e., emotional responses to emotions), perspective taking, and abstraction. Yet when speculating about why moral thinking would thus emerge, his explanation is in terms of avoiding "a potential collapse in social stability" (Prinz 2008, 405), and when considering why some moral systems flourish over others, his explanation is in terms of certain judgments being "essential to forming successful bands and tribes" while others fail to encourage "the level of cohesion needed for survival" (Prinz 2009, 178).

So those who take moral thinking to be a by-product do tend to see morality as having a purpose, but they see that purpose in terms of serving social ends rather than as an evolutionary function. This important difference aside, though, they are generally broadly in agreement with the adaptationists about why moral thinking emerged: because it helped strengthen social cohesion in ancestral communities. (By-product theorists need not say this; they may hold that morality is an accidental by-product of adaptations that has no particular function at all. Such a view would not undermine any of the arguments that I will develop in the remainder of this chapter.)

I have been at pains to highlight this observation because, in my opinion, it can be pressed into the service of a metaethical conclusion. From my metaethical point of view, what all these sketched genealogical hypotheses strikingly have in common is that they can be happily endorsed by someone who is utterly skeptical of moral truths. When it comes to conducting debates about the origins of moral thinking – about whether it is an adaption or a by-product, the result of individual selection or group selection, a genetic or cultural trait, and so on – whether one believes that moral judgments are sometimes true or believes them to be systematically false makes not an iota of difference. The essential feature of the genealogical hypotheses on offer is that our ancestors' moral beliefs were, in some sense, *useful* – the substantive debate concerning in what ways they were useful. But it is no essential feature of any of these hypotheses that any of our ancestors' moral judgments were *true*. This forms the basis of an epistemological debunking argument.

Human Moral Thinking: A Metaethical Puzzle

Let's bring some metaethical actors onto the stage.

First we have the moral error theorist, who takes an attitude toward morality analogous to the atheist's attitude toward religion. The atheist thinks (a) that people engaged in religious discourse generally purport to state truths and express their beliefs, but (b) the world simply isn't furnished with the objects, properties, and relations necessary to render these assertions and beliefs true. Similarly, the moral error theorist thinks that what it would take for a claim like "Stealing is morally wrong" to be true is for stealing to instantiate the property of moral wrongness, but there simply is no such property, so the statement is false. (It doesn't follow, of course, that the error theorist thinks that it's fine to steal. She also thinks that "Stealing is morally good" and "Stealing is morally permissible" are false. And she may be vehemently opposed to stealing on various nonmoral grounds.)

Next we have the moral realist, who agrees with the error theorist that for the judgment "Stealing is morally wrong" to be true, stealing must instantiate the property of wrongness, but the realist disagrees with the error theorist in maintaining that there is such a property (and the realist probably thinks that stealing usually instantiates this property). In addition, the realist holds a view about the nature of this property: that it is, in some fashion to be specified, a mind-independent property – the wrongness of stealing is a fact we discover, not a fact that we create in taking a certain attitude toward it.

Lastly, we have the moral skeptic, who holds that there is no such thing as moral knowledge. Let's accept for the sake of argument (bracketing off some well-known complications) that for S to know that p is for S to have a true, justified belief that p. This tripartite account of knowledge creates space for three kinds of moral skeptic. First, there are noncognitivists, who deny that moral judgments express beliefs. Second, there are those who accept that moral judgments express beliefs but deny that these beliefs are ever true. This is the error theoretic view that we have just encountered (i.e., error theorists are a kind of skeptic). Third, there are those who accept that moral judgments express beliefs but deny that these beliefs are ever justified.

The debunking argument in which I am interested favors the third kind of moral skeptic. The argument is (roughly) that the genealogical hypotheses sketched undermine the justificatory status of moral judgments. Two things must be noted immediately about this argument.

First, although the error theorist has a role to play in the argument (as we shall see shortly), the conclusion of the debunking argument is not the error theoretic view that no moral judgments are true. The conclusion of this debunking argument is compatible with the truth of moral judgments.

Second, since the claim that a belief is unjustified is compatible with that belief's being mind-independently true, the conclusion of the debunking argument is compatible with moral realism. (Analogy: Someone suffering from paranoid delusions may believe that Dr. X is out to get him; but if he has no evidence of this then his belief lacks justification; nevertheless it may be true, and objectively so, that Dr. X is out to get him.) No fewer than six of the chapters in this volume discuss debunking arguments that target moral realism or moral objectivity, and while there do exist such arguments (e.g., Ruse 1986; Street 2006), the one I am focused on is entirely different. Moral *objectivity* plays no role in this argument; it aims to undermine morality *simpliciter*. With these preliminaries sorted out, let's proceed to the epistemological debunking argument.

I said that what is striking about extant evolutionary hypotheses about the origin of moral thinking is that they can be happily endorsed by the error theorist. It is important to see that this observation cannot be generalized. Suppose it turns out that humans have a mechanism dedicated to distinguishing faces from other visual stimuli. (Whether this hypothesis genuinely withstands empirical scrutiny doesn't matter for my purposes; for discussion, see Slater and Quinn 2001.) One way of expressing this hypothesis – though, admittedly, a rather clumsy way – is to say that humans innately have the concept *face*.[2] An advocate of this view owes us an account of why it would have been adaptive for this mechanism to develop in the evolution of the human brain. Such an account would likely mention the importance of social bonding and emotional communication in the very early period of infancy, the stability of the presence of human faces (and their anatomical structure) in infancy, and so on. Whatever the details, any reasonable account of why it was adaptive for our ancestors to have the concept *face* is surely going to presuppose that faces actually existed; it is useful for infants to have a face-seeking mechanism only because this mechanism is likely to put them in perceptual contact with faces. The same thing will go (mutatis mutandis) for evolutionary hypotheses about the human ability to distinguish inanimate from living things (see Simion et al. 2008), to assume object continuity (see Baillargeon 2008), to spot specific dangerous animals such as spiders (see Rakison and Derringer 2008), and so on. The error theorist about the phenomenon in question (though in many cases such a view would be utterly bizarre) would be barred from endorsing the adaptational hypothesis, since such endorsement would entail admitting the reality of the object or state of affairs. Were we to discover that spiders never existed in the ancestral environment – were we to uncover astonishing evidence that spiders evolved in the last few thousand years – then the hypothesis that it was adaptive for our ancestors to develop an innate spider sensitivity would collapse. The evolutionary hypotheses pertaining to moral thinking are different: they explain the origin of moral thinking in a manner amenable to the moral error theorist.

The upshot of this, I maintain, is not that the error theorist is proven correct, but that the epistemological moral skeptic is bolstered. If one believes that x is P, but the correct explanation for one's coming to have this belief is consistent with nothing ever having the property P (i.e., the correct explanation is something the error theorist about P discourse can endorse), then one's belief lacks justification. This is not to say that this lack of justification must be a permanent state of affairs. Beliefs that are unjustified can gain justification (via, say, the gathering of further evidence). In this case, what the non-skeptic needs to do is show that although the error theorist might *think* that she can endorse the evolutionary hypothesis, in fact she is mistaken. But showing this takes argumentation. In advance of a persuasive case being offered, it seems reasonable to grant the error theorist the benefit of the doubt: if it appears that she can endorse a hypothesis, then we should assume that she can endorse it until it is demonstrated otherwise.

Notice that saying this doesn't amount to granting the error theorist any benefit of the doubt *simpliciter*; rather, it is just a matter of accepting what the moral error theorist can and cannot endorse. Can the moral error theorist endorse the proposition that Napoleon lost the Battle of Waterloo? It would certainly seem so. I suppose someone might deny this, but then at least we should demand to see this person's reasoning, and in advance of being persuaded we should stick to the view that the error theorist can endorse the proposition. (Obviously, allowing this in no way implies any sympathy toward the error theoretic view. Allowing that people who believe in ghosts or phrenology can endorse that Napoleon lost Waterloo in no way implies that one is anything other than 100 percent confidently opposed to these ideas.) By comparison, can the moral error theorist endorse the proposition that Napoleon was morally virtuous? Presumably not. So our question is, Can the moral error theorist endorse the proposition that (for example) human moral thinking evolved as a psychological adaptation that strengthened individuals' commitments to reciprocal exchanges? On the face of it, this looks much more like the question of whether she can endorse that Napoleon lost Waterloo than the question of whether she can endorse that Napoleon was virtuous. To the extent that the moral error theorist owes us an account of why humans everywhere employ a massively mistaken conceptual scheme, the evolutionary hypothesis looks like exactly the kind of explanatory answer she will find attractive.

On the one hand, then, the strength of this debunking argument is ultimately quite modest. If we grant the error theorist the benefit of the doubt that she can endorse the evolutionary hypothesis about the origin of human thinking, then we are provisionally granting that our moral judgments lack epistemological justification. On the other hand, this conclusion is the morality-oriented version of a skepticism no less radical than that espoused by the classical tradition. Ancient skeptics such as Pyrrho, Aenesidemus, and Sextus Empiricus all claimed that we lack knowledge, but it is less clear whether they had any argument to support the stronger thesis that knowledge is unattainable. Sometimes, of course, they got carried away by their own rhetoric. The existence of disagreement was central to the classical skeptic's case, and Sextus (our most complete source on Greek and Roman skepticism) discusses this disagreement using phrases that are translated as "unresolvable impasse" (Mates 1996) and "undecidable dissension" (Annas and Barnes 2000). But no argument deployed by Sextus or any other classical skeptic is actually sufficient to establish this stronger claim. Gathering evidence that disagreement is ubiquitous is one thing; gathering evidence that disagreement is *inevitable* is something else. Correspondingly, showing that our moral judgments lack justification is one thing; showing that they are *unjustifiable* is something else.

A Nod to Sextus Empiricus

To interpret skeptical arguments as attempts to shift a burden of proof onto the non-skeptic is to recognize their provisional nature. The skeptic can never rest easy, for she never knows what new putatively justification-establishing argument the non-skeptic may come up with. Such non-skeptical cases will have to be dealt with as they arise. It is not surprising, then, that Sextus combines his general case for skepticism with specific attacks on various schools that make claims of knowledge: "Against the Grammarians," "Against the Logicians," and so on. In the same way, the proponent of the epistemological debunking argument, even if she has succeeded in shifting the burden of proof, must be willing to confront those opponents who believe that justification for moral judgments can be supplied. I will close by taking a page out of Sextus's book and saying a few words specifically against some leading non-skeptical views of moral epistemology. Of course, there is a great deal that could be (and has been) said in general terms against any of these views; I will confine myself to briefly noting for each the potentially undermining impact of genealogical considerations.

Against the Reliabilists

Reliabilists maintain that a belief is justified if and only if it has a certain kind of history, such as being the product of a process that reliably produces true beliefs. Evolutionary genealogical hypotheses of moral thinking provide the big-picture history of moral concepts, and do so in a way that (it would appear) the error theorist can endorse. If beliefs about X are the product of a process that the error theorist about X can endorse, then they are not the product of a reliable process. The reliabilist who wishes to maintain that moral beliefs are justified therefore has work to do. He must demonstrate that, contrary to appearances, the error theorist in fact cannot really endorse the genealogical hypothesis.

Against the Coherentists

Coherentists maintain that a belief is justified if and only if it coheres in a certain way with other beliefs. A person's moral beliefs may cohere quite well together, giving the appearance of justification. However, beliefs about where our moral judgments come from are themselves items that enter into the coherentist mix. Suppose evidence was forthcoming that human moral thinking is due to a genealogical process that definitely presupposes that many of our moral judgments are true (like, say, the processes that give rise to beliefs about faces or spiders). This, I take it, would strengthen the coherence relations that moral beliefs (generally speaking) would have within the doxastic web. However, this is clearly not the evidential situation with which we are actually faced. Rather, the evidence suggests that moral thinking is due to a process that the error theorist appears able to endorse.

Consider an analogy. A parent judges that his daughter's performance at the musical recital is definitely and objectively the best. This belief coheres well with many of his other beliefs: that his daughter was also the best at last year's recital, that his daughter is exceptionally talented in other ways, etc. But then a new belief enters the mix: the parent comes to appreciate that he judges his daughter's musical accomplishments so positively only because of his extreme partiality toward her; perhaps he is provided with some kind of empirical evidence that he would carry on judging her in this way even if her musical talents were meager. It would seem that it's now incumbent on the parent to form yet another belief: a meta-belief concerning what the relation is between his judgment that his daughter is the best and his realization that he thinks this only because of his own partiality. (After all, just leaving the two beliefs sitting there without forming an opinion about the relation between them seems a step away from coherence.) He has a choice. On the one hand he could think, "No, even though I judge her to be best only because I love her so much, she really definitely is the best"; or, on the other hand, he could think, "Hmm, I suppose the fact that I judge her to be best only because I love her so much places a question mark over whether she really is the best." The thing about coherentism is that either option could, with appropriate adjustments to beliefs elsewhere, figure in a total coherent doxastic package. I suggest, however, that the latter option appears to require fewer adjustments elsewhere. Of course, the parent might prefer the former option and then seek out further sources of evidence as to how good his daughter's ability is. Perhaps she really is the best (even though he thinks this only because of his partiality); perhaps the other parents at the recital will agree to this. The point is, though, that to maintain the former option within a coherent doxastic package, the parent has extra work to do.

Against the Intuitionists

The coherentist's traditional rival is the foundationalist, who holds that some beliefs can be non-inferentially justified. If one supposes moral beliefs to be justified on foundationalist grounds, then one holds either that some moral beliefs can be non-inferentially justified or that moral beliefs can be inferred from nonmoral beliefs that can be non-inferentially justified. There are very well-known problems for the latter view; I'll put it aside and focus on the former view – which I shall refer to here as "moral intuitionism." Many moral intuitionists (e.g., Ross 1930; Audi 2004) hold that certain moral propositions are non-inferentially justified in virtue of being (in some sense) self-evident, and it might seem tempting to suppose that there are many self-evident moral beliefs. A lot of moral beliefs seem so obviously true that we are aghast at the prospect of their being denied. ("You don't think torturing babies is morally wrong?!" exclaims the error theorist's opponent in a tone of disbelieving outrage.) Some have argued that our confidence in certain moral judgments is so high that we can dismiss any argument that purports to deny them, even in advance of examining the argument's details, so sure are we that one of its premises or inferential steps must be mistaken (see Enoch 2011).

But genealogical considerations can upset the moral foundationalist's case. Let me make the point using a quick and slightly dirty analogy. Suppose a person is so extremely confident in the proposition that x is P that he is inclined to classify it as self-evident and requiring no inferential justification. Imagine, though, that the only reason he confidently has this belief is that he was successfully hypnotized to hold it. Upon discovering this fact about his own belief-formation process (and supposing the spell of hypnosis now is broken to the point that he can reflect on the epistemic status of the belief), surely the person should at least doubt whether the proposition that x is P has the status of a self-evident truth.

The general problem here is that the very feature that indicates (according to the intuitionist) a belief's self-evident status – such as our according it a very high level of confidence – may be itself subject to a plausible genealogical explanation consistent with an error theory. For example, an evolutionary explanation for moral thinking is not simply going to account for why we have moral beliefs but also probably will, when properly fleshed out, account for why our moral beliefs are entrenched and enjoy a high level of confidence – why, indeed, they seem self-evident. But if evidence is available that the feature that inclines us to classify a belief as self-evident is itself the product of a process that appears consistent with the falsity of the belief, then that feature can no longer be appealed to as a ground for self-evidence.

Against the Conservatists

Coherentists and intuitionists tend to favor epistemic conservatism: the view that the mere fact that a belief is held accords it a kind of prima facie justification. Conservatism is reasonable if one can be confident that the beliefs in question are likely to be true. If I know that humans tend to be pretty good at judging the relative size of nearby objects, and I know that Fred believes that a certain nearby object is larger than another, then it is reasonable for me to accord Fred's belief some prima facie justification just in virtue of his holding it. On the other hand, if I know that humans tend to be pretty bad at judging probabilities when calculating risks, and I know that Sally (who is calculating risks) judges a certain outcome to have a high probability, then it is not reasonable for me to accord Sally's belief any justification in virtue of her holding it. Even if I know that Sally's belief is extremely widely shared in the epistemic community, if I have reasonable grounds for thinking that humans tend to get things wrong in this domain, then the mere fact that the belief is widely held provides it with no positive justification.

In the moral case, we tend to think that our judgments are pretty good – more like Fred's belief than Sally's. The prospect that our basic moral beliefs might be false – not just our own beliefs but also those shared (often emphatically so) by our peers and loved ones – is one so troubling that we tend to shy away from even confronting it. While a person may acknowledge that she has learned many of her moral norms from her community, this is likely to be accompanied by the inchoate thought that surely *someone somewhere* has *somehow* had access to the moral facts. If this inchoate thought were reasonable, then conservatism might provide moral beliefs with a degree of prima facie justification. The genealogical hypotheses discussed earlier, however, starkly reveal what is not reasonable about this thought, for they provide plausible explanations of how it has come about that we are all emphatically confident about moral beliefs in a manner that appears consistent with a moral error theory. One may, of course, embark on the project of showing that, contrary to appearances, the error theorist cannot endorse the genealogical hypotheses – which could amount to showing that humans are pretty good at judging moral matters – but (with apologies for the repetition) this is extra work that needs to be accomplished before conservatism can supply justification in the moral realm.

Any of the epistemic justificatory theories just presented – reliabilism, coherentism, intuitionism, and conservatism – might be correct for all that

I've said here. My efforts have been confined to showing very quickly how within each framework the discovery that a body of beliefs is the product of a process that the error theorist can endorse will undermine the epistemic status of those beliefs. Since plausible evolutionary hypotheses about the origins of moral thinking appear to be endorsable by the moral error theorist, moral believers must shoulder a dialectical burden: they have work to do if they are to maintain their moral beliefs in a non-dogmatic fashion.

Notes

1 It is also unclear whether Darwin thinks the human moral sense is an adaptation or a by-product.
2 I think it is equally clumsy for the moral adaptationist to express the hypothesis by saying that humans innately have the concepts *moral obligatoriness, desert, moral wrongness*, etc. But for some purposes such expressions can be taken as a kind of shorthand for a much more sophisticated hypothesis.

8 Evolution and the Epistemological Challenge to Moral Realism

Justin Horn

Introduction

Moral realism, as I shall understand it, is the view that morality is stance-independent. That is, according to moral realism, the fundamental moral standards are true independently of the attitudes that any human beings have toward them, and independently of the attitudes that human beings would have toward them under idealized conditions of reflection (Shafer-Landau 2003, 15). As a metaethical theory of the nature of morality, moral realism has gained considerable philosophical ground over the last few decades. Recently, however, several authors have challenged moral realism on evolutionary grounds (Kitcher 2005, 2011; Joyce 2006a; Street 2006; Locke 2014). The version of the challenge that I wish to examine centers on the following claim: given the way in which our moral faculties have been shaped by evolutionary processes, those faculties cannot provide us with justified beliefs about any stance-independent moral truths (at least once we are made aware of this evolutionary influence). Insofar as the combination of moral realism and radical moral skepticism is deeply implausible, the evolutionary challenge would, if successful, give us strong reason to abandon moral realism.

Evolutionary arguments in metaethics characteristically rely on empirical assumptions, many of which are, admittedly, somewhat speculative. A complete assessment of evolutionary arguments against realism would require a detailed examination of the evidence supporting various hypotheses concerning the evolution of our capacity for moral judgment. I shall not attempt this difficult task here. Rather, in this chapter, I propose to examine the metaethical implications of one hypothesis that has been endorsed by a number of authors, both philosophers and scientists. The hypothesis is this: evolutionary forces have had a significant influence on the content of our moral judgments (see De Waal 1997, 2006; Hauser 2006; Joyce 2006a; Street 2006). Notice that this hypothesis is stronger than the plausible claim that our capacity to think morally is, in some general sense, the product of evolution. The hypothesis is committed to the stronger claim that evolution has "pushed" us in the direction of making certain moral judgments rather than others. According to the hypothesis, just as the pressure of natural selection played a significant role in bringing it about that tigers have sharp claws (rather than dull ones) and that zebras are speedy (rather than slow), so too did the pressure of natural selection play a significant role in bringing it about that human beings have a very strong tendency to regard certain things as morally valuable or morally reprehensible. Examples of evolutionarily favored moral attitudes might include the widespread positive moral regard enjoyed by activities such as caring for one's own children and reciprocating benefits provided by others, and the negative moral regard commonly held for defecting from agreements or casually harming one's kin. There are, of course, variations in the precise form that such "moral regard" takes in different individuals and cultures. But it is hard to deny that even these vague generalizations get at real patterns in moral judgment, patterns for which there must be some causal explanation. The hypothesis in question holds that some such deep patterns in moral judgment are significantly due to the pressure of natural selection, such that had we evolved differently, we would make moral judgments with very different content.

I hasten to note, however, that the hypothesis is not intended to serve as a complete explanation of why we make all of the particular moral judgments we do. The truth of the hypothesis is consistent with the fact that there are many other significant influences on the content of our moral judgments – influences that include rational reflection, as well as a variety of social, cultural, and historical factors. The claim is only that evolution has been one powerful causal influence on the content of our moral judgments.

To avoid confusion, I should note a few of the consequences of the hypothesis. First, because the hypothesis is not intended to explain all moral phenomena, it is not threatened by the existence of certain moral phenomena for which the best explanation is not an evolutionary one – for example, the changes in moral attitudes toward women and racial minorities that have taken place in the United States over the last 150 years. The following analogy might be helpful here. We have excellent reason to believe that genetic factors have a significant influence on our personalities. This belief is in no way threatened by the existence of personality-related phenomena for which the best explanation invokes nongenetic factors, such as a case in which identical twins have substantially different personalities. Such cases merely show that genetic factors are not the only causal factor contributing to personality development. Similarly, certain moral phenomena may have a non-evolutionary explanation, even if the influence of evolution on our moral beliefs is pervasive.

Furthermore, the claim that evolution has significantly influenced the content of our moral judgments does not entail that we will inevitably make all of the moral judgments that would be adaptive. Nor does it entail that all of the judgments we do make will be adaptive. On the most likely model of the development of our moral capacities, evolution has endowed us with certain very general evaluative dispositions regarding harm, fairness, purity, and the like (see de Waal 1997, 2006; Kitcher 2005, 2011; Street 2006; Haidt and Joseph 2007; Haidt 2012). These dispositions in turn shape the content of our moral judgments significantly, such that if we had a different set of general evaluative dispositions, we would come to very different conclusions about a variety of moral matters. Nonetheless, these general evaluative dispositions do not wholly determine which moral judgments we make; rather, they can be channeled in a variety of ways by culture, learning, and rational reflection. Thus, the foregoing model is consistent with us making moral judgments that are in fact quite harmful to our fitness, such as the judgment that I am required to sacrifice my life for my country.

With these clarifications in place, I will henceforth assume for the sake of argument that the hypothesis is true. The challenge for the realist is to provide some defensible account of the relationship between the putative stance-independent moral truths and the evolutionary pressures that (we are assuming) have shaped our moral judgments. I'll describe three general accounts of this relationship and argue that the realist's best hope lies in establishing what I'll call the indirect tracking account. I'll then argue that any attempt to establish such an account will rely on premises for which the recognition of evolutionary influence will have already defeated our epistemic justification. In light of this defect, I argue that if our initial hypothesis is true, moral realists are saddled with the conclusion that all of our positive moral beliefs – those beliefs that attribute moral properties – are unjustified. Since this sort of radical skepticism about morality is implausible, I conclude that if evolutionary forces have strongly influenced the content of our moral judgments, then we have strong reason to reject realism as a metaethical position.

The Distortion Hypothesis

If we assume that evolutionary forces have "pushed" us in the direction of making some moral judgments rather than others, the realist must give some account of the relationship between this evolutionary influence and the putative stance-independent moral truths. One possibility is that there is no positive correlation between the stance-independent moral truths and those moral judgments that evolution has "pushed" us toward. Let us call this the distortion hypothesis. On the face of it, the distortion hypothesis seems to represent a worst case scenario for the realist. If the distortion hypothesis is true, then evolutionary pressures are no more likely to have pushed us toward true moral beliefs than, say, were we to have based our moral beliefs on a random drawing from a hat containing all logically possible moral judgments.

The hat-drawing method is obviously very unlikely to consistently yield true moral beliefs. But if the distortion hypothesis is true, one considerable influence on the content of our moral judgments is, in all relevant respects, exactly like hat-drawing. If evolutionary influences on our moral judgments are sufficiently deep, and we have no way of correcting this distorting influence, then many of our moral beliefs are very likely to be false.

According to the hypothesis under consideration, evolutionary influences on our moral beliefs are in fact quite deep. Although certainly other factors may influence our moral judgments, the hypothesis suggests that many of our central moral commitments are largely the result of evolutionary pressures. Thus if this hypothesis is true, a realist seeking to embrace the distortion hypothesis without succumbing to skepticism must therefore argue that we possess the tools for weeding out and correcting even deep and widespread errors among such moral judgments. One way to argue for this conclusion would be to postulate a special faculty of moral intuition. It is not clear that this would solve the problem, however. Unless the realist also holds that we can reliably distinguish the outputs of our special faculty of intuition from those moral judgments that have been conditioned by evolutionary forces, the most that a faculty of intuition will achieve is to mix some true beliefs in with those mostly false beliefs that we have due to evolutionary pressures. Without any tool to reliably separate the two, a large percentage of our moral beliefs will still be false.

Some have argued that performing such a separation is relatively easy – we can simply look at the content of a particular judgment to figure out whether it is likely to have been influenced by evolutionary pressure. This is so, it is argued, because "biological evolution would be expected to produce a bias toward favorable evaluations of things that promote one's own inclusive fitness; intuitions that do not imply favorable evaluations of things that promote one's own inclusive fitness are not candidates for being products of this particular bias" (Huemer 2008, 381; see Shafer-Landau 2012, 5–8). This method of sorting is problematic, however. First, since evolutionary change takes time and environments can change rapidly, natural selection can produce organisms with traits that no longer promote their own inclusive fitness. Consider, for example, human tastes in food. As it happens, most of us find ourselves attracted to sweet, salty, and fatty foods. In environments where food is plentiful, these preferences can be quite maladaptive, leading people to consume more calories than would be healthy, and often leading to early death. From the point of view of survival, it would probably be far better for those of us living today if we were to crave whole grains and leafy green vegetables, and to find large quantities of red meat repulsive. Nonetheless, there is a straightforward and plausible evolutionary explanation of our cravings for fats and sugars: in the ancestral environment, in which food was scarce and starving was a very real threat, it was adaptive to be motivated to consume the most calorific foods available. We inherited from our ancestors a palate that helped them survive but that often leads us today to heart problems and obesity. Thus natural selection can produce traits that are, in our current environment, quite detrimental to our own inclusive fitness.

Furthermore, it is not plausible that evolution influenced the content of our moral judgments by directly selecting for some particular judgments over others. On the contrary, the influence of evolution was likely much more indirect: certain general evaluative tendencies were selected for over time, and these evaluative tendencies in turn strongly influence which moral judgments we end up making. Given this sort of influence, it wouldn't be surprising if such dispositions sometimes "misfired" to produce particular judgments that turn out not to be reproductively advantageous, especially given significant differences between the ancestral environment and our own. Consider just one (admittedly speculative) example. It clearly promotes one's own inclusive fitness to ensure that one's own offspring – and to a lesser extent the offspring of one's close relatives – survives to reproductive age. If, in the ancestral environment, most of the small children that one came across were closely related to oneself, a standing disposition to be gentle toward all small children and refrain from harming any of them would be highly adaptive, and could in principle be selected for. In our present environment, we often come across small children to whom we are not related. Perhaps, in some circumstances, harming them would promote one's inclusive fitness. Nonetheless, there could very well be an evolutionary explanation for why we are strongly disposed to regard harming small children as forbidden, even when it would be advantageous to harm them.

To be clear, I am not attempting to offer a complete and accurate explanation of our attitudes toward small children. Rather, the example demonstrates that the mere fact that a judgment does not currently promote one's own inclusive fitness is consistent with the hypothesis that the judgment is substantially the product of natural selection. Furthermore, to the extent that evolutionary influence on our moral judgments took place largely at the level of very deep and general evaluative dispositions (for example, by inclining us to regard certain very general features such as harm, fairness, loyalty, and purity as having positive moral relevance), it will be very difficult to find substantive moral judgments that are plausibly entirely isolated from such dispositions, even when we consider judgments that happen to be detrimental to our fitness in the present environment.

Even if we cannot identify exactly which moral beliefs have been shaped by evolutionary forces, one might think that the widely endorsed method of reflective equilibrium would allow us to correct for any potentially distorting effects that evolution has had on our moral judgments (Rawls 1971). Perhaps by "testing" our judgments about moral principles against our judgments about particular cases and vice versa, while also seeking coherence between our moral judgments and background theoretical considerations, we can root out even deep moral errors generated by evolutionary pressure.

If the distortion hypothesis is true, however, it is doubtful that such a method will be of much help. The method of reflective equilibrium essentially involves simply working back and forth between judgments about moral principles and judgments about particular cases, adjusting each in light of the other until an adequate degree of coherence is achieved. If the set of initial moral judgments with which we begin inquiry is sufficiently corrupt, however, such a process of mutual adjustment is unlikely to be promising as a way of arriving at a stance-independent moral truth. As Sharon Street points out, if the distortion hypothesis is correct, then this sort of reasoning will simply involve "assessing evaluative judgments that are mostly off the mark in terms of others that are mostly off the mark" (Street 2006, 124). Thus, it seems that if the distortion hypothesis is true, a large percentage of our moral beliefs are very likely to be false, even if we prune them so as to bring them into reflective equilibrium.

The distortion hypothesis is therefore an unattractive option for any realist who believes that some of us are reliable moral judges. For this reason, realists might be encouraged to notice the following peculiar feature of that hypothesis: it would be very difficult to establish that the hypothesis is actually correct. To establish that there is indeed no positive correlation between the moral truth and the evaluative judgments that were selected for, it seems that we would need to compile a rough list of some moral truths and then compare the moral truths to those evaluative judgments favored by natural selection. (Elliott Sober makes a similar point in Sober 1994, 107.) Only by having some information about the contents of each list could we provide evidence that these contents were not correlated. If we possessed the information required for this task, however, then clearly any skeptical argument would be very hard to get off the ground. After all, in such a situation, we would already have a rough list of at least some moral truths. Much like global skepticism, it seems that the distortion hypothesis is not one that can be coherently asserted with confidence.

Yet it remains a troubling possibility, for an obvious reason. If the distortion hypothesis is correct, very many of one's moral beliefs are likely to be false, since there is no correlation between those moral beliefs that natural selection has pushed us toward (and thus, many of the moral beliefs that humans tend to have) and the moral truth. It seems to be a very plausible epistemological principle that if one has undefeated reason to think that one's beliefs in a domain have a high probability of being false, one cannot be justified in holding those beliefs. Thus, if realists have reason to believe that there is even a fairly high probability that the distortion hypothesis is correct, realism faces a serious epistemological challenge. Of course, I have not yet given any reason to believe that there actually is a high probability that the distortion hypothesis is correct. At this point, the thing to notice is merely that the distortion hypothesis could, in principle, threaten to undermine the justification of our moral beliefs even if it cannot be firmly established as correct.

The Direct Tracking Hypothesis

Realists might want to forestall this skeptical possibility by arguing that the evolutionary influence on our moral beliefs has been largely benign. One way of doing so would be to suggest that evolutionary pressures have pushed us toward the stance-independent moral truth, precisely because it was adaptive for our ancestors to grasp the moral truths in question. Let us call this the direct tracking hypothesis. Such a hypothesis, if true, would not only save the realist from epistemological objections based on evolutionary grounds but also would in fact provide the realist with a powerful tool for defending our general moral reliability.

This hypothesis is unacceptable on scientific grounds, however. In particular, it is inferior to a competing hypothesis, which Sharon Street calls the adaptive link account. According to the adaptive link account, "tendencies to make certain kinds of evaluative judgments rather than others contributed to our ancestors' reproductive success not because they constituted perceptions of independent evaluative truths, but rather because they forged adaptive links between our ancestor's circumstances and their responses to those circumstances, getting them to act, feel, and believe in ways that turned out to be reproductively advantageous" (Street 2006, 127).

The main problem with the direct tracking account is that the most promising explanations of the evolutionary influence on the content of our moral beliefs simply needn't make any reference to the existence of moral facts. Indeed, it's not clear how postulating such facts would contribute anything to such an explanation. In contrast, consider the best explanation of the origins of our capacity for detecting mid-sized physical objects. Any acceptable explanation of our perceptual abilities will invoke the fact that non-veridical perceptions of mid-sized physical objects (say, predator or prey) would tend to be detrimental to the fitness of an organism. If an organism tends to form beliefs to the effect that it is being chased by predators when this is not so, it will end up wasting a lot of valuable time and energy running and hiding. Still worse, if an organism tends not to form beliefs that it is being pursued by a predator on those occasions when it is in fact being pursued, that organism's genes are likely to be swiftly removed from the gene pool. In short, when it comes to avoiding predators, the truth of one's perceptual beliefs is of paramount importance.

In contrast, it is not at all clear how the truth of one's moral judgments can play any analogous role in an evolutionary explanation of our moral abilities. Other things being equal, it seems it would be adaptive for an organism to believe that it ought to take care of its offspring, and maladaptive to believe that it ought to kill them. But the adaptiveness (or lack thereof) of these judgments would remain exactly the same if it were to turn out, quite surprisingly, that we have a fundamental moral obligation to kill our own offspring. In morality, it simply does not seem that beliefs are ever adaptive in virtue of being true. Thus, we should expect selection for moral judgments that form adaptive links between circumstances and behavior, regardless of whether such beliefs are true or false.

One should note the limitations of the preceding remarks. I have not argued that the mere fact that stance-independent moral facts play no role in scientific explanations justifies eliminating them from our ontology. The present claim is much more modest. Given that we can explain everything worth explaining about the evolutionary influences on moral judgment without postulating moral facts, considerations of parsimony give us a reason to prefer the adaptive link account to the direct tracking account. Thus, we may conclude that while the direct tracking account would save the realist from a skeptical conclusion, it is unacceptable on scientific grounds.

Indirect Tracking and Preestablished Harmony Explanations

I've argued that both the distortion hypothesis and the direct tracking hypothesis are unpromising ways for the realist to reconcile the notion that we have justified beliefs about stance-independent moral truths with the (putative) fact of evolutionary influence on the content of our moral judgments. There is another possibility, however. It may be that there is a strong correlation between the stance-independent moral truths and those moral judgments that were selected for, such that the moral judgments that were selected for are mostly true, but were not selected for because they were true. I say "mostly" because the realist needn't insist that evolutionary pressures have pushed us toward the truth in every case to surmount the epistemological challenge. As David Copp points out, the realist can resist a skeptical conclusion, provided that "our beliefs tend to do well enough in tracking the moral truth that rational refection can in principle correct sufficiently for any distorting influence" (Copp 2008, 194). The position we must consider, then, is one that accepts the adaptive link account as an explanation of why certain moral judgments were selected, while still holding that the moral judgments selected are close enough to the truth. I think that this view, which I will call the indirect tracking hypothesis, is the most promising avenue for the realist.

How might one defend the hypothesis that true moral beliefs were not selected for because they were true, but that nonetheless evolutionary influences have pushed us toward mostly true moral beliefs? The most promising explanation appeals to the widely accepted principle that any moral facts that exist supervene on natural facts: natural facts fix the moral facts in the sense that, necessarily, any two states of affairs that are exactly alike in all natural respects must be exactly alike in all moral respects. According to the evolutionary hypothesis presently under consideration, evolutionary forces have pushed us toward the acceptance of moral beliefs that are appropriately related to certain natural facts (namely, facts about survival and reproduction). If the natural facts that our moral beliefs tend to track are systematically related to the moral facts, this opens the door for a "pre-established harmony" explanation of the correlation between the moral judgments that were selected for and the realist's stance-independent moral truths (Enoch 2010; Skarsaune 2011).

Suppose, then, that the realist accepts that certain moral beliefs were selected for, as described by the adaptive link account. The realist might proceed to argue that these moral beliefs are (mostly) true, because the features that moral judgments were selected to track either constitute or closely correlate with moral features. We can explore how such a strategy would work by considering a simple form of naturalistic realism: hedonistic utilitarianism.

The hedonistic utilitarian might admit that moral beliefs were selected not for their truth, but for their tendency to motivate individuals to behave in ways that increased reproductive success. But the utilitarian might then claim that the moral beliefs that evolution has conferred on us are for the most part reliable. The utilitarian needn't simply see this as a convenient coincidence, but could argue for it as follows. Pleasure is intrinsically good and pain is intrinsically bad. Given this, one can imagine why natural selection would, to a considerable degree, favor true moral beliefs rather than false ones. After all, pain is typically an indicator of bodily harm, so organisms that tend to view pain as bad would tend to survive longer than those who do not. Likewise, pleasure is often an indicator of bodily benefit (or in the case of sexual pleasure, of reproductive success), and therefore organisms that see pleasure as good would tend to have greater reproductive success than those that do not. Thus, while evolutionary forces may have led us astray in some cases (for instance, the widespread belief that we have only very weak obligations to distant strangers), it is no accident that it has given us mostly true moral beliefs.

I use utilitarianism as an example, but it is important to note that this sort of explanatory strategy could in principle be used for a wide variety of normative theories. It needn't be limited to reductive accounts, or even to naturalist accounts. Any view that claims that the moral facts supervene on natural facts could in principle tell this sort of story, by first linking certain natural features of the world with moral features, and then arguing that it was (for the most part) adaptive for our ancestors to regard those natural features as good, even though the explanation of why this is adaptive makes no reference to the truth of their judgments.

The nonnaturalist realist David Enoch adopts this sort of strategy to respond to Street's Darwinian Dilemma. According to Enoch, what I have called the indirect tracking hypothesis can be adequately supported if we merely accept that "survival or reproductive success (or whatever else evolution "aims" at) is at least somewhat good" (Enoch 2010, 430). This claim is not intended as a reductive account of what goodness is; it is merely a rough and ready claim that in most circumstances, survival has value. Enoch argues that if survival has value, and viewing survival as valuable was selected for, then evolution might have left us with mostly true moral beliefs, even if the truth of these moral beliefs plays no role in the explanation of why they were selected for.

It is not clear that such a modest assumption is sufficient to explain the correlation Enoch aims to explain. The claim that survival is at least somewhat good is compatible with the claim, for instance, that the beauty of nature is of far greater value and that we are all obligated to sacrifice our own survival to maximize natural beauty. Likewise, Enoch's normative claim is compatible with the view that while survival is good, this goodness is outweighed by the goodness of excruciating suffering. An indefinite number of logically possible, internally coherent ethical systems are compatible with the claim that survival is at least somewhat good, and many of these systems differ dramatically from our moral intuitions in far-reaching, systematic ways. Thus, even assuming that survival is somewhat good, the realist still needs an explanation of why our system of intuitive moral judgments (which incorporates this assumption) approximates the stance-independent moral truth, while all other such internally coherent sets incorporating it do not.

In general, though, indirect tracking accounts seem attractive because they have the potential to provide an explanation of a correlation between those moral beliefs favored by natural selection and the stance-independent moral truth, and all this without giving up the scientifically preferable adaptive link hypothesis. Nonetheless, such accounts suffer from a serious defect. When presented with a claim linking the moral to the nonmoral, we are entitled to ask what evidence or justification is on offer for the claim. The realist answer, it seems, will typically rely on substantive normative ethical views. This was clearly the case in the previous utilitarian example, as well as the case of David Enoch's more modest bridge principle. In a similar vein, Erik Wielenberg attempts to vindicate our moral judgments in the face of evolutionary influence by assuming the normative claim that there are "moral barriers" that surround all creatures with sufficient cognitive capacities. These cases do not seem to be exceptions to the rule. As realist David Brink writes, "determination of just which natural facts and properties constitute which moral facts and properties is a matter of substantive moral theory" (Brink 1989, 177–178). The problem is that invoking substantive normative ethical views at this point in the dialectic begs the question, since the reliability of these views is exactly what is at stake.

The question for the realist is whether evolutionary influences have left us with the capacity to form moral beliefs that (at least roughly) track a stance-independent moral truth. Supposing we have ruled out the direct truth tracking account, we are left with two options: the indirect tracking hypothesis or the distortion hypothesis. If the distortion hypothesis is correct, then most of our intuitive moral judgments are false. If the indirect tracking hypothesis is correct, then a large number of our intuitive moral judgments are true, at least enough such that rational reflection could (in principle) weed out the bad apples. The trouble for the realist is this: how do we figure out which of these two possibilities obtains?

If we are at all unsure, it simply will not do to invoke substantive normative judgments at this point. Consider the following analogy. Suppose you discover that you've been brainwashed by a cult leader, who has given you all sorts of supernatural beliefs, which are based on visions he experienced while taking a brand new Miracle Drug. Further suppose that you are genuinely unsure whether Miracle Drug visions are a reliable guide to the supernatural truth, and you are trying to ascertain whether or not this is so. Clearly it would not do to "test" the beliefs that the leader formed when using Miracle Drug against your own convictions about the supernatural. After all, you know that your beliefs about the supernatural are the result of the cult leader's brainwashing, so of course his supernatural beliefs will pass this "test," whether Miracle Drug visions are reliable or not. (For a similar point, see Copp 2008, 197.)

Analogous things could be said about the evolutionary influences on our moral beliefs. If we are trying to determine whether evolutionary forces have pushed us toward the moral truth (as the indirect tracking hypothesis says) or not (as the distortion hypothesis says), it will be of no use to "test" the moral beliefs that would be selected for against our intuitive moral judgments. For we know (or so we are supposing) that our moral judgments have been heavily shaped by evolutionary forces. For this reason, the moral beliefs that have been selected for would be very likely to pass this test, even if the distortion hypothesis were correct.

Is there some way of establishing the indirect tracking hypothesis that does not rely on any first-order normative views? This would be nice for the realist, but it does not seem promising. If we accept some version of Hume's dictum that moral claims cannot be established by arguments that invoke no moral premise whatsoever, we must admit that any attempt to establish the indirect tracking hypothesis will rely on normative judgments. I've argued that relying on normative judgments to establish the truth of the indirect tracking hypothesis over the distortion hypothesis begs the question. Thus, any argument for favoring the indirect tracking hypothesis over the distortion hypothesis will be question begging.

Moral Realism and Skepticism

I've argued thus far that arguments for the indirect tracking hypothesis are question begging, while the distortion hypothesis cannot be coherently asserted confidently. One might be tempted to conclude that without any way of resolving which hypothesis is the correct one, the epistemological challenge to moral realism flounders. After all, given everything that I've said, perhaps the indirect tracking hypothesis is correct and our epistemological situation is pretty good. So nothing I've said can be thought to undermine realism.

Furthermore, one might press the following line of argument. Suppose one were to call into question the justification of our perceptual judgments by challenging us to show that they themselves were not distorted in some deep way. One natural reply to such a challenge is to point out that the most plausible account of our basic perceptual capacities will be (to a significant extent) a direct tracking account, according to which the ability of those capacities to yield true judgments was essential to their being selected for. And this reply does seem adequate to vindicate our perceptual capacities to some degree. But notice: we can only establish a direct tracking story about the evolutionary origins of our perceptual capacities by relying on those capacities from the outset. Without the input of sensory observations, scientific theorizing about the nature of evolutionary influence on our perceptual capacities could never get off the ground (see Schafer 2010 and Vavova 2014). And yet we do think that we are justified in believing things on the basis of our senses. So it seems plausible to suppose that our perceptual judgments have some justification from the very beginning.

The moral realist might insist that similar considerations allow us to justifiably accept the indirect tracking hypothesis over the distortion hypothesis in the case at hand. If we are willing to grant some justification to our perceptual judgments from the outset, there seems to be no reason not to allow that our moral judgments enjoy a similar degree of justification at the outset of inquiry. Once we grant this, though, it seems that the realist can rely on her justified moral beliefs to rule out the distortion hypothesis and find in favor of an indirect tracking view (see Schafer 2010).

This is an elegant line of argument, but I think it can be resisted. We should grant the first point: we should, at the outset of inquiry, regard our intuitive moral judgments as having some modest degree of justification. The question we must ask is whether this justification is undercut by the time we face the question of whether to prefer the indirect tracking hypothesis to the distortion hypothesis. And it seems to me the answer is yes.

The first thing to notice is that in seeking an evolutionary vindication of our perceptual judgments, there is never a moment at which we have an explanation of the origins of our perceptual faculties that completely leaves open the question of whether they are reliable in tracking stance-independent facts about our surroundings. We begin with perceptual judgments that have some degree of justification, we do a lot of scientific inquiry, and we wind up with additional reasons to trust our perceptual faculties: our best explanation of their emergence vindicates their (approximate) reliability. But the second thing to notice is that we can imagine things being different. And in such imagined cases, the wrong kind of genealogy of our perceptual judgments could undermine their justification, even while leaving it open whether or not such judgments were actually reliable.

Imagine you were to discover something shocking about your perceptual judgments: they are never caused by external physical objects. Further, imagine you discover this in a manner completely independent of your perceptual capacities – perhaps God directly imparts this knowledge to you. It turns out that all of your perceptions are constantly being caused directly by some supernatural creature. This supernatural creature is akin to Descartes's evil demon, with one important difference: we have no idea whether he is evil. (For some reason, God neglects to tell us this part.) Indeed, we have no indication of the being's intentions whatsoever. Call this creature the Demon of Unknown Intentions (DUI).

With this new and disturbing information about your perceptual capacities, you start to worry about your ability to reliably form beliefs about external physical objects. You reason as follows: on the one hand, it's consistent with your newfound knowledge that the DUI is benevolent and only gives you veridical perceptual experiences. Perhaps you only have the experience of a tree when there is indeed an external physical tree in your vicinity. Perhaps, in fact, the demon is necessarily benevolent, and so couldn't possibly deceive you in any deep and undetectable way. On the other hand, it's also entirely consistent with your newfound knowledge that the DUI is entirely deceiving you. Perhaps, as far as physical reality goes, you are just a brain in a vat, or an eight-armed slimy creature, or perhaps there is no external physical world at all.

What would be reasonable to conclude if (somehow) you were to learn that, as a matter of fact, all of your perceptual experiences were caused by the DUI? You could hope for a kind of indirect tracking explanation. Perhaps you could find some central regularities in the world of your experience and assume that these correspond to physical reality (perhaps relying on our a priori entitlement to trust our perceptual capacities to justify this claim) and then deduce that your initial perceptual beliefs were close enough to veridical to correct any distorting influence through reasoning. For example, you might note that the DUI has made your experiences such that Newtonian mechanics seems roughly true of macroscopic objects. Since, you insist, Newtonian mechanics is roughly true of such objects, the DUI has probably not led you too far astray.

But here this reply seems totally unconvincing. Once you learn that your perceptual judgments are caused by something wholly distinct from any physical objects they seem to report, something that you have no independent reason to regard as a reliable source, the initial justification provided by your perceptual judgments is defeated. Absent any other way of finding out about a world of objectively existing external physical objects, it seems that all of your beliefs about them would be rendered unjustified.

This remains so even if we weaken the case a bit, so that the DUI is not wholly responsible for your perceptual beliefs. Suppose you are informed that the DUI is only one significant influence on the content of your perceptual judgments. Nonetheless, you learn, (a) this influence is such that you have no way of isolating any perceptual judgments that are known to be free of the influence of the DUI, and (b) the influence of the DUI is sufficiently powerful that the following is true: had the DUI influenced you differently, you would make radically different perceptual judgments. It seems to me that learning of even this more modest influence of the DUI on your perceptual judgments has deep skeptical consequences. Continuing to believe that your perceptual beliefs accurately represent an external physical reality in such a case requires trusting that the influence of the DUI has pushed you toward, rather than away from, the truth. But this is exactly what you have seem to have no reason to believe, and no way of figuring out.

In the case of the DUI, what defeats our initial justification for our perceptual beliefs is that we justifiably accept an account of the origins of those beliefs that (a) rules out a direct tracking explanation and (b) gives us no reason to prefer an indirect tracking account to a distortion account. Once we have this, the initial warranted confidence we had in regarding those beliefs as faithful representations of an external physical reality disappears. Yet, if our initial hypothesis about evolutionary influence on the content of our moral judgments is true, then – at least once we come to realize its truth – we seem to be in an analogous epistemic situation. For we will have identified one deep influence on the content of our moral judgments, where the best explanation of the nature of this influence (a) rules out a direct tracking account and (b) gives us no reason to prefer an indirect tracking account to a distortion account. Once we have this in hand, it seems that – at least insofar as we regard our moral beliefs as attempts to represent a stance-independent moral reality – all of our moral beliefs will be unjustified.

9   Evolutionary Naturalism and Valuation

Richard A. Richards

Introduction

The geneticist Theodosius Dobzhansky was surely exaggerating when he proclaimed in the title of a 1973 paper that "nothing in biology makes sense except in the light of evolution." But he was right in that much of what we observe in organic nature makes sense because of what we know about evolution. We can, for instance, explain the anatomy, physiology, psychology, and behavior of organisms across biota based on what we know about evolutionary history and evolutionary processes. That explanatory power surely extends to humans. *Homo sapiens* has the general traits we observe because of its origins in a primate lineage, past population dynamics, principles of inheritance, the operation of natural and sexual selection, and developmental processes. Among these general traits explicable by evolution are the patterns of human social interactions. We can understand, at least in part, why we interact with other humans the way we do, based on our evolved social nature.

In general, we are a highly social species with intense and ubiquitous cooperation. But we could be this way, and cooperate to the degree that we do, only if we had an evolved social psychology with the required capacities. We would need, for instance, the capacity to recognize, accept, and follow the rules of cooperation, as well as the motivation to do so. Moreover, if this cooperation had value in survival and reproduction, then perhaps there is an adaptation story to tell. This is the premise of Michael Ruse's claim, in his *Evolutionary Naturalism*, that morality is an adaptation:

We think we ought to do certain things and that we ought not to do other things, because this is our biology's way of making us break from our usual selfish or self-interested attitudes and to get on with the job of cooperating with others ... in order to make us "altruists" in the metaphorical biological sense, biology has made us altruists in the literal, moral sense.... Morality is no more – although certainly no less – than an adaptation, and as such has the same status as such things as teeth, eyes and noses. (Ruse 1995, 241)

According to Ruse, what makes morality effective are our tendencies to believe that the rules of morality are objective and absolute – they don't depend on our subjective desires or contingent circumstances (Ruse 1995, 254). If so, we cooperate and act ethically to the degree we do so, based on a foundation in our evolved nature.

By itself, this may not seem controversial for those who accept the evolutionary origins of humans, but Ruse goes further, arguing that there are therefore no absolute moral values, and our belief in them is a useful error (Ruse 1995, 248–249). An evolutionary account of human ethics implies that our normative principles – the rules that our ethical systems require us to live by – are contingent on facts about our evolutionary history. If we were social insects instead of primates, perhaps highly intelligent ants, we would have a very different set of normative principles. According to E. O. Wilson, such an ant ethics would likely include rules related to antennal rites, cannibalism, caste determination, colony-foundation rules, colony organization, drone control, larval care, metamorphosis rites, mutual regurgitation, nursing castes, nuptial flights, nutrient eggs, queen obeisance, sterile workers, and more (Wilson 1978, 22–23). The normative principles here, as with actual humans, would be contingent rules of a cooperation game. But we typically don't think of morality or ethics as being contingent in this way. We tend to think, for instance, there is something absolutely and unconditionally wrong about torturing children and something absolutely and unconditionally good in caring for the poor.

Notice that this evolutionary account of morality does not by itself imply that we are wrong about the content of our normative principles or their status. After all, there might be a Supreme Creator who created humans through evolutionary processes to live by some absolute moral principles. But Ruse's evolutionary naturalism seems to rule that out. There are two main commitments here. The first is substantive: because humans are the products of evolution, we understand them (and other living things) through evolution. The second commitment is methodological: the "naturalism" in evolutionary naturalism – the only legitimate method for understanding the world is empirical and through scientific investigation (Ruse 1995, 2–7). This naturalism rules out any appeal to supernatural creators and their absolute ethical principles, or anything else that cannot be discovered empirically through science. And what evolution and scientific investigation seem to reveal are only contingent normative rules of cooperation, based on facts about our evolutionary history – not moral absolutes.

I shall not challenge this conclusion, nor will I investigate the complexities of the argument here. (These are topics for other chapters in this volume.) I think in a broad, rough sense Ruse is correct, and we should thank evolutionary naturalism for revealing unjustified assumptions about moral absolutes. But there is another even more fundamental error that is revealed by evolutionary naturalism – an error in how we think about value in general.

We are inclined to think of value simply as a property of objects and actions. We tend to think, for instance, that some actions have the property of being (ethically) good, or of being (ethically) bad. Torturing innocent children is simply bad. Giving to the innocent poor is simply good. Evolutionary naturalism suggests that this way of thinking about value is incorrect. We should instead think of value as a relation between an action or thing, a subject and an environment. Evolution tells us that values in general, and ethical values in particular, are complex relational facts. This, as we shall see, can serve to explain, clarify, and even correct some of our moral deliberations.

The Property Conception of Value

One standard philosophical approach to thinking about value, perhaps the standard way, is in terms of the simple predication of properties, such as "x is good." We see this way of thinking in the standard locutions, "giving to the poor is good" and "torturing children is bad." G. E. Moore approached value in this way. In his *Principia Ethica* he proclaimed: "The peculiarity of Ethics is not that it investigates assertions about human conduct, but that it investigates assertions about the property of things which is denoted by the term 'good,' and the converse property denoted by the term 'bad'" (Moore 1903, 36). Other philosophers have adopted this way of thinking about value, as we see in the debates about naturalism and nonnaturalism, cognitivism and noncognitivism, intuitionism and realism (Darwal et al. 1992). In one recent textbook, for instance, the author claims that "[m]oral realism commits itself to the existence of external, independent moral properties" (DeMarco 1996, 258). Two critics of Ruse, Rottschaefer and Martinsen, give an evolutionary account, but one that conceives values in terms of properties:

We have adopted a non-reductionist account of the moral realm, according to which the moral properties of persons and things, moral rightness and goodness, are distinct from natural non-moral properties because they supervene on the latter. (Rottschaefer and Martinsen 1995, 398)

John McDowell, in his "Values and Secondary Properties," similarly argues for a property conception of value by drawing an analogy between colors and values. According to McDowell, values, like colors, are real properties of things (McDowell 1988). Here McDowell is following the lead of Moore, who also drew an analogy with color perception.

There is ... no intrinsic difficulty in the contention that "good" denotes a simple and indefinable quality. There are many other instances of such qualities. Consider yellow, for example. We may try to define it, by describing its physical equivalent; we may state what kind of light-vibrations must stimulate the normal eye, in order that we may perceive it. But a moment's reflection is sufficient to shew that those light vibrations are not themselves what we mean by yellow. They are not what we perceive.... The most that we can be entitled to say of those vibrations is that they are what corresponds in space to the yellow which we actually perceive. Yet a mistake of this simple kind has commonly been made about "good." It may be true that all things which are good are also something else, just as it is true that all things which are yellow produce a certain kind of vibration in the light. And it is a fact, that Ethics aims at discovering what are those other properties belonging to all things which are good. (Moore 1903, 10)

As we shall see, the analogy with color has a lot to recommend it, but Moore's analysis of color perception is flawed. How it is wrong is instructive in seeing how the property conception of value is flawed.

The Color Analogy

Since Galileo and Locke, colors have commonly been conceived as secondary properties, distinguished from primary properties such as size and shape. Primary properties have usually been taken to be independent of observers. The shape and size of a basketball, for instance, is the same regardless of who is observing it, or whether anyone is observing it. Secondary properties such as color, by contrast, depend on facts about the observer. The color of the basketball depends on facts about the perceiver, and the conditions under which the object is perceived. Nothing is orange in total darkness, or to a color-blind person. This implies that color (unlike primary properties) has a dual nature, being a property of an object, but also a feature of a subject's experience. The color of the basketball is both part of the basketball and of the subject's experience. Barry Maund begins his book, *Colors: Their Nature and Representation*, with this duality problem:

The status of colour has for long been puzzling. Almost everyone agrees that physical objects have colours: that sunsets are golden or red; that bananas are yellow; that claret is purple; and so on. Everyone agrees that objects are perceived as coloured. Where there is disagreement is over the nature of the colour physical objects have and the nature of the colour we perceive objects as having, and indeed whether the two are the same. (Maund 1995, xii)

Maund and others try to solve this problem within a property framework. Maund argues that color is a "virtual property" – "an intrinsic nonrelational property of physical objects, but one which no object in the actual world has" (Maund 1995, xiii). Other attempts to solve the duality problem conceive color properties as objective, irreducible and supervenient; objective and reducible; dispositional relative to appearance; or dispositional relative to functional role (Maund 1995, 4).

The details and evaluation of each of these solutions are beyond the scope of the discussion here, but there is a simpler way to accommodate the dual nature of color. Treat color as a relation between an object and a perceiver. As Evan Thompson argues, color is always relative to a perceiver and a photic environment (Thompson 2000, 179–180). For those who lack color vision, nothing is ever colored, and in pure red light everything is red. What this implies is that colors are not simple properties predicated of objects. The standard logic of color attribution in the statement "x is color y" is therefore misleading about the nature of color perception. A more accurate statement would be relational and could be schematized as follows:

Relational Color Schema: w is color x for y in context(s) z.

Here w is an object, x is a color term, y is a perceiver, and z is a description of, or reference to, a particular context or set of contexts.

The advantages of such a relational schema are obvious when we start thinking about variability in color perception. Normal human vision is trichromatic, based on the presence of three kinds of photoreceptor cones, each with a different absorption spectrum. But there is variation in human color vision. There are the anomalous trichromats who have the standard three kinds of cones, but cones with abnormal absorption spectra. There are dichromats who have only two kinds of cones and are unable to distinguish some colors distinguishable by trichromats. And then there are the monochromats who lack color vision altogether. On the other hand, there are tetrachromats, individuals who have four kinds of cones and who can distinguish more colors than trichromats (Coren et al. 2004, 102–108). This suggests that color vision among humans is highly variable and dependent on the subject. What may seem green to a normal trichromat may not seem green to a dichromat who lacks the green cone, or to a monochromat who lacks color vision altogether. So for the normal trichromat, the statement, w is green, may seem true, but it would seem false for the dichromat or monochromat. Color vision is even more variable across fauna. Tetrachromacy is common among vertebrates – birds, fish, amphibians, and reptiles – and some are sensitive to light in the ultraviolet wavelengths as well (Thompson 2000, 175). All of this implies that an object can have different colors for different perceivers – even under the same viewing conditions.

Whether or not it is possible to accommodate this variability in color perception by a simple property schema, w is color x, the relational schema has some clear advantages. It implicitly recognizes that there are colors because of how the visual systems of organisms interact with the world. And it explicitly recognizes all of the factors relevant to color perception: the object itself, the nature of the perceiver, and the photic environment. Thinking about colors as properties (even causal dispositional properties) simply predicated of objects misrepresents the nature of color perception (Richards 2005,

275–276). Once we recognize this relational structure of color perception, the analogy with value becomes helpful. This is most apparent from an evolutionary starting point.

Evolution and Value

What evolution tells us about values is that they, like colors, are best understood as relations. We can see how by first looking at value and natural selection. In his *Origin of Species*, Darwin tells us that natural selection preserves and adds up that which is "good," and rejects (eliminates) that which is "bad." And it does so in relation to conditions of life:

It may be said that natural selection is daily and hourly scrutinizing, throughout the world, every variation, even the slightest; rejecting that which is bad, preserving and adding up all that is good; silently and insensibly working, whenever and wherever opportunity offers, at the improvement of each organic being in relation to its organic and inorganic conditions of life. (Darwin 1859, 84)

The idea here is that goodness or badness is relative to a variation and a condition of life. And by extension, an organism is good or bad relative to some condition of life, as Ernst Mayr explains:

The "goodness" of the new individual is constantly tested, from the larval (or embryonic) stage until adulthood and its period of reproduction. Those individuals who are most efficient in coping with the challenges of the environment and in competing with other members of their population and with those of other species will have the best chance to survive until the age of reproduction and to reproduce successfully. Numerous experiments and observations have revealed that certain individuals with particular attributes are clearly superior to others during this process of elimination. They are the ones that are "fittest to survive." (Mayr 2001, 119–120)

As indicated in the last sentence here, the crucial idea is that of fitness. An organism is fit: it is good, insofar as it functions well with respect to survival and reproduction in some environment. And a trait or variation is fit: it is good, insofar as it functions well, making an organism fit.

The logical structure of the fitness conception of goodness is apparent in the ways we talk about fitness. We don't say that feathers are simply good (or simply bad). Rather they are good relative to particular ways of functioning in particular contexts or environments. Feathers, for instance, are sometimes good with respect to their insulating power in cold climates (think penguins). Other times they are good for locomotion in aerial environments. But it is also easy to see that feathers would be bad in other contexts, such as for tapeworms in gastrointestinal tracts. The basic fitness concept implies that a trait is good or bad for an organism with respect to some way of functioning in an environment. If so, we can schematize this goodness or badness as follows:

Simple Valuation Schema: w is good (or bad) with respect to x for y in z.

Here w is a trait; x is a way of functioning (relative to factors relevant to survival and reproduction); y is an organism, or group, or kind of organism; and z is a context/environment, type of environment, or range of environments. There is also a derivative, comparative schema:

Comparative Valuation Schema: v is better (or worse) than w with respect to x for y in z.

Here v and w are alternative traits, and judgment is relative to the comparative goodness of traits. We can also use these schemata to indicate the goodness or badness of organisms as a whole – in terms of their phenotypes: Phenotype w is good with respect to x (survival and reproduction) for y in z. There is also a comparative schema here: Phenotype v is better than phenotype w with respect to x (survival and reproduction) for y in z. Notice that the value of behaviors or objects is also best understood in terms of these value relations. Run and jump behavior in prairie dogs can be good with respect to avoiding predation on the prairie. A rock can be good with respect to cracking nuts or defending against predators, for humans in Pleistocene environments. A gum wrapper can be good for male bowerbirds in enticing females, and a shell can be good for hermit crabs with respect to its shape and size for providing shelter in marine environments.

There is value (positive or negative) in all of these cases, because the things with value (whether it be traits, phenotypes, or objects) matter in some way or other. (I am adopting the terminology of Railton [2003] here.) The most obvious way to matter, as we see with the concept of fitness, is relative to survival and reproduction. Traits, behaviors, or objects that help an organism survive are clearly of some positive value here. Those that reduce chances of survival and reproduction are clearly of negative value in that respect. But mattering isn't limited to actual survival and reproduction. Something can have value if it might matter for survival and reproduction in some cases. Pleasure and pain, for instance, matter in many ways, including the implications they have for the reinforcement of behaviors. If so, the pleasure social creatures get from the company of others, the enjoyment carnivores get from play, and even the satisfaction philosophers get from solving philosophical puzzles all matter and all have value (positive, negative, or neutral). In general, anything that has some effect, or potential effect, on the well-being or well-functioning of a subject can have a positive or negative value for the subject. What has value for a subject is, in the end, an open question, answerable by empirical investigation into the natures of the relevant subjects and their environments.

Values

What is important here is that this relational schema explicitly represents all of the important factors in valuation. But the semantics of valuation, how we instantiate the variables in the schema, also allow us to make important distinctions. Something may be good with respect to one kind of functioning, but not with respect to another. Long necks may be good for giraffes in browsing, but they present disadvantages for cardiovascular functioning. Something may also be good in one environment but not another. Long necks may be good in a savannah setting, but not in a tropical jungle. We can also express the different degrees of generality in the goodness of traits. A trait may be good for many different subjects, or kinds of subjects, or just one or a few. Homothermia, for instance, is good for all those warm-blooded creatures that benefit from its effect on physiological processes. Or a diet might be beneficial for a narrow, small group of organisms, such as we see with those pandas that rely on an exclusive diet of bamboo leaves. Something might also be good or bad in a wider range of contexts or environments. The feeding mechanisms of mice and rats, for instance, work well in many different environments, while the white coloration of arctic foxes works well only in snowy environments.

Moreover, and most significant for purposes here (as we shall see), something might be good or bad at multiple organizational levels of subjects. A behavior might be good or bad for a single ant, an ant colony, an ant species, several species in symbiosis, or even an entire ecosystem. Similarly, an action may be good or bad for a single human, a family, a community, a nation, or the entire human species. Once we recognize this, we can see also that something may have conflicting value, benefiting an individual while harming a family or community, or harming an individual while benefiting a community.

What is also important is that while value is subjective, in that it is always in relation to some subject, it is also objective. Whether something is good or bad is not just a matter of opinion – whether someone thinks it is good or bad. It is a factual and empirical matter, supported in the strongest possible manner by consequences for reproduction, life, and death. So just as no biologist would deny the objective facts about fitness, no one should deny the objective facts of valuation. And insofar as statements of this mattering represent the relations between a subject and context or environment, there will be truth values in a straightforward way. If something really does matter in the manner asserted to a subject in a context, the value statement will be true. It will be false otherwise. The implication of this way of thinking about valuation is that there is no simple fact/value distinction. Instead there are value facts and non-value facts. The former are simply those facts about mattering that have the relational structure described here.

At the beginning of this chapter, I claimed that values are complex relational facts. We can now see precisely what that means. On this naturalistic stance, valuation is the fundamental idea, as the source of value in the mattering for subjects. Values are the patterns of valuation or mattering, brought out in the relational schemata. The mattering of opposable thumbs or trichromatic vision for humans in their environments, for instance, are values. This sense of "value" should be distinguished from the use of that term to refer to beliefs or attitudes about value. There are, after all, facts about how opposable thumbs and trichromatic vision functions that are independent of any beliefs or attitudes. And something may be valued in belief or attitudes without actually having that value – mattering that way – in reality. The bottom line is that there are facts of mattering that are objective and independent of beliefs or attitudes.

Notice that there may be individual, group, species, or superspecific value facts depending on our interests. If something matters to me in a particular way, then it is a personal value. If it matters to those in a community, it is a community value. If it matters in a particular way to the individuals of an entire species, it is a species-wide value. And if it matters in a particular way to the individuals of multiple species, it is a supraspecific value.

Ethical Value

The account here has so far been about valuation and value in general, and not specifically about ethical value. Perhaps evolutionary naturalism can give us some guidance in thinking about what constitutes ethical valuation. If our ethical practices, moral sentiments, and social psychology are adaptations to help us cooperate, then what is at issue is a particular type of mattering: social mattering. If so, then we can understand ethical value semantically, in terms of the instantiation of the variables in the value schema. To be socially good, a trait or action must function well in social contexts – environments where there is interaction with other individuals. We can restrict the variables to reflect this type of mattering. As with value in general, there is a simple and a comparative schema.

Simple Social Valuation Schema: w is socially good (or bad) with respect to x for y in social context(s) z.

Comparative Social Valuation Schema: v is socially better (or worse) than w with respect to x for y in social context(s) z.

Here v and w are traits (behaviors, actions, sentiments, desires, motives, etc.), x is some kind of social mattering, y is a social being, and z is a social context.

We can apply these schemata to some obvious cases of ethical value. Lying is socially bad by creating distrust among the family members in family contexts. Similarly, cheating on a spouse is bad by creating distrust in a pair bond. And in both of these cases there may be additional social effects – the emotional distress of family members or mates at the violation of a valued social obligation. Giving assistance to the poor will have some obvious social mattering and in a variety of ways, perhaps in affecting the overall well-being of a community, satisfying expectations for members of that community, and enhancing the reputation of the giver – all within a particular social context. Stealing has obvious social mattering: violating assumed property rights, costing the victim, affecting behavior, and creating distrust. In social mattering and ethical value there may sometimes be obvious implications for survival, reproduction, and perpetuation of one's genes. A parent who mistreats his or her child – by leaving him in a hot car, neglecting her nutrition, or endangering him in some other way – is widely regarded as morally blameworthy. But social mattering in many cases may not have such obvious evolutionary implications. It might even extend beyond human subjects to those creatures with which we have or might have a social relationship – our pets.

Evaluating the Relational Conception of Value

There are some obvious complications here. First, the distinction between social and nonsocial mattering may be unclear. In part this may be due to disagreement about what social relations we have. Those who believe in a personal God, for instance, might believe that this implies a social relationship, complete with social obligations and prohibitions. Those who believe in Mother Nature and that the earth is like a living thing might also believe that there is a social relation with nature, complete with social obligations and prohibitions. And for those who believe that their relations with their pet dog, cat, bird, or fish are social, and not simply property relations, there will likely be consequences for beliefs about social mattering and ethics. Or, some mattering may also just be more clearly social than other ways of mattering. The mattering related directly to functioning within families and communities is clearly social. But what about recreational drug use in contexts that don't seem to create risks for others? Is there social mattering here? Answering this question would require more space than is available here, but we can see why it is sometimes hard to be precise about what counts as social mattering and ethical valuation.

A second complication is that with social mattering, as with valuation in general, there will be conflicts in valuation. Something might be good in one way, but bad in another. Something might benefit one person while harming another. Something might be good in one social context and bad in another. It isn't obvious how to weigh these different ways of mattering. On what grounds do we say one way of mattering is more important than another? There may be obvious cases, where life and death are at stake versus narrow and superficial social status. One social context may be more relevant simply because it is more common. But there are more problematic cases. We may prefer social benefit for one person over another person, or for one group of persons of another group. But why should we favor the benefit of the first group over the latter? Decisions in these cases may be simply political, favoring the goods of one group over the goods of another group.

There are many advantages to this way of conceiving value in general, and ethical value in particular. Some I cannot address here (see Richards 2005), but three seem to be most important. First, this approach gets valuation right. Something has value precisely because it matters to some subject. It is good or bad for some subject in some way. This is a simple, straightforward, and obvious way for something to be good or bad. Second, this conception of value makes sense of our judgments that implicitly recognize its relational nature. We condemn some behaviors by some people in some contexts, but not in others. Certain kinds of sexual behavior, for instance, are permissible among consenting adults in the privacy of their own homes. But the same actions may not be permissible with children or in public contexts such as restaurants and churches. This is at least in part because of the contrasting ways these actions typically matter in each instance. They are good in some contexts and for some subjects, but not in others. Moreover, we often explicitly recognize conflicts in value – that an action might be good in some respect, but bad in another. Taking from the rich to give to the poor may be good for the poor, but not for the rich. But it might also be good for the poor in one way – providing resources – but bad in another way – encouraging dependency.

The third advantage of this approach is that it can help us think more clearly in our moral reasoning and deliberations. Most obviously, it eliminates the error in thinking about value as a property of things. Actions don't simply have the property of being morally good or bad. They are good or bad for subjects in particular contexts. The recognition of this fact may disincline us from extending moral judgments beyond their appropriate domains. If deliberate misrepresentation of the facts is only ethically bad in some contexts, and we are conscious of what those contexts are, we won't make errors by assuming it is always bad. When an actor "lies" onstage in reading her lines, for instance, that deliberate misrepresentation has a different value than when a witness lies in court and under oath. To attribute the property of being wrong or bad to all cases of deliberate misrepresentation simply gets the valuation of it wrong in many instances. Similarly, killing another human may be good with respect to self-defense, but it is bad in other respects and in most other cases. The property conception cannot easily account for this variation in value. If lying is simply bad, surely it must be bad in all cases.

The relational nature of value also raises worries about the commensurability of value. If goodness were a simple property predicated of different things, then we might be able to compare goodness as we can the simple weight or volume of objects. Just as one object is simply heavier than another, one action could be simply better than another. But if valuation is relational, as argued here, it isn't clear how to compare different instances of goodness or badness. There are different ways of being good, different subjects for which something is good, and different contexts in which something is good. Only by bracketing their relational nature will it be possible to compare goods or bads. We might say, for instance, that "killing innocent babies is worse than telling a lie to a child." But that is misleading. What we should say instead is something like "killing innocent babies is worse for those babies than telling that lie is for that child in that context." And even then, the respect in which each has disvalue is different, and may not be commensurable.

Notice that this relational theory of value does not automatically tell us what normative rules to adopt. It doesn't necessarily tell us that we shouldn't lie, that we should recycle, or that we shouldn't torture children or kittens. But it does tell us how to determine the value of a particular normative rule, by reference to subjects and contexts. Some rules will generate little or no value for anyone, or negative value for many subjects in many contexts. Some rules will enhance cooperation benefiting everyone involved, while others will reduce cooperation, perhaps engendering conflict and harm. In short, some rules will make the cooperation game work better, while some rules will make it work worse. So insofar as we benefit from the cooperation game, we will benefit from those rules that enhance cooperation and be harmed by those that make cooperation more difficult. This value of particular moral principles would then provide at least some reason to adopt or reject those principles.

The details of a normative ethics are beyond the scope of this chapter, but given the complexity and conflicts in valuation, it is likely that moral deliberations and ethical reasoning will be complicated and messy. We cannot simply compare and add up the different ways of mattering for different subjects in many contexts the way we might if goodness were a simple property like weight or volume. Mattering is relative to subjects, and something can be good or bad at different levels of inclusiveness of subjects – a single person, a pair bond, a family, a community, a nation, a species, and even a higher-level group. Since there will often be conflict in mattering at these different levels, there will tend to be conflict in optimal normative principles. One principle may benefit one subject (or subjects) and harm another, or benefit one level and harm another level. Evolutionary theory has long recognized this fundamental conflict of interest, and the relational conception of value reveals it for our moral deliberations. But also the relational conception of value brings common goods as well as conflicts of interest to the foreground and makes us explicitly consider both in our deliberations.

Evolutionary Naturalism and the Logical Error

The analysis here implies that when we predicate goodness or badness as a simple property of things, we are making an error. Things are not simply good or bad; they are good or bad in relation to ways of mattering, subjects, and contexts. To assert that x is good is therefore making a logical error, mistaking a relation for simple predication. But why would we so often make this error about value? One obvious utility of such an error is that it is simpler and faster to think of value as a property of things. What we often need to know is whether something is good or bad in some specified way for ourselves. "That snake is bad" is faster and in many relevant situations more useful than "that snake is bad with respect to the harm its venom could do to me in hiking contexts." Simple predication may result in the better avoidance of poisonous snakes. Similarly, there may be practical value in thinking of a particular person as simply "good" or "bad," rather than as having relational value in particular contexts. A person that is "good" for me in some respect may be a person I should seek for company and shared activities. A person that is "bad" for me may be a person I should avoid for particular reasons in particular contexts. A simple rule to seek out people who "are good" in this way and avoid those that "are bad" may be more useful than rules to seek out people who are good in specified ways, for specified subjects, and in specified contexts. Similarly, it may be better in motivating behavior to think of an action as simply good or bad. The person who spends time and effort trying to determine all of the ways an action may be good or bad for various people and in various contexts may be less likely to make a timely decision relative to value for herself or himself in a particular context, or for most people and in most contexts. In effect, the person who thinks of values as complex relations may be more likely to become a victim of a snake's venom or a person's dishonest intentions, and he or she may be less likely to behave in socially constructive ways.

If property thinking has this value in daily living, then perhaps there is some explanatory adaptation hypothesis. Perhaps we think of value in property terms because that way of thinking has some genetic basis and was an advantage in survival and reproduction. I won't try to develop such an adaptation hypothesis here. But whether or not there is an evolutionary explanation for the logical error in thinking of value as a simple property predicated of things, evolutionary naturalism at least reveals this error for what it is. One might agree, however, that evolutionary naturalism implies that this property thinking is in error, without accepting that it is in fact an error. One might do so by rejecting evolutionary naturalism. A theist might argue that there are good reasons to believe in a transcendent God, or a Platonist might argue that there are good reasons to believe in transcendent forms. Given the successes of naturalism though, I don't find these sorts of arguments promising. But that is a discussion for another time and place.

10 Evolutionary Ethics: A Theory of Moral Realism

Robert J. Richards

Introduction: Darwin's Theory of the Evolution of Morality

In *The Descent of Man and Selection in Relation to Sex* (1871, 1: 72-73), Charles Darwin produced a theory of moral conscience and an explanation for ethical behavior. His theory had several elements. To develop a conscience that recognized morally right and wrong action, an organism had to have (1) a repertoire of social instincts – those urging parental care, group cooperation, altruistic response; (2) sufficient intelligence and memory to make practical judgments of a complex sort; (3) language to codify behavior and communicate desires, requests, and other information to conspecifics; and (4) habits to help shape behavior. Darwin thought the crucial feature of his account to be the social instincts, those innate expressions of cooperative and altruistic response. But he knew there was a problem in providing a natural-selection explanation for behavior that seemed to benefit the recipient rather than the agent. He believed he had the answer:

It must not be forgotten that although a high standard of morality gives but a slight or no advantage to each individual man and his children over the other men of the same tribe, yet an enhancement in the standard of morality and increase in the number of well-endowed men will certainly give an immense advantage to one tribe over another. There can be no doubt that a tribe including many members who from possessing in a high degree the spirit of patriotism, fidelity, obedience, courage, and sympathy, were always ready to give aid to each other and to sacrifice themselves for the common good, would be victorious over most other tribes, and this would be natural selection. (Darwin 1871, 1: 166)

There are many details of Darwin's proposal that would require extended discussion: for instance, though he based his conception of human community selection on altruistic response in the social insects – in which natural selection operated not on the individual but on the entire hive or nest – one might ask whether the social instincts in humans were produced by group selection or individual selection. Michael Ruse and I disagree about Darwin's answer to this particular question – profoundly. Ruse argues that selection in ancient clan-groups was roughly equivalent to kin selection as we now conceive it, and thus a kind of individual selection, while I regard Darwin as having generalized his notion so that it applied to human tribes in which there was no expectation that all individuals would be related, thus a matter of group selection (Richards and Ruse, 2016). Ruse believes that as proto-human communities expanded, individuals began to aid one another under the expectation of reciprocity. He assumes that this original attitude would become cemented into our genes by a process of individual selection, since such an attitude would be individually beneficial. Ruse forgets that Darwin called reciprocal altruism a "low motive" (Darwin 1871, 1: 163), being at root selfish. Darwin claimed his theory of community selection removed "the reproach of laying the foundation of the most noble part of our nature in the base principle of selfishness" (Darwin 1871, 1: 98). Strong evidence, I believe, that he advanced a theory of group selection. Moreover, if it were only a matter of individual selection, the most advantageous trait would be one in which the bearer feigned cooperation while defecting. But in the group selection scenario, feigning cooperation while defecting would be deleterious for the whole group and the group would be selected against. Thus in the long run, the advantage goes to groups in which most of its members are sincere altruists. There is further evidence that Darwin was a group selectionist; but for the purposes of this chapter, the question of the exact character of Darwin's conception – whether supporting group or individual selection – is not immediately crucial. Both Ruse and I agree that Darwin was fundamentally correct when he held that human beings have evolved to act altruistically toward one another, even if sometimes failing to do so.

Despite our differences about the target of community selection, Ruse and I both endorse Darwin's assumption of an innate attitude toward cooperation and toward helping those in need. The empirical evidence for such an attitude is fairly strong, but we are both interested more in the philosophical consequences of this assumption. So in this chapter, I will simply assume that something like the Darwinian theory of morality is our best account of impulses and judgments we have come to call moral. That is, I will assume we have evolved to be altruists, that altruism evokes a special attitude of approval, and that analysis would show that fundamental moral principles – e.g., the Golden Rule, the Categorical Imperative, etc. – amount to altruism.

A further feature of Darwin's account has Ruse's and my allegiance: that the altruistic instinct needs the guidance of rational considerations. We may feel obliged to act for the community good, but only a rational analysis based on experience and acquired knowledge can determine for us who are members of our community and what really does serve the community good. The Hippocratic physicians who administered to victims of the Athenian plague treated their patients by bleeding them and giving emetics, which could only hasten their deaths. However, the physicians did so under a mistaken medical belief about the efficacy of those treatments; they served their patients at considerable risk of becoming infected themselves and many did become infected. The development of medical science over the centuries has supplied effective treatments for plague victims. Contemporary doctors who harbor comparable altruistic instincts now have a rational strategy for successfully requiting those instincts. We would today, nonetheless, look back and judge those ancient physicians to be moral heroes, since their motives and intentions were altruistic, even if their beliefs were invincibly defective.

Most critics who object to an evolutionary ethics would, I think, be willing to grant these assumptions, since the potent philosophical objections do not lie along the lines of empirical adequacy. The deeper question I would like to explore in this chapter is the validity of the most powerful objection raised by philosophers to the general theory of human moral evolution, namely, that it fails to be an account of authentic moral behavior. The objection has been developed in two distinct but related ways: that an evolutionary theory of morality commits the naturalistic fallacy and that it fails to be objective, that is universally binding, and by default must embrace only a subjective relativism. These specifications of the general objection are connected.

Objections to an Evolutionary Theory of Morality: The Naturalistic Fallacy

G. E. Moore claimed, in his classic formulation of the naturalistic fallacy (1902), that it was a mistake to identify the moral good with any natural property. He particularly had in mind Herbert Spencer's identification of "more evolved" with "ethically better" (Moore 1903, 48-54). Such identification would be liable, Moore argued, to the open-question objection: we could always ask whether this altruistic act, or this more evolved behavior, was, after all, really good, and the answer, yes or no, should enlighten us. If moral goodness were simply identified with the natural trait of altruistic impulse, the question would amount to, is this altruistic act really altruistic? The answer would seem hardly to enlighten us. Moore believed we could intuitively perceive the moral good as a nonnatural property accruing to different actions in particular contexts; it was a real property, the perception of which provided reason to approve of actions to which it accrued or supplied reason to pursue such actions. Moore's objection to Spencer's moral theory wasn't that it failed to be a realistic theory, but rather that it was not the right kind of realism.

Thomas Henry Huxley introduced the objection of the naturalistic fallacy even before Moore formally named it. Huxley's lecture "Evolution and Ethics" (1893) was also directed at the ethical theory of Spencer, his erstwhile friend, but it told equally against Darwin's theory of moral evolution. Huxley recognized a disjunction between the biologically fit and the morally fit: "Social

progress means a checking of the cosmic process at every step and the sub-
stitution for it of another which may be called the ethical process; the end of
which is not the survival of those who may happen to be the fittest, in respect
of the whole of the conditions which obtain, but of those who are ethically
the best" (Huxley 1893, 9: 81). Huxley conceded that our so-called moral
sentiments had evolved no less than our aggressive sentiments; but this outcome, simply considered, furnished no moral rules, no moral goals. "Cosmic
evolution," he cautioned, "may teach us how the good and the evil tendencies
of man may have come about; but, in itself, it is incompetent to furnish any
better reason why what we call good is preferable to what we call evil than
we had before" (Huxley 1893, 9: 80). In nuce, Huxley deployed the open question: Why ought I act on this altruistic attitude rather than on that vengeful
attitude? Huxley's analysis reduces to the more common way of stating the
naturalistic fallacy, namely, that one cannot justify an "ought proposition, a
norm, by appealing to a fact – no "ought" can be derived from an "is."

The naturalistic fallacy has lost its sting. Several considerations mitigate,
really abolish, its force (Richards 1986, 1987, 1993; Richards and Ruse 2016).
Take the simple assertion that facts cannot justify norms. In science, favorable experimental outcomes, thus empirical facts, are used in the justification
of natural laws, and laws carry normative force – e.g., this copper, if dunked
into sulfuric acid, ought to produce cupric sulfate. And counterfactually, if all
copper and sulfuric acid suddenly disappeared from the world, it would still be
true that if sulfuric acid were added to copper it should produce cupric sulfate.
So if facts cannot justify norms, science as we have come to understand it
must collapse. Justification is not always or even usually based on a deduct-
ive relationship. Of course, we feel no moral outrage if a well-justified natural
law fails on further testing. The normative implications for natural laws are
indeed different from moral principles that mandate altruism.

What work does the concept "ought" perform? In itself it has minimal
semantic meaning; it connects a particular causal matrix or property with an
expectation. It achieves more definite semantic content from the particular
causal matrix or property that serves as the basis for the expectation. So there
is no moral ought or instrumental ought devoid of a particular basis. Consider
the kind of "ought" the moral imperative is in practice. It cannot be a categorical "ought," as Kant believed, that is, an "ought" without a basis. After all, we
don't admonish a two-year-old with a categorical "ought-not" about swiping
cookies, but recognize that the individual must be in a certain condition (i.e.,
human, of sufficient age and mental capacity) to be morally admonished. So
the structure of the real, moral imperative is, if you are a human being, with
the full capacities of such – the factual basis for the moral judgment – you
ought not lie, cheat, or steal. In light of the evolutionary assumption I've
made, we would say, since you are a human being who has evolved to have

altruism as an essential component of your natural fabric, then you ought to act altruistically. The ought carries moral weight, as opposed to, say, only instrumental weight, because of the evolved capacity for altruism, which we have identified (as part of the granting of empirical fact) as the moral inclination. This evolved capacity, I have premised, comes with a very strong attitude of approval for altruistic behavior and a comparable negative attitude when a violation of such behavior has occurred. Because the evolutionary process has deeply embedded the altruistic capacity in the human group, we are warranted in the expectation – a kind of prediction – that in the requisite circumstances an individual ought to act altruistically. Such an expectation, of course, is modified by other circumstances. When, for example, a known thief is about to strike again, we recognize that this criminal is likely to ignore his better impulses. If those impulses are altogether absent – say, in the case of a profound psychopath – would we think such a person, if we were aware of his character, *ought* not steal? We would certainly want to be protected against the psychopath in our midst; but if such a person were simply, say, born that way – born with a kind of insanity or profound mental deficiency – could we hold such an individual morally responsible? We may want to hold the psychopath legally responsible and be protected against the dangers he or she might pose, but on careful consideration I do not believe we could hold him or her morally responsible; for we cannot morally demand something of one who is quite incapable of the action, or of one who sincerely does not recognize a moral requirement.

At this juncture, let me examine two kinds of objection to the sort of analysis I've offered of the naturalistic fallacy, that of Moore and that of Richard Joyce. Moore urged an open-question rebuttal to the likes of Spencer's evolutionary ethics. To consider the force of the objection, let us presume not only that we have evolved to approve altruistic acts, that is, recognize them and regard them as special, but also that we carry the attitude as a reflective criterion. So if we reflectively ask ourselves, but is altruism really a special good, then the criterion with which evolution has endowed us would certify our immediate judgment. One could keep asking the reflective question, but if we have evolved in the way supposed, the answer would be boringly the same. This case would be similar to a Kantian immediately perceiving various examples of causal relationships, but then, in a reflective mode asking himself or herself: does every event really have a cause? The answer must be "of course," since every such reflective question evokes the same category of judgment – there is no other way to think about the world of our experience. Would the reflective person learn anything? Initially, very likely; he or she would learn that the first impulse conforms to the criterion that reveals itself as inescapable. This consideration holds for the moral reflection: if we have evolved in the way specified, there is no other criterion to apply, certainly not one that would make sense of our general moral experience. The moral criterion, if generally imposed by evolution, is inescapable. So in this scenario, Moore's open question swings closed.

I have distinguished two factors in a moral judgment – the recognition of community needs and a natural sentiment, a sentiment that has evolved over generations and that approves acting in respect of those needs. Richard Joyce argues that pro-social feelings are irrelevant to moral judgment: the difference between doing something "because you want to do it is very different from doing something because you ought to do it" (Joyce 2006a, 50). To use Joyce's example, I might feel love for someone – that is, I don't want to harm the person – but lack any sense of obligation not to harm her (Joyce 2006a, 51). What then does the "ought" of obligation mean? Joyce suggests it means "inescapability and authority (Joyce 2006a, 62)." I have just given an account of the inescapability of the evolutionary criterion. But whence the authority? The authority of a moral judgment gives it "practical clout," a reason for following a moral precept, such as acting altruistically. Joyce doesn't think we can reduce the authority to a sentiment or feeling – a want – since we presume moral values "bind people irrespective of their desires or interests" (Joyce 2006a, 192). Herein lies his debunking or skeptical attitude: there is no adequate naturalistic account of the practical clout of a moral judgment. Maybe, however, Joyce hasn't tried hard enough.

Moral judgments can be usefully distinguished into two types: our immediate response to someone in distress, which elicits a desire or need to help – the archetypal example of impulsively jumping into a river to save a drowning child – and a more reflective and calmer judgment, when, for instance, I take out my credit card to respond to an Internet plea from Doctors Without Borders. In the first instance, no reasons are given, but the urge to action is immediate, impulsive, and undeniably pushed by feeling; nonetheless, we usually think of someone who has so acted, without calculation of the dangers, as deeply ethical and morally courageous. Such a person displays what an Aristotelian would call the *habitus* for ethical behavior. The altruistic desire, honed over millions of years of evolutionary selection, is of a different caliber than the desire for pistachio nut ice cream. In the second instance – that of taking out my credit card – I have time to reflect on my reasons for so acting: but I've evolved, so our assumption allows, to approve of altruism and weigh it against the set of other desires. Its authority is simply that of the special attitude that makes it stand out against other momentary needs and fluctuating impulses. Another desire may be momentarily stronger – e.g., to use my extra money to buy expensive tickets to a play rather than send it off to doctors who are short of bandages. The inescapability of the altruistic attitude supplies the urgent desire that has become biologically entrenched. Under this scenario, the altruistic attitude forms part of what it means to be human; the subjective desire for pistachio nut ice cream lacks the authority of millennia of selection, and it has no special claim to being part of our humanity. I'll elaborate on this notion of human identity in a moment.

Objectivity of Evolutionary Ethics

This rejoinder to the problem of the naturalistic fallacy leads to the other specification of the general problem of the authenticity of moral response under the Darwinian theory: are moral values given to us in experience the same way objects and events are so given – that is, objectively – or are we coerced by evolution into accepting a faux morality, one that is essentially a subjective preference? Some who advance the so-called debunking argument believe we are subjectively constrained, not objectively compelled. Michael Ruse and E. O. Wilson, as well as Sharon Straight, offer nice examples of this kind of argument. Joyce and Horn furnish other examples in this volume.

Ruse and Wilson provide a succinct set of considerations which imply that the Darwinian moral judgment fails to be objective, and thus not universally binding. Human beings, the authors maintain, "are deceived by their genes into thinking that there is a disinterested, objective morality binding upon them, which all should obey" (Ruse and Wilson 1986). The instinct to reciprocate cooperation, which Ruse and Wilson regard as the result of individual selection, is represented by epistemic rules that have become innate. These innate principles dispose "us to think that certain courses of action are right and certain courses of action are wrong." They give us "the illusion of objective morality." They are, however, entirely subjective, relative to the contingencies of environmental circumstances during the course of our peculiar, human evolutionary history. Ruse and Wilson contend that these rules are simply the "idiosyncratic products of the genetic history of the species and as such were shaped by particular regimes of natural selection." To make this assertion vivid, they contrast our human genetic history with that of "an alien intelligent species evolving rules its members consider highly moral but which are repugnant to human beings, such as cannibalism, incest, the love of darkness and decay, parricide, and the mutual eating of feces" (Ruse and Wilson 1986, 186). This alien species could have been us. So what we take as the universally binding and objective rules of morality could have been otherwise. We prefer feeding the hungry and aiding the poor; they prefer dining on feces and throwing their fathers under a bus. There but for the grace of our peculiar selection history go we.

Ruse and Wilson believe our moral attitudes are the result of the evolutionary process. What they deny is that, precisely because of this process, our ethical rules could meet an ideal of what such rules ought to be, namely objective and universally binding. (One should note that Ruse and Wilson are assuming an objective standard of morality to deny that evolution could provide one. But whence this objective standard?) Thus for the evolutionary theorist, there is no reason to regard the moral rules we adhere to as any different than personal preferences, preferences like that for pistachio nut ice cream. Before responding to this assault on our ordinary moral attitudes, let me make some distinctions relative to the terms "subjective" and "objective." I believe Ruse and Wilson glide over important discriminations.

The pair of terms "subjective" and "objective" could be given an ontological meaning or an epistemic meaning. A proposition, an idea, or an attitude is held, of course, in the mind of an individual, a subject; these mental events are thus ontologically subjective. An event that occurs extra-mentally, in the natural world, would be objective in the ontological sense – a solar eclipse, for instance. Epistemically, however, a subjective belief would be one that is the result of personal idiosyncrasy or prejudice, while an epistemically objective belief – the kind characteristic of scientific propositions – would be one intersubjectively verifiable, capable of confirmation by others acting without bias. Ruse and Wilson slide around between the ontological meaning of these terms and the epistemic.

Ethical propositions are certainly ontologically subjective. After all, they are held by individual subjects – they are held in the minds of individuals. They needn't be epistemically subjective, however. Consider logical propositions (e.g., the principle of noncontradiction, the principle of modus ponens, etc.) or mathematical propositions (e.g., the Pythagorean theorem). Except for Platonists, such propositions are usually taken to be ontologically subjective, but are also considered epistemically objective – that is, they follow from accepted standards that are ultimately intersubjectively verifiable. Ruse and Wilson reject the idea of "morality as a set of objective, eternal verities," "extrasomatic guides," which for their money would constitute an authentic morality (Ruse and Wilson 1986, 186). But they seem to be looking for a set of propositions chiseled in stone and lying on a mountain top – ontologically objective entities. Certainly, moral codes exist in the minds of individuals; they are ontologically subjective. But this does not mean they must also be epistemically subjective. The validity of a given syllogism or even a natural law is objective if justified by accepted standards, which standards are ultimately intersubjectively verifiable. A given argument or proposition can be epistemically objective without being ontologically objective.

Can ethical arguments be epistemologically objective? Consider the history of our species. Most evolutionists will maintain that our rational ability is the result of the evolutionary process. Our ability to assess arguments – say, for the validity of evolutionary theory itself – is not impugned by its origin as an evolved capacity. Those of our hominid predecessors who regarded the saber-toothed cat in their paths as both gentle and dangerous have left a very short line of descendants, while those who evolved cognitively to avoid contradictions have prospered. We have come to hold the principle of noncontradiction because of natural selection. Yet, the application of the principle to establish mathematical or logical systems can hardly be rejected as merely a subjective preference. Mathematical and logical propositions can be perfectly objective without being derived from "extrasomatic guides." The claim that extrasomatic guides are necessary for objectivity is to confuse epistemic objectivity with ontological objectivity. Moreover – and this I'll explore further – the fact that our cognitive faculties have survival value does not at all imply that they are unfit for evaluating truths about the world. Rather they have survival value because, quite generally, they are reliable guides in determining truths about the world.

Ruse and Wilson suggest that an ideal of objective morality cannot be realized, since we could have evolved otherwise than we have. They seem to believe that there could be an alien species, in all essentials like us, but one with a moral code not based in cooperation and altruism. But if that species were intelligent and social, thus like us in that respect, then it would have to be moral like us as well. For what would a social group be like if it were not bound together by ties of cooperation and altruism? Ruse and Wilson's scenario of social aliens that were non-altruistic and noncooperative would be comparable to a story about an intelligent species that did not observe the principle of noncontradiction. What kind of hippogriff-like species could that be? Such an alien species could not be us. Nor could it even evolve as a social species. To put it crudely: early in our evolutionary line a sharp right turn could have been taken, producing a species with very small brains and non-social attitudes. But then, we would not be talking about human beings and their moral judgments. Ruse and Wilson's debunking argument simply fails.

Debunkers sometimes cite an imaginary case proposed by Darwin himself that seems to endorse the utter subjectivity of moral judgments. In the *Descent of Man*, Darwin supposed that if humans had evolved much like hive bees, then unmarried females might regard it "a sacred duty to kill their brothers, and mothers would strive to kill their fertile daughters" (Darwin 1871, 1: 73). This imaginative scenario seems to imply these bee-men would have evolved a moral sense quite different from ours. Thus our human moral sense would appear to be merely a contingent fact of our particular history, quite relative to the ancient environment of our evolutionary descent. These bee-men would exactly mirror Ruse and Wilson's alien species. Two responses seem apposite. First, I believe Darwin's scenario would be just as impossible as that of Ruse and Wilson. It's hard to imagine that if you knew – had rationally determined – your sister would be gunning for you, and your mother was going to stab your sisters, that you could be a member of a stable social community. Bees have a workable cooperative community, since they also have a very dim rational capacity. But second, we need to recognize two features of moral judgment: the moral (altruistic) motive and the rational consideration of how to enact that motive – both are essential for a practical moral judgment, and Darwin thought both necessary for the development of conscience. We might properly judge the Aztec priest, who sought to sacrifice young maidens to make the corn grow and thus save the community – and who did so with the community's approval – perfectly moral but with defective agricultural knowledge. In Darwin's case of the bee-men, if we abstract from the impossibilities of his scenario, we might say that they have the same moral motive as we do – desire to act for the community welfare – but that they have unwarranted beliefs about what would advance the community welfare. They would not be an alien species, but one with us, yet one that needed enlightenment and needed rational knowledge, much like our own ancestors.

Sharon Street (2006) contrives another strategy for the debunking argument, though ultimately bases her approach on ideas similar to those of Ruse and Wilson. She begins with the assumption that realist theories of value must posit truths existing independently of our evaluative judgments. By contrast, the adaptive non-realist theory – which she believes to be the most reasonable theory in light of evolution – simply regards our value judgments as adaptations whose principal function is that of reproductive success. So the issue of realism versus adaptational relativism reduces to the following: do we hold certain values because they are antecedently and independently true, or do we hold them as true because of our contingent evolutionary history and then merely assume them to have been antecedently and independently true? Street argues the latter is the case: they are adaptations that could have been otherwise, and so lack objectivity. The basic structure of Street's argument, then, is comparable to that of Ruse and Wilson.

Street argues that those who adopt the realistic stance concerning moral values face a dilemma. Either they must deny there is any relation of moral values to natural-selective forces or they must posit a relationship. If there is no relation, then any impact of contingent evolutionary forces on our value system, she believes, could only be accidentally enforcing, but more likely distorting. On this horn of the dilemma, any adaptive cognitive response would likely mislead us in regard to any supposed real moral values. Taking the other horn of the dilemma, if there is a relationship, then adherents of this view must regard our evolved capacities as designed to recognize moral truths and to act on them. So on this horn of the dilemma, the adherents maintain that we have evolved rational faculties of evaluation, which allow us to recognize certain evaluative truths, like the value of altruism, and on the basis of recognizing these facts or truths, we judge certain acts as morally appropriate and others as moral transgressions. But Street thinks this latter hypothesis of an evolved evaluative faculty must fail in light of a stronger one, which she calls "the adaptive link account." The adaptive link account simply explains our adoption of certain moral truths because "they forged adaptive links between our ancestor's circumstances and their responses to those circumstances, getting them to act, feel, and believe in ways that turned out to be reproductively advantageous" (Street 2006, 127). She believes the adaptive link account to be superior because it's more parsimonious: it cuts directly to the ultimate cause, namely, reproductive success, without positing the superfluous notion of an evolved general capacity to recognize evaluative truths. At this juncture, Street's argument joins that of Ruse and Wilson, as well as that of Joyce and Horn, and all slide into the slough of implausibility.

Street maintains that a general capacity to recognize factual truths would actually be of "no advantage or even a disadvantage" (Street 2006, 130). She asks: is there any survival advantage to recognizing "truths about the presence or absence of electromagnetic wave-lengths of the lowest frequencies" or, presumably, most of the truths of physics, cosmology, and higher mathematics? Of course not. There might be a selective advantage in directly recognizing the dangers of fire, for example, but she perceives no advantage in a general rational capacity for evaluation. So if a general evaluational capacity carries no reproductive advantage, natural selection could not account for it and thus it could not evolve as part of our repertoire of traits. It's crucial for Street's argument that she dismisses an evolved capacity to make evaluations in many domains. Because, if we have such an evolved capacity, it would show two features of human evolution: that an evolutionary adaptation is not likely to be distorting our truth judgments and, concomitantly, that evaluations might be more than contingent preferences, might be revelatory of facts in the world.

Two simple rejoinders seem decisive in nullifying Street's objection to an evolved capacity for making general evaluations. First, having a large collection of separate and independent instincts, each designed to respond to a particular scenario in the environment – e.g., innate responses to the dangers of fire, flood, lightning, dark Chicago alleys at 2:00 am, as well as the infinite number contingencies that might assail us – would be hugely more costly in evolutionary terms than a general capacity to learn and make evaluative assessments in several domains. But secondly, we do have a rational and evaluative capacity. Where did that come from? Did it fall from a tree? Straight's argument is similar to that of Alfred Russel Wallace, who maintained that natural selection could not account for man's excessively big brain and his capacities for music and mathematics, since these had no survival value. Wallace contended that this kind of objection to a natural-selection explanation made room for and furnished evidence for positing a spirit world, where higher powers would confer these traits on mankind (Richards 1987, 176–184). Is that the implication of Street's own position?

Street's further objection, and an apparently more telling one, concerns the kind of natural entities evaluative facts might be and their use in making evaluative judgments. She asks, "exactly what natural fact or facts does the evaluative fact that one should care for one's offspring reduce to?" (Street 2006, 131) Or comparably, when we judge pains to be bad, "we need make no reference whatsoever to the fact that they are bad; we need only point out how [they] tended to promote reproductive success to take them to be bad" (Street 2006, 151). In these cases, Street's assumption is that we immediately endorse care of offspring and shun pain because these evaluative attitudes have resulted from their aid in the reproductive success of our ancestors. We don't judge these situations to be good or bad because we have first perceived the fact that they are good or bad; rather the perception of an infant in need or the feeling of pain immediately evoke from us an evaluative feeling and that feeling has had reproductive advantage for our ancestors. We make the evaluation, according to Street, and then call it a fact, rather than initially perceiving an independent truth that caring for infants is morally good and pain is physically bad, and then so decide to endorse the one and avoid the other. Thus there is no independently existing moral fact in the world, only our own valuing some attitude or behavior, which we subsequently call a fact. So Street's analysis is similar to that of Ruse and Wilson: we are deceived if we think there are independent moral facts; rather our moral judgments are really about survival.

Street's version of the debunking argument sounds like it furnishes a plausible objection to moral realism, or it does until we examine the two fundamental assumptions on which it is based: that facts in the world stand ontologically apart from us as independent entities and that a trait naturally selected for, and thus part of our evolutionary acquirements, must be epistemically directed only to survival in the first instance. As is the case of the Ruse-Wilson analysis, both of these assumptions confuse ontological objectivity with epistemic objectivity.

Consider my judgment that the ball on the floor is red – a prototypical judgment about a fact in the world. Do I judge the ball to be red simply because I first recognize an independent fact in the world, a red ball, and on that basis make the true judgment "the ball is red"? This is the paradigm Street's discussion suggests. But I think it's the wrong paradigm. Isn't it rather that the experience of the object evokes from me an evaluative attitude – that is, a trait that has been selected for – that leads me to call the ball red? What kind of fact is a red ball, after all? The best physical and physiological theories would say something roughly like this: in the world there are clumps of atoms whose electrons have been excited by electromagnetic energy to give off photons that strike the cones in the retina – the color receptors in the eye – in a certain way; and, further, our neural reaction is an adaptive response that leads us (who possess normal trichromatic vision) to have the sensational quality of red in our visual field. Absent a normal human eye (or trichromatic vertebrate eye), there are no red balls in the world. Most all of an object's qualities that we take to be existing independently in the world of our experience depend on the physiology of our nervous system, a nervous system that has evolved over millions of years. Moreover, being able to make basic color discriminations by way of qualitative sensation has survival value, and so such a response can only be the result of an adaptive link causally forged by evolution, one which unites our qualitative sensations to a certain kind of physiological stimulation that usually stems from a physically independent source. Thus our judging the ball to be red, a quintessential factual judgment, is a function of our evolutionary history, though it's not epistemically about that history, but about the fact of a red ball. Is my judging that torturing innocent children to be morally bad really any different than judging the ball to be red? Both judgments are functions of an evolutionary history and both are immediate, virtually a part of the perception itself and both are typically regarded as factual. If "badness depends in an important sense on our evaluative attitudes" (Street 2006, 151), so does redness. Both are the result of the interaction of independent structures (e.g., atoms excited by a light source or the social structures of cooperation and altruism) with our neuro-mental evaluations. Those evaluations are part of the experience of facts in the world. Moreover, when we reflect and ask ourselves "but is the ball really red?" we presumably do so by applying an acquired internal standard of the quality red, without which the reflective judgment would not be possible. The same is true when we reflectively ask, "but is that act really morally sanctioned?" What assures us that our judgments are objective is that they can be tested and intersubjectively verified. After all, we sometimes discover that the ball only looks red because of the nonstandard character of the environment, just as we might be mistaken in judging the moral character of an act (e.g., not torture but a doctor injecting a vaccine).

We do think it a fact of the world that red balls might be found rolling around on the floor. We also believe the obvious fact that torturing innocent children is morally bad and, consequently, that it should be condemned. But in neither case do we initially assert to ourselves that it is a fact that there is a red ball before me (or that torturing children is bad) and, having established the fact, evaluatively judge there to be a red ball before me (or that child torture is wrong). Both are immediate, perceptive judgments about the world – the world we live in – and both are functions of our evolutionary history. So Straight's argument that the evolutionary process precludes a realistic view of the world either defeats realism not only in morality but in all areas of epistemic concern (e.g., red balls), or her analysis fails for want of understanding what "fact in the world" could mean. I would suppose neither option would be acceptable to her or to the others who advance a debunking argument. That means the debunking argument itself ought to be rejected. That classes of cognitive judgments have an evolutionary genesis is thus no reason per se for rejecting the objectivity of those judgments, and thus their factual character.

Our access to a reality beyond mind can only be indirect. This is the message of realism – critical realism, that is. To make sense of our experience, we have to make a number of assumptions. We test and use those assumptions especially in our commerce with the world. Those assumptions were shaped during the course of the history of science, some assumptions having been rejected by reason of evidence and others added. Gradually these assumptions have become more firmly established and refined through the historical development of reason, of science. Developing science provides ever better evidence for constructing a map of reality. And that map constitutes the factual terrain of the world, at least one part of the factual world. But that terrain is not something we directly perceive. We well understand how the visual system produces color discriminations, and also how this likely arose through the long history of natural selection. In the same way, we understand why certain social behaviors have been perceived as morally good or morally bad, and also why these perceptions likely arose through natural selection operating on our ancestors. Together the extra-mental structures and their representations in our experience constitute the relational facts of our world. The extra-mental structures are postulated by the best scientific evidence we have, such as electromagnetic energy of a certain frequency and the effectiveness of cooperation and altruism for social creatures. As creatures with an evolved nervous system and cognitive apparatus, we perceptively evaluate both, one giving us something as factually red and the other as giving us something as factually morally good (or bad). Our experiential judgments are thus about complex, real structures of the world, one aspect more remote and dependent on our best theories, the other more perceptively immediate. Both aspects are sustained by the criterion of intersubjectivity. They are thus perfectly objective.

Conclusion: The Darwinian Theory of Morality

The Darwinian theory of morality supposes (1) that humans have evolved to be altruists and that fundamental principles of morality – the Categorical Imperative, the Golden Rule, etc. – can be reduced to altruism and (2) that evolution has instilled in mankind the social instincts of caring for the young and acting for the benefit of others, just as it has instilled in the human vertebrate the qualitative sensations produced by various neural impulses. These social instincts, of course, have survival value, as do our qualitative, sensory discriminations. As human beings, we have evolved to be intrinsically social creatures. That means we have wrapped into the very fabric of our being inclinations that are powerful and regarded as special. In the same way, we have evolved a rational, evaluative capacity. Both have had survival value. You might even conclude our ancestors were social long before they became rational. Of course, in the human species, both traits have a distribution of more or less. In some few instances, these capacities may be completely absent. While there may be some imaginative scenarios in which a population, nominally called human, could have evolved into mindless and completely asocial creatures, that could not be our population. As a result of our moral and rational capacities, we make claims and formulate principles based on these capacities, which have the sanction of intersubjective affirmation – that is, they are objective and universal. That simply is who we are as human beings. Aristotle assumed humans were moral animals. Darwin demonstrated it.

11   Moral Mismatch and Abolition

Ben Fraser

Much work on the evolution of morality examines the past adaptive benefits of morality but little considers the current utility of morality. This chapter presents the case that our evolved moral psychology is, perhaps surprisingly, in important respects like our evolved "sweet tooth": useful in ancestral environments but potentially harmful in our much different modern context. In light of this, I aim to make an under-recognized and counterintuitive view, moral abolitionism, more salient and plausible. Moral abolitionism is the view that a prudential cost/benefit analysis favors doing away with morality. That is, we prudentially ought to stop thinking and talking in moral terms and to cease engaging in the practices and institutions underwritten by such thought and talk. Moral abolitionism is clearly a radical and controversial view, so my goal here will not be to settle matters but rather to motivate further discussion.

"Let us hope it is not true, but if it is, let us pray that it will not become generally known," so spoke the wife of the Bishop of Worcester upon learning of Charles Darwin's new account of humanity's origin, the theory of evolution by natural selection (Sagan and Druyan 1993, 274). Why should the introduction of a scientific theory be met with such horror? One reason is that evolutionary theory raises significant philosophical challenges to cherished notions across a wide range of areas, including aesthetics, metaphysics, epistemology, religion, and morality. Many of these challenges date back to Darwin himself. In *The Descent of Man* ([1871] 2004), he argued that elaborate ornamentation in the natural world, such as the male peacock's tail, was explained by animals' aesthetic preferences. This threatened the conceit that humans were unique in possessing a sense of beauty. Darwin also worried about the implications of evolutionary theory for our ability to grasp the true nature of the world around us, admitting that "the horrid doubt always arises whether the convictions of man's mind, which has been developed from the mind of the lower animals, are of any value or at all trustworthy." "Would anyone," he wondered, "trust in the convictions of a monkey's mind?" (Darwin 1881).

Darwin was not alone in worrying about evolutionary theory's philosophical implications. Philosophers themselves recognized its significance. Bertrand Russell's *Unpopular Essays* ([1950] 2009) highlighted questions raised by an evolutionary explanation of human origins, such as "When did our ancestors acquire free will?" – a question that is still live today, as in Daniel Dennett's *Freedom Evolves* (2003). Dennett's metaphor of evolution as a "universal acid" is especially evocative: "it eats through just about every traditional concept, and leaves in its wake a revolutionized world-view" (Dennett 1995, 63). The action of this acid has been of particular interest with respect to two elements of our worldview: religion and morality.

Reflecting on the origins of morality and religion raises at least three kinds of questions: (a) empirical questions about the evolutionary explanation for those phenomena; (b) epistemic questions about the implications of that explanation for moral and religious belief; and (c) evaluative questions about the current utility – as opposed to past adaptive benefits – of the evolved beliefs and practices. By way of introduction to the central concern of this chapter, I wish to draw attention to an interesting asymmetry between discussions of the evolution of religion, on the one hand, and on the other, the evolution of morality.

In the case of religion, there is a great deal of scientific work relevant to the first question regarding the evolutionary explanation of religion. While even a brief summary is beyond the scope of the current chapter, this body of work attempts to explain religious beliefs, practices, and institutions in terms of cognitive biases, cultural evolution and group selection, and costly signaling (see, e.g., Boyer 2002; Wilson 2002; Bulbulia 2004; Atran and Heinrich 2010). A great deal of disagreement exists within this literature, but there is a thriving and productive interdisciplinary research project aimed at answering this first, empirical question in the case of evolution and religion.

In the case of morality, a great deal of work has already been done on the first question. This is an interdisciplinary endeavor with relevant findings coming from evolutionary biology and psychology, anthropology, primatology, and experimental economics (Joyce 2006a; Boehm 2012; Richerson and Henrich 2012; Chudek et al. 2013; Sterelny 2014a). Many disputes about detail persist, but it is widely agreed that the human capacity for moral judgment is susceptible to evolutionary explanation.

In the religious case, there is also a good amount of work on the second question regarding the epistemic upshot for religious belief of an evolutionary explanation for religion. Here, the issue is whether the evolutionary explanation for religion somehow undermines religious beliefs, for instance by showing them to be unjustified (see, e.g., Shafer-Landau 2007; Schloss and Murray 2009; Bergman and Kain 2014). Opinions divide on this issue, and there is a lively debate ongoing about the epistemic implications for religious belief of an evolutionary explanation for religion.

In the moral case too, much attention has focused on whether an evolutionary explanation of morality casts doubt on the truth and/or justification of moral beliefs. Evolutionary debunking arguments against morality are a hot topic, and a great deal of progress has been made as early arguments (e.g., Ruse and Wilson 1986) have been elaborated and critiqued (Joyce 2006a; Street 2006; Kitcher 2011; Nagel 2012). Again, much controversy remains, but there is fairly wide agreement that an evolutionary explanation of morality presents at the very least a challenge to be answered by those wishing to maintain that we hold generally reliable beliefs about an objective realm of moral fact.

To the third evaluative question concerning the current utility of religious beliefs and practices, a negative answer has been forcefully pressed by the current cadre of New Atheists: Sam Harris (2004), Daniel Dennett (2006), Richard Dawkins (2006), and Christopher Hitchens (2007). According to this line of thought, whatever past adaptive benefits religious beliefs, practices, and institutions secured for our ancestors, modern humans are now better off without, given the harms these things generate in contemporary contexts. A different take on this question has it that religion is a positive influence on human social life, in virtue, for example, of facilitating social cohesion (Wilson 2002). As with the previous two questions, there is a substantial discussion of the third evaluative question in the case of evolution and religion.

In the case of religion, then, all three questions – empirical, epistemic, and evaluative – have received considerable attention. In the case of morality, however, while the first two questions have been much discussed, the last evaluative question has been relatively neglected. Only a handful of moral philosophers have given serious consideration to the costs of morality (Mackie 1977; Hinckfuss 1987; Garner 1994; Burgess 2007). Nobody yet has taken a specifically evolutionary perspective on the evaluative question. Here, I aim to rectify that.

This chapter will present two related theses: moral mismatch and moral abolitionism. According to the former, our evolved moral psychology inclines us toward attitudes and behaviors that, although adaptive in ancestral environments, are harmful in modern contexts. This is perhaps a surprising and implausible claim, so before going any further, some brief remarks by way of motivation (and provocation) are probably in order. Why think morality is currently harmful? For one thing, we are groupish. Moral norms are significant markers of group identity as well as powerful generators of in-group bias and out-group hostility (Cohen et al. 2006; Haidt and Graham 2007; Leach et al. 2007; Rutland et al. 2010; Seabright 2010). For another, we are vengeful. Our taste for tit-for-tat retribution can tempt us into costly, mutually destructive retaliatory cycles (Trivers 1971; Axelrod 1984; Wright 1994). We are also prone to objectify morality, and seeing moral matters as fixed independently of individual or social attitudes makes us more intolerant of diversity and less able to cope with conflict (Goodwin and Darley 2008, 2010, 2012; Wright et al. 2014). When it comes to morality, then, it is not entirely outrageous to suppose that our evolutionary history influences not just what but also how we think, in ways that have surprisingly high but under-recognized costs in modern contexts.

The moral mismatch hypothesis will be offered in support of the second thesis, moral abolitionism. This is the moral analogue of the New Atheist attack on religion. Moral abolitionism is the claim that, whatever its past benefits, morality is something we are now better off without. Like the moral mismatch hypothesis, moral abolitionism likely strikes one as implausible if not obviously false. So, the chapter from here on will proceed as follows. First, moral abolitionism will be presented and an attempt made to motivate its discussion and defense. Then, the moral mismatch hypothesis will be presented and supported. In particular, the chapter will offer evidence that our commitment to moral objectivity is a case of mismatch. Finally, the mismatch hypothesis will be recruited in support of moral abolitionism. No pretense to definitiveness is made; rather, the aim is to highlight a currently under-discussed kind of question about evolution and morality and to motivate discussion of some interesting and currently under-explored views, mismatch and abolition.

To understand the moral abolitionist's view, it may be helpful to creep up on it by degrees. To begin with, everyone is surely familiar with calls to give up specific moral beliefs. Take beliefs about slavery. Opponents of slavery urged the rejection of the belief that people of a certain color were inferior beings and that enslaving those people was thus morally permissible (or even required, for their own good). Moral abolitionism makes a stronger claim than merely that we should give up some specific moral beliefs. Perhaps less familiar, but still prominent in the history of moral philosophy, are calls to give up on using specific moral concepts. Jeremy Bentham, for instance, derided talk of "natural rights" as nonsense and was in favor of abandoning the notion. Moral abolitionism makes an even stronger claim than that we should give up employing some particular moral concept. We should give up each and every one of our specific moral beliefs and throw out all the items in our moral conceptual toolkit: the notions of moral right and wrong, duty, value, and so on. We should stop publicly using moral language and refrain also from making moral judgments privately, internally. According to the moral abolitionist, we should "stop making moral judgments," period (Garner 2007, 499).

The moral abolitionist's proposal is, clearly, radical. Before trying to motivate or defend it, though, an immediate question that needs answering is just what is involved in making moral judgments (which I will henceforth shorten to "moralizing"). For the abolitionists I will consider, the key feature of moralizing was described by John Mackie (1977): it is a commitment to "objective prescriptivity." The idea here is best gotten at by contrast.

When we make judgments of prudence, these judgments provide us with reasons for action only contingently, depending on our desires, goals, and interests. When we moralize, says Mackie, we make judgments that purport to provide reasons non-contingently, independently of agents' desires, goals, or interests. The supposed authority of a moral judgment to command action is not "contingent on any present desire of the agent to whose satisfaction the recommended action would contribute as a means" (Mackie 1977, 69). Neither, says Mackie, does the authority of moral judgments depend on our subscribing to some institution, "our choosing or deciding to think in a certain way" (Mackie 1977, 16). Morality in this respect contrasts with etiquette. If you "opt out" of the game of table manners, then although the demands of etiquette still apply to you, they cease to provide any reason for you to act in a particular way. But, Mackie thinks, one cannot opt out of morality in this way; even if one resolutely refuses to care about or endorse any moral demands, such demands nevertheless purport to provide genuine reasons for you to act. According to Mackie, then, when we moralize, we make judgments that are supposed to be objectively prescriptive in this sense: their power to command action is not dependent on agents having any particular desires, or on them choosing to follow a particular set of rules.

While Mackie himself was not an abolitionist, his idea that moralizing involves a commitment to objective prescriptivity is common ground among the three main moral abolitionists: Richard Garner, Ian Hinckfuss, and John Burgess. According to Garner, moralizing involves accepting the notion of "moral bondage," the idea that morality issues in "external and authoritative demands" (Garner 1994, 57). According to Burgess, "ethical vocabulary … carries an implication, or at least an inevitable suggestion, of a claim to objective and absolute … backing for one's standards" (Burgess 2007, 437). According to Hinckfuss, moral judgments about rightness, value, and so on are supposed to be "absolute" – not relative to social or cultural standards – and "objective," meaning not fixed by individuals' subjective feelings (Hinckfuss 1987). Garner, Hinckfuss, and Burgess agree with Mackie and each other about what moralizing involves; they also take this feature of moralizing to be key when building the case for abolitionism. It is the commitment to objective prescriptivity that is problematic, enough so that we should stop moralizing.

Moral abolitionism is a radical and prima facie implausible thesis. Put off for a moment the question of evidence; first, let us ask why we should consider such an outré view at all. There are reasons to take moral abolitionism seriously. Some are theoretical and internal to philosophy. Others are practical and of relevance well beyond the field.

First and perhaps most naturally, moral abolitionism should be of interest to moral error theorists. Error theorists think that moral discourse is somehow systematically mistaken. Either moral judgments are uniformly untrue – they aim to describe the world but fail due to faulty assertion or assumption – or moral judgments are all unjustified and furthermore, unable to be justified. Whichever version of error theory one opts for – truth or justification – a "what next?" question looms. What should we do with our erroneous moral thought and talk, and the practices and institutions built on those things? One obvious option is to consider abolishing the erroneous discourse and dismantling those practices and institutions. Error theory is in fact the starting point for the abolitionism of Garner, Hinckfuss, and Burgess. But, error theorists who are undecided about how to proceed are not the only ones who should care.

Fictionalists should also care. The fictionalist agrees with the error theorist that moral thought and talk is systematically in error, and further claims that such thought and talk should be retained as a useful fiction (see, e.g., Joyce 2005; Nolan et al. 2005; Olsen 2011). Fictionalists have tended to focus on building the positive case for morality as a useful fiction. But, for the fictionalist to be correct, it needs to be true not only that retaining a fictionalized morality is beneficial in some ways but also that doing so is beneficial overall. To show the latter, one must consider not only the benefits but also the costs of moralizing. Hence, the strength of the case for moral abolitionism should concern moral fictionalists.

Less obviously, moral abolitionism should be of interest to those who accept normative ethical theories that are, potentially, self-effacing. For a normative theory to be self-effacing is for that theory, taken as a criterion of rightness, to recommend that agents employ an ethical decision-making procedure other than the theory itself (Parfit 1984). Versions of global consequentialism are typically taken to have this feature. Global consequentialists hold that the moral status of everything – acts, rules, character traits, and so on – is determined by consequences for, say, overall happiness. So, when the global consequentialist asks, which moral theory ought one adopt as a guide to action? the answer will be, whichever one would, if adopted, have the best consequences. A possibility here is that the moral theory that would, if adopted, have the best consequences is not global consequentialism. So, global consequentialism is potentially self-effacing. This is a well-recognized possibility. Another possibility, however, is that the best overall consequences are secured by eschewing moralizing entirely, that is, by adopting no moral theory at all. Global consequentialists might think this possibility remote: fair enough. Still, unless they are content to dismiss the possibility out of hand, without evidence, assessing the case for moral abolitionism should matter to such folks.

Moral abolitionism should also interest anyone who thinks that moral reasons are not a special kind of trumping or overriding reason. To say that moral reasons are overriding or trumping is to say that when moral reasons conflict with other kinds of normative considerations – reasons of prudence, law, etiquette, etc. – then what one has all-things-considered reason to do is the moral thing. To deny the overriding aspect of moral judgments is to admit that one's moral reasons could validly give way to other kinds of reasons. If one thinks the latter, the case for moral abolitionism should be of interest. Grant (contra the moral error theorist) that we have some moral reasons. Still, if moral reasons are not overriding, it may be that what we have all-things-considered reason to do is stop attending to those reasons, to stop counting them in our weighing up of reasons – in effect, to stop moralizing.

The last (and I think most straightforward and compelling) way to motivate consideration of moral abolitionism is just to note that it should concern anyone who engages in and is affected by the practice of moralizing. Return to the parallel with religion: one need not be a theologian to take an interest in the New Atheist case or to worry that religious beliefs and practices may be harmful on balance. Similarly, the possibility that moral beliefs and practices are on balance bad for us should concern not just sophisticated metaethicists but anyone affected by moralizing, which (I submit) is virtually everyone.

So much then for the motivation question. Having supplied reasons to take the abolitionist view seriously, the next task is to provide grounds for thinking it might be true. The moral abolitionist's central claim is that a prudential cost/benefit analysis favors getting rid of moralizing over continuing to do it. Four questions are relevant to deciding whether or not the abolitionist is correct:

(1) What are the benefits of moralizing?
(2) What are the harms of moralizing?
(3) What would be the benefits of abolishing morality?
(4) What would be the harms of abolishing morality?

While all four questions would need to be answered to fully assess the case for moral abolitionism, here I will focus on question (2). I will show how moralizing can impose costs in terms of our straightforward prudential interests. I will do so by developing a moral mismatch hypothesis.

This section gives a general statement of the form of an evolutionary mismatch hypothesis, then illustrates this with a pair of nonmoral examples and formulates the moral mismatch hypothesis. It then focuses on a specific element of our evolved moral psychology – our tendency to objectify morality – describing both its plausible past benefits and the costs it imposes in modern contexts.

An evolutionary mismatch is "a situation whereby an organism that evolved in one environment develops a phenotype that is harmful to its fitness or wellbeing in another environment" (Lloyd et al. 2011). It thus involves four elements: a trait (call this T), at least two environments – the old (E1) and the new (E2) – and some relevant costs (C) of T in E2. A textbook case of evolutionary mismatch is the peppered moth. In this case, T was color, specifically, the light coloration that predominated among these moths in E1, the pre-industrialization forests around Manchester in the United Kingdom. The transition to E2 occurred as industrial pollution darkened the trunks in these forests. In this new environment, T was selectively disadvantageous. In this case C is a fitness cost: light-colored moths were more conspicuous and now faced greater predation risk. (For more detail on this case, see Kettlewell 1973; Grant 1999; Cook 2003).

The peppered moth case provides an example of evolutionary mismatch in a nonhuman organism, where the cost of the mismatched trait is a fitness cost. To further elucidate the idea of evolutionary mismatch, consider now a likely all-too-familiar human case: our sweet tooth. Humans generally prefer sweet foods to bitter ones; call this preference T. In ancestral human environments where efficient energy capture was at a premium (E1), it is not hard to ascribe an adaptive benefit to possessing T, since sweeter foods tend to be more energy dense. In modern First World human nutritional environments (E2), however, energy sources are so abundant that possessing T can lead to over-consumption, obesity, and associated health costs. Notice, here we can state C not in terms of fitness (although there may well be fitness costs to T in E2) but rather in terms of welfare or well-being. A sweet tooth can produce effects that run counter to the interests of the individual, not the interests (loosely speaking) of her genes.

Having run through some examples of evolutionary mismatch and identified the general form of mismatch hypotheses, it is now clear what a moral mismatch hypothesis would look like. It would need to specify a trait T, describe relevantly different environments E1 and E2, and identify the costs C of T in E2. Schematically, that is clear enough. Turning to details, matters swiftly become complex.

Take the trait first. What is T in a moral mismatch hypothesis? Defining "morality" is far from straightforward (Gert 2012) but I will not, nor need I, do anything so ambitious here. I will remain neutral on what is necessary and sufficient for something – a belief, judgment, utterance, system of rules, etc. – to count as moral, or indeed whether any such conditions can be stated (for skepticism on this point, see Stich forthcoming). Moral abolitionism is fundamentally a practical claim that is grounded in how we in fact moralize and what benefits and costs this activity generates and imposes. Accordingly, I will focus on features of our actual moral thought and talk and consider their evolutionary explanation and current effects.

One important distinction here is between the content and the form of our moral judgments. It is plausible that selective forces have influenced what we think is morally right and wrong, good and bad. Sharon Street, for instance, claims that natural selection has strongly shaped the content of human morality, favoring judgments that promote survival and reproduction, to the point that "our moral judgments are thoroughly saturated with evolutionary influence" (Street 2006, 114). A distinct claim is that evolution has affected how we think about moral matters, in particular, whether we see them as objective. For example, Ruse and Wilson suggest that we have been selected to see morality as objective, because "human beings function better if they are deceived by their genes into thinking there is a disinterested objective morality binding upon them" (Ruse and Wilson 1986, 179).

There may well be scope to develop a moral mismatch hypothesis with reference to the content of our moral beliefs. I will set aside that possibility here, in favor of focusing on the form of moral judgments. In particular, I will develop the idea that our commitment to moral objectivity is a case of moral mismatch. So, T in the moral mismatch hypothesis I am tabling here is the tendency to objectify morality, in the sense described by Mackie and outlined earlier.

Turning to the question of which environments are relevant to stating and assessing the moral mismatch hypothesis, the key idea is that human social life has undergone a radical and evolutionarily recent shift. From about seventy-five thousand until roughly ten thousand years ago, our Pleistocene ancestors were cooperative hunter-foragers living in small, mobile multi-family groups of approximately twenty to thirty individuals. About ten thousand years ago, during the Holocene revolution, human social life underwent a dramatic change: with the advent of agriculture, group size greatly increased, as did social complexity, and groups became more sedentary. (For more on this, see Sterelny 2014b and references therein.) Now, few readers will need to be told, human social worlds are very large, complex, and interconnected. So, the key change when it comes to filling in the moral mismatch hypothesis is this: our ancestral environment E1 featured relatively small, homogenous and independent social worlds, whereas in our modern environment E2 social worlds are relatively large, heterogeneous and interdependent.

Having filled in T, E1, and E2, the moral mismatch hypothesis now needs detail regarding C, the costs of objectivist moralizing in modern social environments. Until fairly recently, there was little empirical investigation probing the metaethical views held by ordinary folk, let alone study of the ways in which folk metaethical views manifest in behavior. Happily, there is now a growing body of work on "the psychology of metaethics" (Goodwin and Darley 2008, 2010, 2012). This literature is particularly concerned with whether the folk are objectivists about morality and whether and if so how this metaethical stance matters. Such work is promising in relation to the case for moral abolitionism. Below, I will concentrate on some landmark studies by Goodwin and Darley and on follow-ups to those studies.

Goodwin and Darley start with a definition of "objective." Since the aim of their work is explicitly to engage with metaethical debates in philosophy, they follow ethicist Geoffrey Sayre-McCord (1986) in saying that to see morality as objective is to

hold that the truth conditions of moral claims are mind-independent in the sense that a moral claim can be true without reference to the subjective states of the individual making the judgement, and without reference to the conventions of any group of people who are making the moral judgement. (Goodwin and Darley 2010, 163)

At this point, it is worth noting that the feature of the folk view targeted by Goodwin and Darley is not precisely the same as that which concerns the moral abolitionists, namely, the notion of objective prescriptivity as spelled out by Mackie. Objectivism as just defined concerns the truth conditions of moral claims: moral claims are true (if/when they are) independently of individual subjective states and group conventions. Objective prescriptivity concerns the supposed special authority of moral demands. Objectivity does not equal objective prescriptivity. One point to emphasize, then, is that the empirical case for moral abolitionism needs work specifically on objective prescriptivity. But, even though current experimental work doesn't exactly target objective prescriptivity, it nevertheless turns up some interesting findings about folk objectivism and, even more interestingly for the moral abolitionist, links objectivism of this sort to practical matters concerning disagreement and conflict (as will be detailed later in this chapter). So, it is worth looking here, even if we should look elsewhere too.

Goodwin and Darley tested for objectivism among the folk using the following experimental setup. Subjects were given a collection of statements on various topics: ethics, geography, music, science, etc. For each statement, subjects were asked (a) to rate their level of agreement, then (b) if they thought there was a correct answer about whether that statement was true or false, and (c) whether, if a classmate disagreed with them over the truth of that statement, then either that person or the subject must be wrong.

Notice that one could sensibly answer yes to (b) but no to (c) only if one thought that the truth of the relevant statement was somehow mind-dependent. For example, in the case of statements about music, one might answer in this way if one thought that which musical styles are better than others depends on an individual's tastes. If one thought that the truth of the relevant statement was not mind-dependent, as most people presumably do in the case of statements about geography, then one should answer yes to both questions (and indeed subjects overwhelmingly showed this pattern of response on such items). Answering yes to both questions for a given statement was thus taken as evidence of being an objectivist about that statement.

Goodwin and Darley found that, far from the folk being univocally objectivist about moral statements, there was variation both among individuals in
whether they considered moral statements to be objective and across moral
statements in how likely they were to be considered objective. That is, some
subjects more so than others treated moral statements as objective (specifically those who self-reported as believing that right and wrong depend on the
existence of a divine being), and some moral statements more so than others
were taken to be objective (specifically the negatively valenced ones, such as
statements about the wrongness of stealing a wallet, as compared to positively
valenced statements about the goodness of donating part of one's income to
charity).

A finding of folk "pluralism" when it comes to moral objectivity has also
been reported by Jennifer Cole Wright and colleagues (Wright et al. 2013,
2014). Other work by Hagop Sarkissian and colleagues (Sarkissian et al. 2011)
has suggested that the folk only give the appearance of being objectivists in
some cases; they are really relativists all the time, and this tacit metaethical
relativism can be brought out by the right probes (specifically, probes asking
about disagreements between culturally very dissimilar disputants, in which
cases subjects are willing to say that neither party need be mistaken). In sum,
empirical investigations paint a complex picture on which ordinary moral
agents are not universally objectivist about morality. Being an objectivist
does, however, affect one's attitudes and behavior in cases of moral disagreement. Moreover, its effect is relevant to the case for moral abolitionism.

Goodwin and Darley report that objectivism influences how "open" or
"closed" subjects are in reaction to disagreement. "Openness" and "closed-
ness" are measured by comfort with proximity to a disagreeing other (probe
question: how willing would you be to have this person as a roommate?), the
personality attributions made (how likely are you to say the disagreeing other
is "not a moral person"?), and by self-reported rigidity of view (would it be
possible for you to give up your initial belief?). Goodwin and Darley found
that those who treated a moral issue as objective gave more "closed" responses
to disagreement on that issue: they showed more discomfort with disagreeing
others, were more likely to say the disagreeing other was "not a moral person," and were less likely to admit the possibility of giving up their own belief.

Jennifer Cole Wright and colleagues have also investigated the practical upshot of a commitment to moral objectivity and have reported results
in line with Goodwin and Darley's findings. Wright et al. (2014) tested the
links between objectivism about a moral issue and attitudinal and behavioral
responses to disagreement on that issue. They found that being an objectivist
about the issue predicted greater intolerance across a range of measures: willingness to date, work with, or live near a disagreeing other and willingness to
help such others even in minor ways, such as by making change or running

a trivial errand. Again, metaethical stance appears to matter, and in a way in line with the abolitionist's claim that moralizing can play a harmful role.

The moral abolitionist's central claim is that a prudential, cost/benefit analysis favors getting rid of moralizing over continuing to do it. Four questions must be asked here:

(1)   What are the benefits of moralizing?
(2)   What are the harms of moralizing?
(3)   What would be the benefits of abolishing morality?
(4)   What would be the harms of abolishing morality?

All four questions would need to be answered to fully assess the case for moral abolitionism. Previously, I have focused on question 2, attempting to show how moralizing – especially the kind of objectification of morality that is if not universal then at least common among folk moralizers – can impose costs in terms of our straightforward prudential interests. Now, granting that moralizing is in this respect at least costly, why think of this as evidence that our moral psychology may be a case of evolutionary mismatch? Here is the thought. Objectifying moral demands (recall, this is our T in the moral mismatch hypothesis) has been theorized (Joyce 2006a) and recently experimentally shown (Young and Durwin 2013) to boost motivation to behave as one believes morality demands. In small-scale ancestral social worlds (E1), this extra motivation plausibly provided fitness benefit by boosting intragroup cooperation. In modern social worlds (E2), we are confronted far more often with diversity, including moral diversity, in virtue of the greater heterogeneity within groups and the increased interdependence between groups. Objectivist moralizing, insofar as it makes us more intolerant of diversity and less able to manage conflicts (C), can be harmful under these circumstances.

Our tendency to objectify morality may well have been adaptive in small-scale, relatively homogeneous and independent ancestral social worlds, but now operates in a highly diverse and interdependent modern social context, leaving us more prone to conflict and less able to manage it. If something like this is true, moral abolitionism gains a measure of plausibility.

To the extent that moral abolitionism has been made more plausible, a further question becomes more pressing: if moralizing is abolished, what could take its place? Moralizing is indeed important in our cognitive lives both as individuals, in the form of moral deliberation, and collectively, in the form of moral education and persuasion. Moralizing provides an important means of shaping our own and others' behavior, and eliminating it would leave a void. The abolitionist owes some story about how that void might be filled. I will end with a suggestion, a couple of examples, and an outline of a general strategy.

Humans are subject to many cognitive biases (for an overview, see Baron 2007). One kind of bias is susceptibility to framing effects. A framing effect occurs when choices that are logically equivalent are nevertheless perceived differently due to the way in which they are presented. A classic example comes from Tversky and Kahneman (1981), who presented subjects with a hypothetical choice between different public health policies and found that subjects' choices were heavily influenced by the language used to describe the policies (and not merely the expected outcomes of implementing them). Framing effects can influence choice and, thus, provide a lever with which to manipulate behavior.

Shifting to a more directly relevant case, suppose we wanted to encourage people to be concerned about the welfare of more kinds of things: more ethnic groups than just their own, say, or more species, or perhaps even more kinds of entities, such as ecosystems as well as biological individuals. Notice, one need not be making any *moral* judgments in encouraging people to care in these ways. One can simply recognize the prudential benefits of people extending concern to other cultural groups, the natural world, etc. Plausibly, and all else being equal, this would result in less inter-cultural conflict and increased environmental conservation. Assuming those things are in fact prudentially valuable, the abolitionist need make no appeal to moral considerations. That said, one way to encourage people to widen their circle of concern, and so to promote the prudential values just mentioned, may be to exploit a specific sort of framing effect: inclusion-exclusion discrepancy.

Work in moral psychology (Laham 2009) has found that people's decisions about how widely to draw their circle of concern, given a list of candidate entities, are influenced by whether they are asked to cross off things they think should not be on the list or to circle things they think should be on the list. Subjects asked to "rule out" items are said to be operating under an "exclusion" mindset; they end up with longer final lists than subjects operating under an "inclusion" mindset, who are asked to "rule in" items. In effect, an exclusionary "rule out" process leads subjects to be concerned about the welfare of more kinds of entities than does an inclusionary "rule in" approach. So, should we want to encourage people to give consideration to the welfare of a wider range of others, we might do this in part by trying to put them in an exclusionary mindset, in effect, challenging them to say why those others should not be considered.

Similar work has been conducted on subjects' stated willingness to adopt pro-environmental behaviors (McDonald et al. 2014). Should we want to encourage people to take shorter showers, say, or buy "greener" products, then fostering an exclusion mindset to exploit inclusion-exclusion discrepancy may be one surprisingly effective way to go. Obviously these are merely a couple of examples. They suggest a general strategy, though: identify and exploit cognitive biases to influence behavior. Much more would need to be said, but the moral abolitionist is not without resources when it comes to replacing morality.

To end, I will reiterate my aim as stated at the outset of this chapter: not to settle matters but to raise neglected issues about the practical value of moralizing, to make an under-recognized and counterintuitive view – moral abolitionism – more salient and plausible, and to spark further discussion.

Part III

Against Debunking Arguments

12   Moral Realism and Evolutionary Debunking Arguments

Russ Shafer-Landau

The Nature of Evolutionary Debunking Arguments

Moral realists conceive of moral truths as stance-independent: true, but not in virtue of the attitudes taken toward their content by any actual or hypothetical agent (or set of agents). Suppose, for instance, that there is some non-derivative moral norm that prohibits deceit. For the realist, this prohibition isn't made true by virtue of having been endorsed by anyone – not even the most enlightened ideal observer. Though the verdicts of such an observer may be infallible, these verdicts are not constitutive of moral truth. If moral realism is correct, there is a battery of moral truths that are conceptually and explanatorily prior to the edicts of even the most ideal appraiser.

It has always been a challenge for realists to explain how we might gain access to a realm of robustly objective moral truths. Recently, this challenge has been made more acute by the development of a line of criticism that relies on the following assumption: that much of our moral outlook originates in natural selective pressures. Such pressures operate so as to produce adaptive behaviors and beliefs. But why think that adaptive moral beliefs are also true? There is no good answer to this question, if proponents of evolutionary debunking arguments (EDAs) are correct.

Though EDAs are relatively new on the philosophical scene, they are but the latest in a long line of genealogical critiques of our moral beliefs. Despite the variety of specific forms that such critiques have taken, all instances of the form share a common structure. They first allege an empirical claim about the causal origins of our moral beliefs, and then proceed to raise doubts about the reliability of beliefs formed on that basis. These doubts are said to be powerful enough to render our moral beliefs unjustified or unwarranted (I will treat these as equivalent for purposes of this chapter).

One approach to appreciating the nature and potential disruptive power of EDAs is by considering close cousins and seeing how the debunking strategies that are the focus of this chapter are meant to be distinct from them. In the first case, consider an error theoretic argument that seeks to undermine the possibility of moral knowledge by arguing first for the claim that there are no moral truths, and therefore no moral knowledge. By contrast, the EDAs that have been developed over the past decade have an exclusively epistemological aim. They are not designed to settle questions about the metaphysics of morality, but rather typically assume the existence of moral truth – if only for purposes of argument – and then proceed in an attempt to show that evolutionary considerations make it unlikely that our moral beliefs manage to track it.

Second, while some EDAs target only a subset of first-order moral claims for skeptical treatment (typically, those based on deontological seemings; see Singer 2005, Greene 2008, Singer and de Lazari-Radek 2012), the ones I am interested in are focused instead on moral realism. Their proponents argue conditionally: if moral realism is true, then we can have no justified moral beliefs or moral knowledge. Since realists are loath to accept this sort of skepticism – and since almost everyone else is, too – this conditional is meant to serve as the major premise of a modus tollens inference. If moral truth is stance-independent, it's not clear why we should think that our current moral beliefs, influenced so heavily by adaptive pressures, are managing to track the truth.

Third, EDAs are to be distinguished from all wholly a priori arguments for moral skepticism. EDAs are crucially reliant on an empirical premise – the allegation of actual causal influence by natural selective pressures on our moral faculties – whose tenability is essential to that of the EDA in which it is situated.

Fourth, EDAs differ from other a posteriori skeptical arguments by virtue of being a species of etiological critique – they point specifically to a certain kind of doxastic history as the primary source of skeptical concern. Rather than trying to make moral skepticism attractive by pointing out, say, the pervasiveness of actual moral disagreement, proponents of an EDA will offer evidence of some salient fact about the origins of our moral faculties,[1] and then argue that suspicions about these origins should transmit themselves to our moral beliefs.

Fifth, unlike skeptical arguments that seek to show that moral beliefs cannot be even presumptively justified, the role of EDAs is more limited – at their strongest, they will serve as undercutting defeaters. Proponents of EDAs allow that our moral beliefs are defeasibly justified. After all, some of these beliefs appear to have a great deal of intrinsic credibility, are endorsed by almost everyone, and will find a place within a maximally coherent doxastic network. Considerations of evolutionary doxastic origins are meant to show that whatever presumptive justification our moral beliefs enjoy is in fact defeated.

Finally, EDAs are, as Katia Vavova (2015) has put it, targeted – they are meant to be directed exclusively at our moral beliefs, rather than at all of our beliefs. In other words, any successful EDA will not overgeneralize; its success will target only the credibility of our moral beliefs, leaving the epistemic authority of other kinds of belief intact. Proponents of EDAs have often failed to honor this aim, in some cases developing arguments whose soundness would undermine the justification of belief in premises that play a crucial role in the skeptical arguments on offer.[2]

Just How Threatening Are EDAs?

Those who advance EDAs intend their critiques to yield an inescapable moral skepticism – once again, on the assumption (left unstated and taken for granted in what follows) of moral realism's truth. The mandated suspension of judgment about our moral beliefs is meant to be permanent; there is no way of resuscitating their good standing. This is owing to the assumption that ethics is in some important sense autonomous: if the justification of a given moral belief is legitimately called into question, that justification can be restored only by means of enlisting at least one other moral belief or moral seeming. There can be no direct epistemic movement from "is" to "ought"; if there could be, then we might be able to rely on various, well-confirmed empirical beliefs to restore the justification that is undermined by an EDA.

The thought, then, is that a successful EDA will undercut the justification of all of our moral beliefs. If that is so, then it is illegitimate to rely on any of them when seeking to reestablish their positive epistemic status. Although I am not opposed to all forms of epistemic bootstrapping, there is a relevant difference between bolstering one's existing beliefs by means of other, as yet unjustified beliefs or seemings, on the one hand, and trying to do so by means of beliefs that have been shown to be unjustified, on the other. The success of any EDA leaves our moral beliefs and moral faculties discredited and, as such, ineligible to do any justificatory work. It follows that if some version of an inferential barrier from nonmoral to moral claims is plausible, then once an EDA has undercut the justification of our moral beliefs, it will be impossible to restore it. The realist, then, must show that there is something fundamentally mistaken about the debunking strategy.

The hardest question in knowing how to proceed is whether the realist is entitled to rely on any substantive moral claims in evaluating an EDA. If so, then things will go relatively smoothly for the realist. After all, the standard way to establish the reliability of a belief-forming process is to show that it yields true beliefs. If realists are entitled to assume that certain specific moral beliefs are true, then they need only explain how evolutionary pressures may have influenced us to hold such beliefs, et voilà – one can show that such pressures turn out not to have been distorting factors after all.[3]

But one might think that allowing the realist to assume the truth of any specific moral beliefs is illicitly question begging. Suppose it is. Still, there is another general path for realists to take to immunize realism from doubts raised by EDAs. Perhaps realists can show that EDAs are not all they are cracked up to be – that their critical force can be sapped once we appreciate certain vulnerabilities, ones that do not require the realist to assume the truth of any moral belief.

This is cryptic as it stands; I hope all will become clear in due course. For now, however, consider a general epistemic point that applies to all defeaters: if a consideration D undercuts the justification attaching to belief B, then we can restore that justification without having to offer any further support for B. All we need do is to undercut the probative force of D. We can do that, in principle, without piling on further evidence for B. In the moral case, this means that if an EDA challenges our moral beliefs, we needn't enlist other moral beliefs to restore justification, so long as we can isolate a flaw in the EDA – and do so without already presupposing the truth of some moral beliefs.

Leave EDAs aside for the moment and consider a simple example of the general point. Suppose I round the corner and see a man running from a bank with a gun and a huge sack with dollars flying out of it. I form the justified belief that what he is doing is immoral. Then someone tells me that what I've just witnessed was a movie scene being filmed. Now the justification for my original belief is defeated. But suppose that I later discover that the person who told me this was a co-conspirator, trying to deceive me and prevent me from calling the police. Then the defeater has been defeated, and whatever justification my belief originally enjoyed is now restored, even though the "defeater-defeater" is not itself a moral belief.

The lesson here is that even if there is an inferential barrier from nonmoral to moral claims, we may be able to resist the skeptical implications of an EDA if we can show that the threat that it represents is less grave than it might initially appear. We can do that by criticizing either or both of its central premises. In other words, we can subject its empirical premise to critical scrutiny and argue that the causal influence on our moral faculties is less extensive than debunkers have alleged. Or we can take issue with the epistemological premise and argue that, even if selective pressures are as great as debunkers have claimed, this influence does not warrant the skeptical conclusions drawn by debunkers. I think that both strategies are promising. Let's turn our attention to each of these options.

The Empirical Premise

Proponents of EDAs aim to undermine the justification of all moral beliefs (again, on the assumption of moral realism's truth). But it is highly implausible to suppose that all of our justified moral beliefs are the causal product of what I call doxastically discriminating evolutionary pressures, i.e., those that incline us to form beliefs with certain contents rather than others.[4] So at most, only some of our justified moral beliefs are the product of doxastically discriminating evolutionary origins. And now, a dilemma: either we are able to discern which of our presumptively justified moral beliefs have such origins or we are not.

The first horn means smooth sailing for the moral realist. For even if evolutionary influence amounts to taint in the case of our moral beliefs, we could rely on the set of untainted beliefs to critically assess and thence to purify the others.

So suppose the second horn is true: we are unable to determine which of our moral beliefs has been formed as a result of doxastically discriminating evolutionary pressures. The debunker's assertion is then that we can be confident that at least some of our moral beliefs can be attributed to selective pressures, though we know not which. But in this case, the allegation of evolutionary influence threatens to become untestable. Absent a specific story about how selective pressures actually did work to form our moral dispositions and beliefs, the debunker is left with an unsubstantiated allegation of unspecified influence, which cannot be enough to warrant a skeptical attitude toward our moral beliefs. Yet any more specific account will provide us with the means to determine, at least in principle, which moral beliefs owe their content to selective pressures and which do not. And then we are back to the first horn.

Here is a related worry about the empirical premise of an EDA. All such arguments have to provide some basis for moving from this claim – (1) Evolutionary pressures have exercised a doxastically discriminating influence over some but not all of our moral beliefs – to this claim: (C) All of our moral beliefs are suspect.

Suppose, for the moment, that we can supply at least initially plausible intermediate premises that, along with (1), entail (C). If I am right, then the inference from (1) to (C) depends on

Ignorance: We are unable to distinguish the moral beliefs that are the product of doxastically discriminating evolutionary forces from those that are free of such influence.

If Ignorance were false, then EDAs would pose no threat to the justification of those moral beliefs that we can identify as having been free of doxastically discriminating evolutionary influence. Wholesale moral skepticism would thereby be avoided.

Consider the support that debunkers provide for their empirical claim that selective pressures have exercised a doxastically discriminating influence on our moral beliefs. The evidence on offer consists almost entirely in the fact that endorsement of various moral platitudes appears to be adaptive. Some might object that this isn't really evidence of any causal connection – we need more than the correlation here to infer that we hold the moral platitudes we do because doing so has been adaptive.[5] If the objection can be sustained, this would be devastating to the debunking effort, since it would undermine support for the empirical premise of an EDA. But suppose that the objection can be met. Success on that front requires that the following principle, or something very like it, is true:

Causal Influence: If a set of cognitive and conative commitments was likely to have been adaptive in past, then, probably: If we currently have such commitments, then their presence is largely the result of doxastically discriminating selective pressures.

So far as I can tell, EDAs are sound only if Causal Influence (or some quite similar principle) is true. So there is a lot at stake for the debunker in vindicating this principle. But apart from a couple of efforts in this direction (Joyce 2006a; Kitcher 2011), debunkers have for the most part presupposed its truth, rather than doing the work needed to support it. So the first point to make here is this: the empirical premise of an EDA, no matter how it is specified, will rely on the truth of Causal Influence, or something very close to it. Until that principle or a close cousin is vindicated, we have not yet been presented with a compelling debunking story. And if we bear in mind the dialectic, it follows that we don't as yet have a defeater for our moral beliefs. The possibility that Causal Influence will someday be vindicated is not enough, right now, to cast doubt on any of our moral beliefs.

Suppose I am wrong about that. Indeed, let's assume the truth of Causal Influence for purposes of argument. But once we do that, we can see that the following principle seems to enjoy exactly the same amount of support:

Causal Immunity: If a set of cognitive and conative commitments was not likely to have been adaptive in past, then, probably: If we currently have such commitments, then their presence is not (or not very substantially) the result of doxastically discriminating selective pressures.

Causal Influence appears to be no more (or less) plausible than Causal Immunity. So if debunkers are prepared – as they must be – to embrace Causal Influence, then they ought to accept Causal Immunity. But if we do that, then we have a path to identifying a set of highly justified moral claims that are immune from doxastically discriminating evolutionary influence. The contours of the path will differ depending on the specifics of the etiologies advanced by debunkers. Absent any such specific stories, their claim of doxastically discriminating evolutionary influence on our moral beliefs is just speculation. But as soon as the story becomes less speculative and the details are filled in, the realist can use them to pursue this path and identify a set of justified moral beliefs that are immune from such influence.

The Epistemological Premise

Debunkers seek to establish this central epistemic claim: (E) If moral realism is true, then we are not justified in thinking that evolutionary pressures have left us with reliable moral faculties and largely true moral beliefs.

There are two ways to defend (E). The first, more ambitious path requires a defense of

Distortion: If moral realism is true, then evolutionary pressures have (probably) served a distorting role when it comes to our moral beliefs and moral faculties.

A second path requires defense of a weaker principle, one that takes no stand on whether evolutionary pressures (are likely to) have been distorting ones. The debunker must instead argue for

Incapacity: Moral realists are incapable of offering viable reasons to believe that evolutionary pressures have been morally reliable ones.

I don't believe that debunkers are able to vindicate Distortion. And what I take to be their best effort to sustain Incapacity also falls short. If I can make out these criticisms, then (E) will be left without adequate support, in which case the skeptical threat posed by EDAs vanishes. Because debunkers allow for the pro tanto justification of our moral beliefs and are offering a purported defeater of them, skepticism is called for only if the defeater is vindicated. And the defeater is vindicated only if we should be convinced that (E) is true. If I am right, that requires a compelling case for either Distortion or Incapacity. Let's take these in turn.

The clearest way to establish that a causal influence is a distorting one is by means of an argument like this:

1. Causal factor C influences agents to form beliefs that P.
2. P is false.
   Therefore,
3. C is a distorting factor with regard to beliefs that P.

But this path is barred to debunkers. When it comes to moral beliefs, they cannot plausibly assume or argue for any instance of premise (2). If moral realism is true, then once we identify some moral beliefs as false, we are thereby entitled to believe that their contradictories are true. With that entitlement in hand, moral skepticism is easily defeated.

So the debunker must convince us that evolutionary pressures are likely to have been distorting without herself taking any stand on which moral beliefs are true and which false. The most straightforward way to do this would be to argue that evolutionary pressures are, quite generally, distorting factors, joining such states as drunkenness, agony, and deep depression as equal opportunity distorters. But no debunker is alleging that evolutionary pressures, across the board, serve as distorting factors.

To substantiate Distortion, the debunker needs more than just to raise the possibility that evolutionary forces are distorting. The debunker needs to provide some positive reason for thinking this. (After all, just about any cause might be a distorting one.) But establishing a claim of unreliability presupposes a commitment to a standard of truth by which to determine reliability. Once possessed of that standard (whatever it may be), we have all we need to insulate beliefs that meet it from any skeptical force an EDA might otherwise have. If we had no idea at all about what qualified as morally right or wrong, morally good or bad, then we would not be in a position to determine that selective pressures are likely to have played a contaminating role in shaping our moral faculties.[6] But once we have some idea – i.e., once we are entitled to confidence in the truth of at least some beliefs in a domain – then we have all we need to resist wholesale skepticism about that subject matter.[7] So even if evolutionary pressures are often morally distorting, vindicating that claim will provide realists the materials to identify a set of moral beliefs whose truth we can be justifiably confident of. That is enough to block wholesale moral skepticism.

We can put the basic point in terms of a dilemma: Debunkers either do or don't assume wholesale ignorance of the content of morality for purposes of developing an EDA. If they don't, then whatever substantive moral claims they are presupposing are immune to EDAs, thereby undermining hopes for a wholesale moral skepticism. If they do, then the charge of likely distortion cannot be made to stick. For such a charge either mistakenly assumes that evolutionary pressures are quite generally distorting, or it presupposes a standard of truth to substantiate the charge of distortion, and realists can utilize that standard to identify a set of moral beliefs whose justification can then withstand any skeptical force an EDA might otherwise have.[8]

Debunkers might retreat at this point to an effort to vindicate Incapacity. They might begin with a concession: it is possible that adaptive dispositions are also, for the most part, morally reliable ones. But until we have good reason to think this possibility a reality, it is best to suspend judgment on the matter. And, say debunkers, the realist is unable to supply any such reason. We know that selective pressures act so as to yield adaptive behaviors, including adaptive doxastic practices. What we don't know – and can't know, without assuming the truth of some moral beliefs – is whether adaptive doxastic practices also incline us to believe the moral truth, realistically construed. But since the debunker is calling all moral beliefs into doubt, the realist would be illicitly begging the question were she to try to establish the reliability of our moral faculties by taking some moral beliefs for granted. And if she doesn't help herself to any moral beliefs, then she can't possibly substantiate the reliability of our moral faculties, given the existence of a plausible inferential barrier from nonmoral to moral claims.

I agree that barring the realist from enlisting any moral beliefs is going to make it impossible for her to establish the reliability of our moral faculties. (Of course, barring realists about the external world from enlisting any perceptual beliefs will also make it impossible for them to establish the reliability of our perceptual faculties.) But the moral realist isn't yet illicitly begging any questions by invoking substantive moral beliefs to substantiate the reliability of our moral faculties. The realist is entitled to rely on moral beliefs until such time as the debunker has successfully provided a defeater. Without a very powerful debunking argument on the table, it is the debunker who would be begging the question against realists. She'd be doing that were she to insist that we must suspend judgment about moral matters in the absence of a positive argument for the reliability of our moral faculties, and then denying realists the use of the only possible means (substantive moral beliefs) to craft such an argument.

We can see why the debunker's defense of Incapacity is problematic by constructing a parallel argument to the one offered two paragraphs above and noting its implausibility:

We know that selective pressures act so as to yield adaptive behaviors, including adaptive doxastic practices. What we don't know – and can't know, without illicitly assuming the truth of some substantive arithmetic beliefs – is whether adaptive doxastic practices have also inclined us to believe arithmetic truths, realistically construed. So we should suspend judgment about our arithmetic beliefs.

Since this line of reasoning is perfectly analogous to the debunking one just aired, and its conclusion deeply implausible, we should be suspicious of that initial argument.

The debunker might respond: We can know that adaptive arithmetic beliefs are likely to be true because if they were false, then they wouldn't be adaptive. Whereas we have no reason to suppose that false moral beliefs would fail to be adaptive. So the parallel I've been pushing won't work.

But this response fails. We can rightly believe this claim:

(M) If our arithmetic beliefs were false, then they wouldn't be adaptive

only if we already assume that some of our arithmetic beliefs are true. If we really had to suspend judgment about all of our arithmetic beliefs, then we would be in no position to verify (M).

What the debunker needs is a further argument that can vindicate Incapacity. Specifically, she must provide an independent debunking argument that works to defeat the presumptive warrant enjoyed by our moral beliefs. Only at that point can she justifiably claim that moral realists are forbidden from enlisting any substantive moral beliefs in an effort to establish the reliability of our moral faculties. Let's conclude by reviewing the debunker's best chance to do just this.

The Insensitivity Argument

The EDA that seems most threatening derives from a charge of insensitivity. Ordinarily, we regard a belief-forming process as insensitive to the truth just when it would continue to generate the beliefs it does even if the beliefs were false. But the assumption made by most moral realists (myself included) is that at least the basic moral principles are necessary truths, making a hash of this way of understanding the nature of insensitivity. So perhaps we can formulate what I'll call the Insensitivity Argument like so:

1. If the truth of a belief fails to play a role in explaining why we hold that belief, then that belief is unwarranted.
2. If moral realism is true, then the truth of our moral beliefs fails to play a role in explaining why we hold them.

Therefore:

3. If moral realism is true, then our moral beliefs are unwarranted.

Formulated as I have done, the first premise allows debunkers to spare themselves the difficult task of identifying and then applying a substitute counterfactual to replace the standard one employed when testing for insensitivity (viz., Would we believe p even if p were false?). Instead, premise (1) invites us to focus on the explanatory role that a given truth plays in the formation of a belief. The idea is that if an agent forms a belief in such a way that its truth plays no role in its generation, then the agent has lighted on the truth by accident, thereby preventing her belief from being justified. That is so even if the content of her belief is a necessary truth – even if, given the belief she has actually formed, she couldn't possibly be mistaken.

The defense of premise (2) is pretty straightforward: if we hold the moral beliefs we do because of having been subject to selective pressures, then it is these pressures, rather than the moral truth, realistically construed, that explains why we have the moral beliefs we do.[9] It might be that our moral beliefs are true. But since we don't hold them because they are true – we hold them, instead, because evolution effectively "programmed" us to hold them – it follows that our moral beliefs are unjustified. They are unjustified because even if we have landed on the truth, our having done so is wholly coincidental. If we have the moral beliefs we do because of evolutionary forces, then we would continue to hold these beliefs, regardless of where the truth in ethics lies. This is an intuitive way of expressing the kind of insensitivity that (per premise 2) allegedly attaches to our moral belief-forming operations. Such insensitivity undermines any warrant we might have for our moral beliefs.

Though the first premise, even understood capaciously, may be problematic,[10] let us grant its truth here and direct our critical attention to its second premise. Realists can offer at least two lines of resistance. The first says that even if evolutionary origins of our moral beliefs are distorting, a number of our moral beliefs do not arise from doxastically discriminating evolutionary pressures. The beliefs that emerge in these latter ways are immune to the line of argument advanced on behalf of premise (2) of the Insensitivity Argument, for even if evolutionary influence were distorting, beliefs that have different origins may be perfectly epistemically above board. Indeed, for all the debunker has yet said, those moral beliefs may be held because they are true. This point lines up nicely with the arguments of Part III in limiting the skeptical implications of EDAs. And imposing such limits entirely undermines an EDA's ability to threaten moral realism, for that threat depended on showing that if moral realism were true, then a wholesale moral skepticism would follow.

Here is a second, complementary reply, related to the one just offered. Suppose again that evolutionary etiologies of moral beliefs are distorting. The realist, then, must show how we can correct for such distortion. This of course is a very serious challenge for realists. Answering it requires the development of a positive moral epistemology, one element of which is the articulation of a method of moral inquiry that promises to correct for moral mistakes that arise from unreliable influences. This would be impossible if all of our moral beliefs were subject to distorting influences, or if a great many were, and we were unable to sort the ill-formed beliefs from the better ones. But the debunker has not yet shown that either of these unhappy alternatives is actually the case. Without having done that, though, the debunker has yet to show that realists are unable to offer an explanation of how moral agents can correct for any distorting influence that evolutionary forces exert.

To see why this is important, and how it bears on premise (2), consider a familiar skeptical analogue of EDAs: if moral realism is true, then cultural influences are often morally distorting. We know this because these influences often lead people to have incompatible moral beliefs, some of which must be false if moral realism is true. The realist should grant that some of our moral beliefs are products of cultural influence, and that such influence is sometimes distorting. But this does not yet force us to wholesale moral skepticism, since the realist will offer diagnoses of error and seek to explain why some such influences are likely to have led us astray and others not. These diagnoses will point to general sources of cognitive distortion, ones that apply to nonmoral beliefs as well as to moral ones. They will also, very likely, invoke some substantive moral claims in assessing certain cultural pressures as distorting ones (e.g., those that have led citizens to regard their political leaders as gods, or those that encourage children to inform on parents who privately express antiauthoritarian sentiments).

It is of course possible that these further substantive beliefs, introduced as correctives to certain cultural influences, are themselves products of distortion.

But to assume this from the get-go is to just assume that the realist is incapable of doing what she claims – namely, to be able, sometimes, to identify the moral truth, and to form our moral beliefs on the basis of an adequate grasp of that truth. We reasonably think that some of our moral beliefs arise primarily from cultural influences, others from parental influences, yet others from evolutionary influences. Unless we are assuming from the start that they cannot arise from an appreciation of the truth, from an understanding of the grounds in virtue of which moral facts are what they are, then for all we know, some of our moral beliefs also stem from our grasp of the moral truth. To assume otherwise – i.e., to assume the truth of premise (2) of the Insensitivity Argument – is just to beg the question against the realist.[11]

This is not the end of the debate between realists and debunkers. There may be a better option for debunkers than the Insensitivity Argument, or a better version of that argument than the one I have considered here.[12] But until debunkers manage to clearly present and vindicate such an argument, we lack reason to think that EDAs are going to succeed in their aim of committing moral realism to a thoroughgoing moral skepticism.

Notes

1 I am using talk of moral faculties very loosely, to designate *whatever* mechanisms and processes are responsible for our making the moral judgments we do. So in my use of the term, it is a definitional truth that moral beliefs are the product of moral faculties. I am not committing myself to the existence of discrete "modules" of moral thinking, or to the existence of neural networks, separate from those that govern thought about nonmoral topics, that are devoted exclusively to generating moral beliefs.

2 See Shafer-Landau 2012 for details.

3 This is the strategy deployed by Wielenberg 2010; 2014, Brosnan 201; Copp 2008; Enoch 2011; and Skarsaune 2011.

4 There is no doubt that evolutionary pressures have had some doxastically nondiscriminating influence. We are capable of thinking about morality (and philosophy and set theory and chemistry) in part because of the way that evolutionary pressures have contributed to the development of our cerebral cortex. But results in philosophy or chemistry are not cast into doubt by such generalized influence. EDAs get what force they have from the concern that evolutionary pressures have pushed us to hold certain moral beliefs rather than others; this is the phenomenon that I am calling doxastic discrimination.

5 Cf. Parfit (2011, 535).

6 To my knowledge, there is one exception to the line of argument I am pushing here. That is the view, advanced by Street (2006, 122ff.; 2008b, 208;

2011, 14), which says that we have excellent reason to believe that evolutionary forces are morally distorting influences because the odds that they have inclined us to believe the truth, realistically construed, are vanishingly small. Street's argument for this explicitly assumes that morality could be about anything at all; since that is so, and since evolutionary pressures have pushed us toward one subset of commitments that represent a very small percentage of all possible moral commitments, the chances are extremely slim that they have inclined us in the right direction. I reject the view that morality could be about anything at all (see Cuneo and Shafer-Landau 2014, 2016). And even if there were no conceptual constraints on what could qualify as morality or as a moral judgment, the "low-odds" argument just given would overgeneralize to yield a wholesale nonmoral skepticism, thereby defeating the aims of the evolutionary debunker. For defense of this last point, see Shafer-Landau 2012, 13–14; Berker 2014, section 7; and Vavova 2015, section 3.2.

7 See Vavova 2014.

8 See Vavova (n.d.) for a similar line of argument.

9 One might also defend premise (2) by means of the following argument: (a) The truth of our beliefs play a role in explaining why we hold them only if we can make causal contact with the facts that such beliefs represent. (b) Moral facts (if any there be) are causally impotent. Therefore: (2) The truth of our moral beliefs fails to play a role in explaining why we hold them. Assessing the argument would take us too far afield; suffice it to say that both (a) and (b) are highly controversial and that this argument, whatever its merits, is most certainly not an EDA, since it lacks an empirical premise that postulates evolutionary influence on our moral faculties.

10 The first premise obviously needs lots of interpretation, especially with regard to its central notion of a truth's explaining an agent's belief in it. A capacious understanding of explanation here will assume that explanation needn't be causal and will allow, for instance, that believing on the basis of accurate testimony (rather than firsthand experience or understanding) can count as the kind of case in which the truth helps explain why someone has the belief she does. (Still, there are worries. One example: Do facts about the future help explain why we hold beliefs about it? If not, then premise 1 indicts the justification of all such beliefs.)

11 A complementary and more fully developed version of this reply can be found in FitzPatrick (2015).

12 In Shafer-Landau (2012), I reconstruct five EDAs that have appeared in the literature; Wielenberg (2010) reconstructs four. But my own view is that the Insensitivity Argument, or some version of it, poses the most serious threat to moral realism.

13 Why Darwinism Does Not Debunk Objective Morality

William J. FitzPatrick

Two Perspectives and a Philosophical Question

Long before we develop a scientific understanding of the world and adopt a Darwinian perspective on ourselves as products of biological evolution, we live in that same world as committed moral agents. We embrace certain values and norms in our practical life, revising and refining them over time to reflect our maturing sense of what it is to live well and to be good human beings. And if we are serious about our central moral commitments, we will not tend to think of them as mere matters of taste or as arbitrary or optional personal projects or social conventions we happen to have but could equally jettison without error. Instead, it is natural to view our deepest moral values in the context of striving to get something right – something it matters greatly that we do get right; and insofar as our moral beliefs undergo changes, those changes seem answerable to that aim as we seek improved moral understanding and practice.

Consider a simple example. If you are serious about the value of racial or gender equality, and about related moral norms prohibiting race- or gender-based restrictions on liberty or the right to vote or to receive an education, then you will presumably be inclined to see social changes made from earlier racist or sexist practices to current norms of political and social equality as genuine progress, better realizing important values that it would be a grave mistake to abandon. If someone proposed reversing these changes, you would see this as something to fight against, and not just because it would involve changing a convention you happen to be comfortable with – as if someone were to propose changing which side of the road we drive on or the structure of the academic calendar. While we may resist changes to mere conventions for reasons of personal comfort and convenience, there seems to be something very different at stake in moral cases, giving us very different grounds for fighting for some changes (viewed as progress) and against others (viewed as regress). From this perspective, then, certain values and norms seem to have a special status that merits our attention and commitment as moral agents in a way that mere taste and convention do not.

At the same time, however, we have also learned to think about the world from a more abstract, scientific perspective, detached from the personal perspective of the engaged and committed moral agent, looking instead at the world – including the human parts of it – "from the outside," employing only the concepts and methods sanctioned by empirical scientific inquiry. From this perspective, our values and commitments appear as just more empirical phenomena in the world, to be viewed through the same lens employed by an amoral anthropologist studying the desires, feelings, and practices of a distant tribe. When viewed "from the inside," our core values, such as racial and gender equality, strike us as correct and as grounding moral truths such as that race-based voting restrictions are morally wrong; but this is not even an issue on the radar from the scientific perspective, which does not trade in such categories as moral truth and progress. What we see or infer from that perspective are behavioral and psychological phenomena – people doing certain things and reacting certain ways, driven by various desires, feelings, and judgments they make. This is all studied with a view to discovering causes of a sort available to empirical inquiry without considering philosophical issues of moral truth and justification. It is the sort of inquiry someone could conduct even without being a committed moral agent herself – just as a biologist can study the goings on of a beehive without sharing in the ways and "concerns" of bees, looking simply for scientifically accessible causes of observed phenomena. There is a difference, of course, in that the anthropologist has to be able to understand the complex psychological states of the people being studied, unlike the entomologist. But these states are still approached from a scientific perspective simply as causal upshots of various empirical factors, likely extending back to Pleistocene evolutionary influences, rather than as potential apprehensions of moral truths. In this way, the scientist still views moral agents from a stance much like that of the biologist observing the beehive – an example to which we will return shortly.

The philosophical question all of this raises is, how should we ultimately think about morality once we are capable of thinking about it from both perspectives – from the outside, as scientists, as well as from the inside, as engaged moral agents? As it happens, many are led, without ceasing to be moral agents in their practical lives, to focus so heavily on the scientific perspective that it seems to them necessary for theoretical purposes to discount the view from the inside and what it seems to reveal to us, in favor of a deflationary view of morality limited to the offerings of the scientific perspective alone. This might take such forms as moral subjectivism, relativism, or skepticism, often marching under the banner of "debunking" the view that there are objective moral truths (such as the truth that racial discrimination and persecution are wrong) or that we could know them if they existed. In particular, many are led to this deflating picture of morality as a result of coming to see human beings from a Darwinian perspective: they take the fact of evolution somehow to undermine the usual pretentions of morality.

I believe that this use of science, and specifically the appeal to Darwinism to support such a view of morality, is misguided. More precisely, it is a mistake if the implication is that all of us, regardless of the philosophical views we start out with, should be led by Darwinism to adopt a deflationary view that denies knowledge of objective moral truths. We can take the scientific perspective seriously without neglecting or discounting the "internal" perspective we also occupy as engaged and committed moral agents and ultimately maintain a non-deflationary understanding of morality that allows for the possibility of objective moral truth and knowledge. There are, of course, familiar philosophical challenges that must be answered by such a view, but the point is that the science – and in particular, evolutionary biology – does not show such a view to be untenable. In the next two sections, I will sketch some common skeptical or debunking claims and then, in section four, explain why they are far less potent than they might appear to be.

Darwin's Rational Bees

One reason sometimes given for discounting the appearance of objectivity that morality presents from within the internal perspective is the alleged fact that our moral beliefs are radically contingent and driven by ecological peculiarities stemming from our evolutionary background. Something like this thought was strikingly expressed by Charles Darwin himself in a fanciful reflection (later echoed by E. O. Wilson [1978, 204–206], among others), which puts an interesting spin on our example of bees:

It may be well first to premise that I do not wish to maintain that any strictly social animal, if its intellectual faculties were to become as active and as highly developed as in man, would acquire exactly the same moral sense as ours. In the same manner as various animals have some sense of beauty, though they admire widely different objects, so they might have a sense of right and wrong, though led by it to follow widely different lines of conduct. If, for instance, to take an extreme case, men were reared under precisely the same conditions as hive-bees, there can hardly be a doubt that our unmarried females would, like the worker bees, think it a sacred duty to kill their brothers, and mothers would strive to kill their fertile daughters; and no one would think of interfering. Nevertheless the bee, or any other social animal, would in our supposed case gain, as it appears to me, some feeling of right and wrong, or a conscience. (Darwin 1871, 73)

This passage is puzzling in certain ways, but the basic idea Darwin is getting at is that our moral sense has the shape and content it has largely because of the contingencies of human ecology. We could therefore have had very different moral sensibilities had evolution supplied us with dispositions less geared toward sympathy and cooperation (Ruse 2006). And what is more, if creatures with a very different ecology, such as bees, had developed the capacity for rational judgment, they would have developed a very different moral sense suitable to their ecology, endorsing the killing of brothers by unmated sisters, the killing of fertile daughters by mothers, and so on. And if that is right, then it might seem naive to suppose that there is any objective morality at all – any set of objective moral truths to be grasped. From a scientific perspective, looking at all these possible forms of life and associated moral sensibilities, it might seem that all there is are these different sets of moral beliefs causally conditioned by the ecologies that shape the psychologies of believers – in each case giving rise to the illusion "from the inside" that they are on to something real and true in the realm of value and duty, though in fact there are no objective moral facts on the scene (Ruse 2006).

Alternatively, even if the possibility of objective moral facts is granted, it might still seem that we could not lay claim to any knowledge of them: if our moral beliefs spring from the contingencies of our evolved ecology, and different ecologies would have given rise to innumerably many different moral senses and hence different sets of beliefs, it would seem to be a merely lucky coincidence if our moral beliefs (as contrasted with those of the hypothetical rational bees, say) happened to track the objective moral facts. In the absence of any special reason to think that the contingencies of our primate evolutionary development coincidentally happened to push our beliefs toward the objective moral facts, we seem to have no grounds for confidence that our moral beliefs are true. Or at least this would seem to be so unless the moral facts were radically relativized to ecological circumstances by just being reduced to facts relevant to Darwinian biological fitness within each form of life – hardly an attractive prospect to most defenders of objective morality.

Epistemically Defective Processes and the "Best Explanation" of Our Beliefs

Picking up on the second general worry just mentioned, a central debunking theme focuses on two related questions about the explanation of beliefs: What best explains our moral beliefs? And what best explains our (metaethical) belief in moral objectivity? In both cases, the strategy is to provide scientifically based explanations that somehow undermine our justification for those beliefs. To see how this works, consider a parallel with the belief, held by many grieving families during World War I, that spiritualists were communicating with the dead (Ruse 2006, 22–23). One way to undermine justification for that belief would be to show that it was caused in believers by mere wishful thinking and emotional stimulation of the imagination, which are epistemically defective processes (i.e., unreliable for producing true beliefs). This hypothesis is supported, for example, by the discovery that in some cases the supposedly dead soldiers that people thought were offering reassurances from beyond the grave were in fact alive and suffering in a military hospital (Ruse 2006, 23). Clearly the things people were taking as evidence of spiritualist communication were produced in these cases by extraneous factors such as overactive imagination stoked by grief and wishful thinking – and sometimes by charlatanism. And given the lack of other evidence for life after death and contact with the dead, as well as good reasons for doubting such things, this suggests that such psychological causal factors provide the true, exhaustive causal explanation for the belief, thus undermining any warranted confidence in its truth.

Similarly, then, if it could be shown that the sole source of our belief in moral objectivity lies in some equally epistemically defective process(es), then presumably we should likewise lose any confidence we might have had in the objectivity of morality (Nichols 2004). For example, suppose it could be shown that our belief in moral objectivity is caused simply by the disgust and anger we feel toward violations of moral norms, or by our desire to punish such transgressions, where these responses can be traced to the adaptive role played by such dispositions in our Pleistocene ancestors. This might be supported, for example, by studies indicating that even artificially induced, unrelated disgust and anger can increase perceptions of objectivity for moral judgments (Nichols 2004); and this effect might be explained as an adaptation helping to reinforce the motivational force and hence adaptive benefits helping to reinforce the motivational force and hence adaptive benefits of making moral judgments in cooperative social contexts. If that is all correct, it might seem to debunk the belief in moral objectivity, showing such objectivity to be an illusion that is with us because of its contribution to the functioning of moral judgment in promoting social coordination among Pleistocene hunter-gatherers (Ruse 2006, 21).

Other arguments focus on our various first-order moral beliefs themselves. Some seek to debunk them, appealing to Darwinism to undermine our justification for our moral beliefs, leaving us with moral skepticism (Joyce 2006a); others allow that our moral beliefs are largely justified but try to show that they wouldn't be if moral truths were objective (thus debunking not our moral beliefs but objectivism or realism about those beliefs) (Street 2006). We'll focus on the former, ignoring various complications (see FitzPatrick 2008, 2014, 2015 and Shafer-Landau 2012 for detailed discussions). Consider the explanation of the belief that murdering or betraying one's comrades is wrong. Suppose it could be shown that the best explanation of this moral belief appeals only to extraneous causal processes and factors having nothing to do with the truth of the content of the belief. Upon discovering this fact, would we not lose all warranted confidence in this moral belief, since it would require a sheer coincidence for it to be true (which we would have no grounds for thinking has occurred)? Suppose you discovered that your belief that racial discrimination is wrong is simply the causal upshot of a pill someone slipped you, which causes such a belief to arise in a way that is utterly independent of any moral facts; indeed, the pill was chosen randomly from a group that also included pills that would cause the opposite belief to arise (cf. Joyce 2006a, 179). Surely upon discovering this fact about the etiology (causal origins) of your belief, you should lose all confidence in its truth: even if there is a moral fact about racial discrimination, you would not be in a position to know it, since your belief would, at best, merely happen accidentally to be true. What some have argued, then, is that we are in roughly this situation with respect to our moral beliefs, not because of magic pills but because of pervasive evolutionary influences on our moral beliefs (Joyce 2006a; Street 2006 argues that we would be in this situation if our moral beliefs were answerable to objective moral facts).

The basic argument goes roughly as follows. Why, after all, do human beings have the capacity and disposition for moral judgment and motivation in the first place (much as we all have a capacity for language)? The best answer seems to be that this is an adaptation that evolved in the human lineage because it promoted social coordination and cooperation, ultimately enhancing biological fitness for ancestral humans who engaged in moral judgment and practice within their social groups (Kitcher 2011). And this plausibly involved the evolution of particular emotional and cognitive dispositions influencing the direction and content of moral judgment in ways that were adaptive to Pleistocene hunter-gatherers. That is, our capacity and tendency (1) to experience characteristic emotions such as sympathy, parental love, jealousy, resentment and guilt, and (2) to make judgments condemning incest, unfairness, cheating, and in-group harming or praising loyalty, reliability, and cooperation all traces back to biological adaptations in the sphere of human psychology (Barkow et al. 1992; Buss 2005). Suppose, then, that our moral judgments can best be explained as just the causal upshots of such evolutionary influences (though of course the particular forms they take will reflect contingent cultural developments and conditioning as well). If so, then we have a problem.

On this picture, the psychological mechanisms giving rise to our moral beliefs emerged as adaptations for solving problems of social exchange faced by our Pleistocene ancestors, and they were sculpted by Darwinian forces that rewarded modifications in these psychological traits ultimately for one thing: maximizing their possessor's (or kin's) genetic contribution to succeeding generations in that ancestral environment. Now this may well have resulted in some dispositions we now intuitively associate with a plausible, true morality, such as tendencies to cooperate with coalition partners or to care for one's children. But it can equally be expected to have resulted in a variety of nastier tendencies as well, such as a disposition for selective cheating and for aggression toward weaker outsiders, or more generally for tribalism and xenophobia. The crucial point is that these psychological dispositions influencing moral belief were shaped by Darwinian, genetically oriented processes having nothing to do with moral facts as such, since in causally shaping moral beliefs, they operated insensitively to the truth of the content of those beliefs, simply rewarding belief-forming dispositions according to their effects on hunter-gatherer genetic propagation. So if our moral beliefs are merely causal upshots of such "morally blind" processes (dressed up with cultural embellishments along the way), then we would seem to be no better off than if our beliefs resulted from random belief pills. Upon discovering that our moral beliefs can best be explained, and explained completely, in terms of extraneous causal influences operating insensitively to the truth of the content of those beliefs, it seems we would be in no position to lay claim to moral knowledge, as it would be only a lucky accident if our beliefs turned out to be true. We would be stuck with moral skepticism (Joyce 2006a).

Responses to the Darwinian Deflationary Arguments

In the previous two sections we have seen appeals to Darwinism to cast doubt on the plausibility of objective morality, or on our justification for the belief in moral objectivity, or on our justification for our moral beliefs, allegedly undermining the possibility of moral knowledge (or at least knowledge of objective moral truths). In the remainder of this chapter, I will provide reasons for skepticism in turn about whether these appeals to Darwinism to deflate morality or moral knowledge really carry much weight in the end (see FitzPatrick 2008, 2014, 2015 for further development of some of these points).

On Ecological Contingency and Darwin's Rational Bees

To begin with Darwin's hypothetical example, the first point to notice is that with the emergence of genuine rational capacities of the sort necessary to turn these hymenoptera into buzzing moral agents would come also the possibility for achieving critical reflective distance from their evolutionarily given instincts – just as we find in human beings. Precisely because of our reflective capacities, we are not merely slaves to the genetic imperatives built into our evolved human nature, destined to act out the instinctive psychological tendencies stemming from our evolutionary heritage. That heritage might include strong tendencies toward tribalism or racism, or a disposition in males to philander (as found in other species for obvious Darwinian reasons). But with the emergence of rationality and moral agency we acquire also the ability to step back from such evolved dispositions and to subject them to evaluative reflection in light of our best culturally developed conception of what is truly worthwhile and appropriate in life, and how it is ultimately good and right to live – choosing, for example, to honor a faithful marriage or to treat even distant strangers with kindness, justice, and generosity, as we seek to live well. We obviously engage in autonomous exercises of this general capacity for reflection, reasoning, and choice in countless spheres of life – autonomous in the sense of not being micromanaged by evolutionary influences, but instead involving reasoning in accordance with standards internal to various culturally developed modes of inquiry. This is how we exercise our (evolved) mental capacities when doing mathematics, physics, and philosophy (as in the present debate), and there is nothing preventing us from doing the same in the course of ethical reflection – a point we will come back to.

The same, then, would be true for the rational bees. Their rational capacity and moral agency would not merely transform them into creatures that now make moral judgments in the course of still slavishly carrying out the Darwinian genetic imperatives for hymenoptera. Perhaps this is all we would find in unreflective and tradition-bound bees, who might be found thundering from bee pulpits about "sacred duties" to kill brothers or fertile daughters, their moral senses being mere expressions of their evolved psychologies. But we could equally expect more reflective bees (spurred on by the philosophers among them, after sufficient cultural development) to raise the question whether they really had good reason to do everything that their evolutionary history has disposed them to do in the service of maximizing their genetic propagation. They might reflect, in light of their ongoing cultural and individual experience and innovation, about the potential goods of rationalized bee life that might be available to them if they resisted certain impulses and acted differently and about what principles of action can truly be justified in light of those possibilities – just as we do when we reflect and reject tribalist principles in favor of more rationally consistent and expansive principles we come to see as properly justified.

We must be careful, then, not to exaggerate the contingency of our moral beliefs by supposing that they simply reflect our evolved ecological situation and would therefore automatically be radically different across rational species. That assumption overlooks the role played by autonomous moral reflection and reasoning, which reduces the ecological contingency of moral beliefs. Of course, it would also be naïve (or at least overly Kantian) to suppose that deep ecological differences across diverse species would be canceled out completely by the power of rational reflection. But this is not a problem for moral objectivists because any plausible version of objectivism will allow that the moral facts depend heavily on relevant facts and features of life – not evolutionarily given genetic imperatives, but facts relevant to possibilities of relationship and well-being, for example, affecting what is

fitting and right and worth pursuing. So those of us who believe in objective morality should not expect the moral facts to be wholly invariant across
radically different forms of life, but should instead embrace a reasonable
pluralism reflecting the different possibilities for a good life available to
rational creatures of different biological kinds. And the ineliminable variation in moral belief across such species needn't reflect anything more than
a reasonable sensitivity to these differences in the moral facts; in particular,
it needn't be as radical as what Darwin imagined, and so needn't lead us to
despair of the idea of (appropriately pluralistic) objective moral truths or of
coming to know the ones that pertain to us as human moral agents – such as
the truth that it is wrong for us to organize our societies in ways that embody
racial and gender discrimination. There could well be such truths, and our
arriving at them would not require any miraculous coincidence involving
an ecological lottery. It would require only our successful employment of
evolved mental faculties that have been trained and refined in cultural contexts through experience and reflection – just as with higher mathematics,
physics, or philosophy – to discover, in this case, the moral wrongness of
racist or sexist practices.

What Really Best Explains Our Beliefs?

Some will be prepared to grant this possibility in principle but will still claim
that we are unjustified in believing it to be true. Perhaps it just seems like too
much of a coincidence that (1) evolution should have given us a capacity and
disposition to value things and to judge things right and wrong, for Darwinian
reasons having nothing to do with actual moral value or duty, while (2) it
also turns out that certain things really do matter and certain actions really
should or shouldn't be done. I suspect that this simple thought underlies much
Darwinian skepticism about objective value and morality, especially when
combined with a scientistic prioritizing of the "external" perspective discussed
earlier, where all that is visible to inquiry and relevant to methodology are
empirical psychological and behavioral phenomena and scientifically accessible causal factors. What use, from this perspective, are actual moral properties
and facts? Isn't it simpler and more parsimonious just to drop them from the
picture and to see morality as just so much evolved psychology and behavior,
stemming from biological and cultural causal factors having nothing to do
with moral facts, dragging behind it a clever illusion of greater significance
that we have now finally seen through?

This line of thought finds more precise expression in the claim that scientists are now in a position, thanks to Darwinism, to argue that the "best
explanation" of our moral beliefs – and similarly of our belief in the objectivity of morality – appeals in every case only to scientifically accessible causal

factors that have nothing to do with the truth of the content of those beliefs as such. That is, the claim is that we can give a "complete explanation" of our moral beliefs that nowhere presupposes their truth in explaining the origins of the beliefs, just as we can give a complete explanation of belief in spiritualistic communication with the dead that nowhere presupposes communication with spirits. This is then taken by debunkers to undermine our justification for our moral beliefs, as it would seem to be a mere coincidence if they turned out to be true, since we arrived at them for reasons unrelated to their truth; moreover, if our moral beliefs can be completely explained without appeal to moral truths, then it would seem superfluous to posit such truths in any case (Joyce 2006a).

Now one problem with this line of argument is that it is just not the case that epistemic justification is undermined merely by showing that a belief can be explained in a way that doesn't presuppose its truth. We have justified beliefs about the future (e.g., that ocean levels will be significantly higher a century from now, based on current evidence of climate change and our best models), as well as justified false beliefs (where we have excellent evidence but the belief turns out to be false), despite the fact that neither of these are explained by appeal to the truth of the beliefs in question (Nichols 2004). So as it stands, the argument proves too much. But this problem can be avoided by modifying the argument so that it claims only the following: epistemic justification is undermined when a belief can be completely explained in a way that neither presupposes its truth nor appeals to an agent's taking account of considerations that constitute good evidence of such truth. This avoids the previous counterexamples since the explanation of our justified beliefs about the future does appeal to an agent's taking account of considerations that are evidentially related to the truth of the belief contents in question, despite not being caused by such truths, and similarly with cases of justified false beliefs. And the claim in the present case will be that the causal factors that fully explain our moral beliefs – whether evolutionary influences or sociological conditioning – do so in a way that neither presupposes moral truths nor involves agents' recognizing good evidential relations: they are merely extraneous, "morally blind" causes on a par with the magic belief pill.

There remains a much deeper problem with this whole line of argument, however, and it can be put quite simply: no one who starts out believing in objective moral truths, based on philosophical reflection that takes seriously the "internal" perspective they occupy as engaged moral agents, has compelling reason to accept such strong claims made in the name of science about the "best explanation" of our moral beliefs (or of our belief in moral objectivity) in the first place. At most, what the science shows us is that there are various extraneous causal factors that have plausibly had some degree of influence on some moral beliefs, which is a far cry from establishing that the complete explanation for all of our moral beliefs appeals solely to such extraneous causal factors. From a purely scientific, external perspective such explanations may naturally appear complete: what could be missing? But as soon as we reoccupy the internal perspective of the engaged moral agent, we see exactly what may be missing from the scientist's explanations – namely, our coming to at least some of our moral beliefs through grasping good reasons for thinking them to be true, as by understanding certain features of an action to be wrong-making and hence good reasons for thinking that the claim that the action is wrong is true. (This would be precisely the recognition of evidential relations denied in the previous debunking claim.) These good reasons are not visible to the scientific perspective, which does not deal with such things as moral truths, truth makers, and right- or wrong-making features of actions, and instead sees only empirically accessible causes. But such good reasons are intelligible and salient from within a morally engaged perspective, as we think critically about racism or sexism, for example (i.e., the wrong-making features of racist or sexist policies, which are good reasons for thinking such policies to be wrong), just as good reasons for believing a mathematical theorem are visible and salient from within a mathematically engaged perspective, leading us to hold our mathematical belief for good reasons and not merely as a result of extraneous causes.

Just as you have good reasons to offer in support of your mathematical beliefs, which are what would be cited in any charitable explanation of those beliefs (e.g., we cite the mathematical proof that you take to establish a certain theorem, in explaining why you believe the theorem to be true), we have what are plausibly good reasons to offer in support of many of our moral beliefs. And it is similarly reasonable to seek to explain these beliefs by appeal to our reasons – the considerations we take to show that the contents of these beliefs are true – rather than looking straightaway for mere extraneous causes of the beliefs. This does not, of course, guarantee that what we took to be good reasons really are good reasons (any more than in the mathematical case). We could be mistaken in any given case, and it could even turn out that there are never any good reasons for believing anything in the domain of morality because there are no moral truths at all. But that is certainly not something that is established by Darwinism or any other part of science.

Once we take seriously the significant autonomy with which we can approach ethical reflection, and the fact that our exercise of that autonomous reflection within the perspective of engaged moral deliberation involves our arriving at beliefs via judgments about certain considerations constituting good reasons for drawing certain moral conclusions (just as in math, science, or philosophy), it is a wide open question whether we might at least often be basing our beliefs on genuine recognition of good reasons rather than merely being caused to have them by morally blind causal pushes and pulls of the sort available to scientific observation. This is a philosophical question that cannot be settled one way or the other by scientific investigation, and how we answer it will depend on what strikes us as the best explanation for our beliefs, which in turn will depend on how confident we are in the reasoning that supports our moral beliefs.

Those who think they have independent philosophical grounds for doubting the soundness of our moral reasoning and who also believe that the scientific perspective – with its empiricist methodologies and the prioritizing of parsimony as an explanatory virtue – is the only reliable way to learn about the world will naturally claim that their explanations of our moral beliefs in terms of extraneous causal factors are the "best explanations," standing ready to do debunking work. But many of us instead begin from a morally engaged perspective within which we find certain moral arguments – or claims about right- or wrong-making features, or about good reasons for certain moral conclusions – to be compelling. We thus begin from the conviction that at least many of our moral beliefs are based on our recognition of such good reasons rather than merely being caused in us by extraneous factors. From this perspective, we will not find the parsimonious scientific causal explanations to be the "best explanations" at all – any more than you would find such an explanation of your mathematical or philosophical beliefs to be the best explanation of them. It would be insulting to your mathematical or philosophical competence if we were to purport to explain (completely) your mathematical or philosophical beliefs in a way that just bypassed the reasons you offer as good reasons for them from within your mathematically or philosophically engaged perspective, instead appealing directly to "mathematically or philosophically blind," extraneous causal factors (having to do with mere psychology or sociology). But in the same way, it will strike many of us as insulting to our moral competence to purport to explain (completely) our moral beliefs in a way that just bypasses the reasons we offer as good reasons for them within the perspective of moral deliberation, as if it could just be assumed that these are never really good reasons and we believe what we believe only due to extraneous causal factors (such as evolutionary or sociological factors).

The crucial point here is that we have no obviously compelling reason to accept such purported explanations of why we believe what we do: They may seem like the best explanations from an exclusively scientific point of view for those who already doubt the existence of good reasons in the moral sphere, but for the rest of us, they are not the best explanations at all. They are in fact seriously defective explanations, leaving out the most important part of the story of why we believe what we do – i.e., our recognition of good reasons for these beliefs. If I am convinced through moral and philosophical reflection that I have good reasons for believing slavery to be wrong, and

I take these to constitute a correct reason-giving explanation for why I believe what I do, I am obviously not going to grant that an explanation simply in terms of my upbringing or sociological causes – making no mention of my recognition of good reasons as such – is the "best explanation."

All of this applies equally to the best explanation of my belief in moral objectivity: someone may purport to explain away this belief by appeal to extraneous causal factors, such as a desire to punish, but if I am convinced through moral and philosophical reflection that I have good reasons for believing in objective moral facts – as by recognizing that the wrong-making features of slavery do not depend on mere convention or on our subjective responses – then I am not going to be moved to accept this undermining explanation of my belief, nor should I be. Of course, it is possible that I am deluded. But nothing forces me to concede that the debunking explanation is in fact correct as the complete explanation for my belief. I am far from being in the compromised position of believers in spiritualist communication with the dead after the debunking evidence: there is nothing from Darwinism or social psychology approaching that sort of debunking of our belief in moral objectivity. There may be scientific evidence that some people's belief in moral objectivity is influenced to some degree by various extraneous causal factors, just as there is such evidence regarding some moral beliefs, but again that falls far short of showing us that our beliefs in this domain are all entirely due to such factors and not at all to our recognition of good reasons for holding those beliefs. (It is no accident that vastly many more reflective, educated people continue to believe in moral objectivity than continue to believe in spiritualist communication with the dead, or in witches, and so on.)

What we have, then, is a philosophical standoff, even in light of all the best science. The moral objectivist needn't deny anything real science tells us about how and why humans first got our moral faculties or how evolutionary influences plausibly contributed to certain dispositions in some of our moral thinking. But embracing what we learn in this regard from a scientific perspective on the world does not in itself require us to renounce the philosophical possibilities that loom large from within the morally engaged perspective we also occupy as moral agents in the world. While we can respect the perspective of those who are skeptical about those possibilities and take a purely empirical approach to moral judgment and practice, we remain equally entitled to hold on to a belief in moral reality and truth, in which case we are never led down the debunker's path because we never grant the overreaching claims about what really "best explains" our moral beliefs.

It is true that in holding out for moral objectivity we must grant some coincidence in the fact that evolution designed us to be valuers for Darwinian reasons having nothing to do with real value as such, and yet it turns out that some things really do matter, i.e., merit our valuing. For example, evolution "designed" us, for merely Darwinian reasons, to value our children, and yet it also happens to be true that our children's welfare matters (as does the welfare of distant strangers, which evolution did not "design" us to value). But there is really nothing very surprising in this combination of facts unless we start out with the default expectation that the world should be an inherently valueless place. Perhaps many do start from that expectation, but this is more a metaphysical prejudice than a scientific result. If value is an inherent feature of the sphere of sentient creatures and their flourishing then there is nothing particularly surprising in the fact that evolution, which is also concerned with such creatures and their survival and reproduction, should give rise to psychological capacities that turn out to enable us, with culturally developed training and experience, to grasp value in the world.

The debunkers could of course turn out to be correct. They do not, however, win by default, and they cannot legitimately use science to provide decisive leverage in favor of their position, since the science itself fails to provide adequate support for the very strong and generalized explanatory claims on which they rely. Darwinism leaves ample room for genuine knowledge of objective moral facts, and the debate among those who believe in moral reality and objectivity and those who instead adopt some deflationary conception of morality is a matter for continued philosophical exploration.

14 Debunking Arguments: Mathematics, Logic, and Modal Security

Justin Clarke-Doane

Genealogical Debunking Arguments

In the précis of a recent book, Richard Joyce writes,

> Nativism [the hypothesis that moral concepts are evolutionarily innate] offers us a genealogical explanation of moral judgments that nowhere ... presupposes that these beliefs are true.... My contention ... is that moral nativism ... might ... render [moral beliefs] unjustified.... In particular, any epistemological benefit-of-the-doubt that might have been extended to moral beliefs ... will be neutralized by the availability of an empirically confirmed moral genealogy that nowhere ... presupposes their truth. (2008, 216)

Such reasoning, falling under the heading "Genealogical Debunking Arguments," is now commonplace. The hypothesis is that there is an explanation of our moral beliefs that fails to imply their truth.[1] Because this thesis was pressed in Gilbert Harman's *The Nature of Morality* (1977), I will call it Harman's Thesis. The key assumption is that knowledge of Harman's Thesis defeats our moral beliefs.[2] I will call this Debunkers' Assumption. Debunkers do not claim that knowledge of Harman's Thesis "rebuts" our moral beliefs, or gives us direct reason to believe that they are false. They claim that it "undermines" our moral beliefs, or gives us reason to no longer believe their contents, without giving us direct reason to believe that those contents are false.[3]

Some philosophers (Street 2006) suggest that (knowledge of) Harman's Thesis undermines our belief in moral realism, not our moral beliefs.[4] Moral realism is, roughly, the view that typical moral sentences are true or false, and that some are true, interpreted at face value, independent of human minds or languages (for details, see Clarke-Doane 2012, Sec. 1). But "debunking" explanations of our ordinary beliefs do not lead us to reject realism about their subject matter. Suppose that Tom tells us about what he claims is a novel species of bird. We then gain knowledge that Tom told us this because he is a pathological liar. Then our beliefs about the novel species seem undermined, not our belief that facts about the bird are independent of our beliefs.

Harman's Thesis and Coincidence

Harman's Thesis is relatively uncontroversial (but see Sturgeon 1985). It is questionable whether the evolutionary explanation advanced by evolutionary debunkers is adequate for explaining their origin. But there is some explanation of our moral beliefs, and prima facie it will be neutral as to their truth.

Debunkers' Assumption, by contrast, is quite controversial. How could Harman's Thesis undermine our moral beliefs? Street suggests the most promising answer. She writes:

[T]he realist must hold that a massive coincidence took place ... that as a matter of sheer luck, [causal] pressures affected our ... attitudes in such a way that they just happened to land on ... the true [moral] views.... [T]o explain why [we] ... make the [moral] judgments that we do, we do not need to suppose that those judgments are true. (2008a, 208–209)

Street's suggestion is apparently that, if the truth of our moral beliefs is not implied by their explanation, then their reliability would be "coincidental" in a sense that is undermining (assuming that moral realism is true).[5] Why would this be? Street offers the following intuition. If Harman's Thesis (and, hence, the previous antecedent) were true, then the forces responsible for our moral beliefs would have "nothing to do with" the moral truths (Street 2006, 121).

The background principle is due to Field. "[O]ur belief in a theory [is undermined]," he writes, "... if it would [appear to] be a huge coincidence if what we believed about its subject matter were correct" (Field, 2005, 77).[6] Street's argument can be understood as an application of Field's principle, given the following assumption. If it is no coincidence that our beliefs of a kind, F, are true, then the truth of our F-beliefs is implied by their explanation.[7] I will call this the Necessity Assumption. Whether the Necessity Assumption is defensible will be discussed later in this chapter.

Mathematics, Logic, and the Sufficiency Assumption

An important question surrounding genealogical debunking arguments is the extent to which they generalize. Debunkers frequently allege that they do not work equally against our mathematical beliefs (see, for example, Crisp 2006, 17; Gibbard 2003, Ch. 13; Sinnott-Armstrong 2006, 46; Sosa 2002; and Street 2006, 160, fn. 35). Joyce goes so far as to write:

[T]he dialectic within which I am working here assumes that if an argument that moral beliefs are unjustified or false would by the same logic show that believing that 1 + 1 = 2 is unjustified or false, this would count as a reductio ad absurdum. (2006a, 182, fn. 5)

He assures the reader,

There is ... evidence that the distinct genealogy of [mathematical] beliefs can be pushed right back into evolutionary history. Would the fact that we have such a genealogical explanation of ... "1 + 1 = 2" serve to demonstrate that we are unjustified in holding it? Surely not, for we have no grasp of how this belief might have enhanced reproductive fitness independent of assuming its truth. (2006a, 182)

There are two questions. The first is whether the truth of "1 + 1 = 2" is "a background assumption to any reasonable hypothesis of how this belief might have come to be innate" (and similarly for other elementary arithmetic beliefs).[8] It might seem to be. Consider a creature, A, who believes that 1 + 1 = 2 and a conspecific, B, who believes that 1 + 1 = 0. Creature A seems to have an advantage over B. For instance, in the presence of 1 lion to the left and 1 lion to the right, A will be less likely than B to walk out from behind the bushes and get eaten. Any explanation of this fact would seem to need to appeal to the premise that 1 + 1 really does equal 2, and not 0.

This appearance is suspect, however (Clarke-Doane 2012, Sec. 3). The sentence "1 + 1 = 2" is about numbers. It says that the plus function maps the number 1 onto itself and to 2. More relevant is the (first-order) logical truth that if there is "exactly one" lion to the left, and "exactly one" lion to the right, and no lion to the left is to the right, then there are "exactly two" lions to the left or right (where the phrases "exactly one" and "exactly two" are abbreviations for constructions out of ordinary quantifiers and the identity sign, and do not refer to numbers). In symbols, $[\exists x[\text{lion}(x)$ & $\text{right}(x)$ & $\forall z[(\text{lion}(z)$ & $\text{right}(z)) \rightarrow x=z]$ & $\exists x[\text{lion}(x)$ & $\text{left}(x)$ & $\forall z[(\text{lion}(z)$ & $\text{left}(z)) \rightarrow x=z]$ & $\sim\exists x[\text{lion}(x)$ & $\text{right}(x)$ & $\text{left}(x)]] \rightarrow \exists x \exists y[(\text{lion}(x)$ & $[\text{right}(x)$ v $\text{left}(x)])$ & $(\text{lion}(y)$ & $[\text{right}(y)$ v $\text{left}(y)])$ & $\sim(x=y)$ & $\forall z[(\text{lion}(z)$ & $[\text{right}(z)$ v $\text{left}(z)]) \rightarrow (x=z$ v $y=z)]]$. The truth of "1 + 1 = 2" is not, in fact, "a background assumption" of this explanation.[9]

It might be objected that this strategy is not sufficiently general (Braddock et al. 2012). Logical truths like the previous one are about concrete objects, such as lions or cliffs. But to explain the general fact that creatures who believed that 1 + 1 = 2 had an advantage over those who did not, it seems insufficient to cite a (first-order) logical truth about lions, say. The fact that if there is "exactly one" lion to the right, and there is "exactly one" to the left, and no lion to the left is to the right, then there are "exactly two" lions to the left or right does not explain the general fact at issue. One way to address this problem would be to ascend to second-order logic. Creatures who believed that 1 + 1 = 2 had an advantage over creatures who believed that 1 + 1 = 0 because, for any properties F, G, and H, if there is "exactly one" F that is G, and there is "exactly one" F that is H, and no G is an H, then there are "exactly two" Fs that are G or H. But while this would seem to show that we can explain the usefulness of the arithmetic belief that p without assuming that p, knowledge of second-order logic is often thought to be comparably mysterious as knowledge of arithmetic. Two other options would be to appeal to schemas, rather than truths per se (take second-order logical truths like the previous one and remove the second-order quantifiers), and to appeal to mereological principles.

The second question is whether if the truth of "1 + 1 = 2" is "a background assumption to any reasonable hypothesis of how this belief might have come to be innate" (and similarly for other elementary arithmetic beliefs), then the truth of our mathematical beliefs is "no coincidence." The contention that it is "no coincidence" follows from the converse of the Necessity Assumption, which I will call the Sufficiency Assumption. The Sufficiency Assumption says that if the truth of our F-beliefs is implied by their explanation, then their truth is no coincidence.

The Sufficiency Assumption is immediately suspect. If it were correct, then it would be trivial to show that the truth of our logical beliefs is "no coincidence," since every logical truth is implied by every explanation at all.[10] In particular, for any logical truth that we believe, p, the explanation of our belief that p is the case implies that p. But even if we can show that the truth of our logical beliefs is "no coincidence," this is not trivial (Clarke-Doane 2015, 86, fn. 14; Forthcoming A, secs. 5 and 6).

Why might debunkers be tempted to believe the Sufficiency Assumption? Because they have confused the challenge to explain the reliability of our beliefs (i.e., the challenge to show that their truth is "no coincidence") with the challenge to explain their justification (Clarke-Doane 2015).[11] If the truth of our mathematical beliefs were implied by their explanation, and we had empirical evidence for that explanation, then, arguably, we would have empirical evidence for their contents. Those contents would be implied by an empirically confirmed theory. This would seem to generate an explanation of the justification of our mathematical beliefs that even an empiricist could accept (Quine 1951, sec. VI). But it would not seem to generate an explanation of the reliability of our mathematical beliefs. It would leave it mysterious as to why our mathematical beliefs are reliable symptoms of the mathematical truths (Field 1989, 26).

Such considerations illustrate the obscurity of Street's suggestion that the forces generating our moral beliefs "have nothing to do with" their truth. In one sense, the forces generating our logical beliefs obviously "have something to do with their truth." Their truth is a consequence of their genealogical explanation. But even if that explanation is evolutionary, this does not show that we were "selected to have true logical beliefs," or selected to have logical beliefs at all.

The Necessity Assumption

If the Sufficiency Assumption is dubious, then what could recommend the Necessity Assumption? In particular, why think that if Harman's Thesis is true, then the truth of our moral beliefs would be "coincidental"? Recall that the relevant sense of "coincidental" is such that, if the truth of our moral beliefs appears to be coincidental, then this undermines those beliefs.

It might be thought that Harman's Thesis gives us ("direct") reason to doubt that our moral beliefs are sensitive in the sense that, had the contents of our explanatorily basic moral beliefs been false, we would not have believed them – where explanatorily basic F-beliefs state the conditions under which an F-property is instantiated.[12] (The parenthetical "direct" is needed, since if Harman's Thesis undermines our explanatorily basic moral beliefs, then it gives us reason to doubt that those beliefs are actually true. If our explanatorily moral beliefs are actually false, then the closest world in which their contents are false is a world in which we still believe them.[13]) Debunkers frequently suggest that Harman's Thesis gives us such reason. Ruse writes:

> You would believe what you do about right and wrong, irrespective of whether or not a "true" right and wrong existed! The Darwinian claims that his/her theory gives an entire analysis of our moral sentiments. Nothing more is needed. Given two worlds, identical except that one has an objective morality and the other does not, the humans therein would think and act in exactly the same ways. (1986, 254)

Harman's Thesis does not "rebut" our moral beliefs. Hence, if it gives us reason to doubt that they are sensitive, then it gives us reason to doubt that, if our explanatorily basic moral beliefs are true, then they are sensitive. But the contents of our explanatorily basic moral beliefs are widely supposed to be ("metaphysically") necessary, if true, and beliefs in necessary truths are vacuously sensitive on a standard semantics. Moreover, Harman's Thesis does not seem to give us reason to believe that the contents of our explanatorily basic moral beliefs would be contingent. Hence, Harman's Thesis does not give us reason to doubt that those beliefs are sensitive on a standard semantics.

Maybe Harman's Thesis gives us reason to doubt that our explanatorily basic moral beliefs are sensitive on a nonstandard semantics incorporating "impossible worlds"? Even if it did, it is hard to see how this could undermine all of our moral beliefs. For virtually any supervenient property F, it seems that had – per impossibile – the contents of our explanatorily basic F-beliefs been false, we still would have believed them. In particular, it seems that had atoms arranged chairwise failed to compose a chair, we still would have believed that they did (Clarke-Doane 2015, sec. 3). If such counterfactuals do not undermine our non-explanatorily-basic ordinary object beliefs (such as that we are sitting in a chair), then analogous counterfactuals do not undermine our non-explanatorily-basic ordinary moral beliefs. In fact, the previous counterfactual does not even seem to undermine our explanatorily basic moral beliefs. For virtually any metaphysically necessary truth p, it appears that had – per impossibile – it been the case that ~p, we still would have believed that p. In particular, as Field notes, it seems that "we would have had exactly the same mathematical ... beliefs even if the mathematical ... truths were different" (2005, 81). Unless debunkers advocate global skepticism about necessary truths, Harman's Thesis cannot undermine our moral beliefs by giving us reason to doubt that they are sensitive.

Perhaps, then, rather than giving us reason to doubt that our explanatorily basic moral beliefs are sensitive, Harman's Thesis gives us (direct) reason to doubt that they are safe. That is, it gives us reason to believe that we might easily have had false explanatorily basic moral beliefs.

Again, not even debunkers claim that Harman's Thesis rebuts our explanatorily basic moral beliefs. So, if the thesis gives us (direct) reason to doubt that those beliefs are safe, then it gives us reason to doubt that if they are true, then they are safe. Given that Harman's Thesis does not give us reason to believe that the contents of our explanatorily basic moral beliefs might easily have been false, it must give us reason to believe that we might easily have had different explanatorily basic moral beliefs. However, debunking arguments themselves demonstrate that Harman's Thesis does not do this. Consider the most austere interpretation of Street's suggestion that "among our most deeply and widely held judgments, we observe many ... with exactly the sort of content one would expect if the content of our evaluative judgments had been heavily influenced by selective pressures" (2006, 116). Suppose that we were evolutionarily "bound" to have the explanatorily basic moral beliefs that we have "for reasons that have nothing to do with their truth." Then we could not easily have had different explanatorily basic moral beliefs. Hence, given the necessity of the contents of our explanatorily basic moral beliefs, we could not easily have had false explanatorily basic moral beliefs.

I am not suggesting that we could not easily have had different explanatorily basic moral beliefs, or even that Harman's Thesis proves that we could not have. My point is that Harman's Thesis is consistent with, and sometimes even taken to suggest, this. So, it certainly does not give us reason to believe that we could easily have had different explanatorily basic moral beliefs.

To sum up, Harman's Thesis cannot undermine our moral beliefs by giving us reason to doubt that our explanatorily basic ones are sensitive or safe. Note the irony. A tentative sign that a realist about an area F can argue from the truth of her explanatorily basic F-beliefs to their sensitivity and safety is that there is a genealogical debunking argument aimed at her F-beliefs.[14]

Modal Security

Could Harman's Thesis undermine our moral beliefs, but not by giving us reason to doubt that they are sensitive or safe? That is, could Debunkers' Assumption still be true? The answer depends on the following (Clarke-Doane, 2015, 2016, 2017).

Modal Security: If information E undermines all of our beliefs of a kind F, then it does so by giving us reason to doubt that our F-beliefs are both sensitive and safe.

The intuition motivating Modal Security is that there is no such thing as a "non-modal underminer" (obviously there is such a thing as a non-modal defeater, namely, a rebutter). Whether this is true is uncertain. However, if it is true, and if the Sufficiency Assumption is false, then whether the truth of our beliefs from an area F is implied by their explanation is independent of whether their truth is "coincidental" in any sense that could be undermining.

Conclusion

I have discussed the structure of genealogical debunking arguments. I have argued that they undermine our mathematical beliefs if they undermine our moral beliefs. The contrary appearance stems from a confusion of arithmetic truths with (first-order) logical truths, or from a confusion of reliability with justification. I concluded with a discussion of the cogency of debunking arguments, in light of the previous argument. Their cogency depends on whether information can undermine all of our beliefs of a kind, F, but not by giving us reason to doubt that our F-beliefs are modally secure.

Notes

1 "Moral beliefs" is always shorthand for token moral beliefs in what follows.
2 Strictly speaking, the key assumption is that it defeats our moral beliefs that are not logical truths, such as "either Hitler is wicked or it is not the case that Hitler is wicked." I will ignore this complication in what follows.
3 See, for instance, Joyce (2006a, 181). See Pollock (1986, 38–39) for the distinction between rebutting and undermining (or "undercutting") defeaters.
4 I will not continue to add the qualification "knowledge of," but this is always intended.
5 I will not continue to add the parenthetical qualification.
6 See also the introduction to Field's *Realism, Mathematics, and Modality* (1989).
7 Since any belief will presumably have multiple explanations, a more exact statement of the Necessity Assumption would say that if it is no coincidence

that our beliefs of a kind, F, are true, then the truth of our F-beliefs is implied by some explanation of them. I ignore this complication in what follows.

8 In fact, few evolutionary psychologists would claim that this belief is innate (and Joyce explicitly oversimplifies in the relevant passage). They would claim that the capacity for quantitative reasoning is innate. (Thanks to an anonymous referee for suggesting that I clarify this.)

9 The psychological literature is mostly orthogonal to Joyce's argument. But see Butterworth (1999) and Dahaene (1997).

10 P is a consequence of Q just in case there is no way to make Q true and P false. But there is no way to make any logical truth, T, false. So, for any Q, T is a logical consequence of Q.

11 I am not saying that there are no connections between reliability and justification, simply that the concepts are different. A lucky guesser is reliable, but not justified, whereas a brain-in-a-vat is justified, but not reliable.

12 This notion is similar, but not identical, to that of Nozick (1981).

13 I will not consistently add this qualification in what follows. But it is always intended.

14 An argument that our moral beliefs are sensitive and safe suggests an argument that the (objective) probability that they are true is high. For any explanatorily basic moral truth p, presumably $Pr(p)=1$, given that such truths would be necessary. Also, it may be that $Pr(\text{we believe that } p) \approx 1$. But, then, $Pr(p \text{ and we believe that } p) \approx 1$. Since (p and we believe that p) implies (our belief that p is true), we have $Pr(\text{our belief that } p \text{ is true}) \approx 1$.

15 Evolution and the Missing Link (in Debunking Arguments)

Uri D. Leibowitz and Neil Sinclair

Introduction

What are the consequences, for human moral practice, of an evolutionary understanding of that practice? By "moral practice" we mean the way in which human beings think, talk, and debate in moral terms.

The first type of answer holds that moral practice – or some feature of it – is vindicated by an evolutionary account of its origins. One might argue, for example, that evolutionary considerations support substantive moral claims. Richards, for instance, holds that an evolutionary account of moral practice justifies the claim that each of us "ought to act for the community good" (1986, 289; compare Casebeer 2003). Alternatively, evolutionary theorizing might vindicate moral practice insofar as it makes true or justifies a particular view of the nature of that practice. Harms (2000), for instance, argues that the fact that moral practice is an adaptation supports a version of moral realism – the view that moral judgments are maps of a moral reality, made true independently of human attitude. Finally, evolutionary theorizing might vindicate moral practice insofar as it helps demonstrate how actual human beings have sufficient reason to go on engaging in that practice (Campbell 1996).

The second type of answer to the initial question is less optimistic. It takes moral practice – or some of its features – to be debunked by evolutionary understanding. There are several different types of debunking, and several possible targets. First, evolutionary considerations might show particular moral theories to be false or unjustified. Here "moral theories" cover both substantive first-order moral theories (such as utilitarianism) and second-order metaethical theories of the nature of moral practice (such as realism). Greene (2008) and Singer (2005), for example, take evolutionary accounts of the origins of moral intuitions to place pressure on first-order deontological theories. Street, by contrast, argues that "realist theories of value prove unable to accommodate the fact that Darwinian forces have deeply influenced the content of human values" (2006, 109). According to both, evolutionary considerations are a threat to the validity of moral theories. A different type of debunking has a more down-to-earth target: moral judgments of the ordinary folk, who may embrace no clear theory. According to the first instance of this type, evolutionary considerations show certain folk moral judgments to be false (or incoherent). This is truth-debunking (Joyce 2013). Ruse (1986, 254), for example, argues for the falsity of moral judgments based on the explanatory redundancy of the supposed moral facts they concern. According to the second instance of this type, evolutionary considerations show folk moral judgments to be unjustified, not knowledge, or otherwise lacking in epistemic status. This is epistemic debunking. Joyce, for example, argues that "we have an empirically confirmed theory about where our moral judgments come from … which forces the recognition that we have no grounds one way or another for maintaining those [judgments]" (2006a, 211). The most interesting types of truth- and epistemic-debunking have global targets: they urge that all moral judgments are false or unjustified. It was the possibility of such global debunking that so alarmed many of Darwin's contemporaries. In what follows, we focus on global epistemic debunking arguments.

Evolutionary Explanations

Evolutionary debunking arguments (EDAs) begin with the claim that moral practice can be given an evolutionary explanation. An evolutionary explanation here is (roughly) an account of how the possession of a trait, or another trait that is in some way connected to it, increased the relative evolutionary fitness of some ancestral population. The "relative evolutionary fitness" is a direct measure of the ability of individuals of a population to pass their genes onto the next generation and is always relative to the competition and to a given environment.

When it comes to evolutionary explanations of moral practice, there are several distinct phenomena one might take evolution to explain. First is the capacity to think, talk, and debate in moral terms. Call this the "capacity hypothesis." Second is a population-wide tendency to make particular types of moral judgment, distinguished by their contents. Call this the "tendency hypothesis." Whereas the capacity hypothesis explains our possession of moral concepts, the tendency hypothesis explains common patterns in the way these concepts are deployed. Neither hypothesis entails that the relevant trait will develop independently of appropriate environmental conditions. Joyce (2006a, 108–142) defends the capacity hypothesis. According to his account, the capacity to make moral judgments can aid the reproductive fitness of an individual by providing a motivational bulwark against weakness of will. By categorizing certain paths of action under moral labels (e.g., as right or required), the individual is more inclined to pursue those paths and less inclined to eschew them for short-term gain. Where the paths of action thus preferred are prosocial behaviors (e.g., keeping a promise), pursuing them leaves the individual better able to reap the reproductive benefits of cooperation. Further, the capacity to make moral judgments can aid the evolutionary fitness of a population by enabling members of that population to negotiate, test, refine, and sustain cooperative social structures. By contrast, Street adopts the tendency hypothesis. She notes that "one enormous factor in shaping the content of human values has been the forces of natural selection" (2006, 114). For Street, the explananda of evolutionary explanations are the "deep and striking patterns, across both time and cultures, in many of the most basic evaluative judgements that human beings tend to make." Street goes on to list particular moral judgments that seem to admit this kind of explanation. For example: "it is clear how beneficial (in terms of reproductive success) it would be to judge that the fact that something would promote one's survival is a reason to do it" (2006: 115).

One feature of these evolutionary hypotheses is worth emphasizing. Both can be viewed as offering a type of functional explanation – that is, they specify a selected-for effect of a given trait (Wright 1976). According to Joyce's hypothesis, the capacity to make moral judgments was selected for insofar as past instances of that type had the effect of stabilizing mutually beneficial cooperative structures. According to Street's hypothesis the tendency to make particular moral judgments was selected for insofar as past instances of that type had the effect of producing (or making more likely) survival-promoting behaviors. The important feature of both hypotheses is that the selected-for effect concerns the actions and behavioral tendencies of agents who possess the trait (and perhaps a derivative effect on the actions and tendencies of other agents). Call these "downstream motivational effects." Just as importantly, it seems that according to these hypotheses the effect for which moral traits were selected was not the effect of successfully tracking explicitly moral facts. These moral traits were selected for not because they somehow enabled agents to "hook up" with a realm of moral facts, but simply because they prompted fitness-enhancing behaviors. More controversially, it is often held that the effect for which moral traits were selected does not involve those traits tracking facts of any kind – moral or otherwise.[1] In both ways, the evolutionary explanation of moral traits seems importantly different to the evolutionary explanation of our perceptual capacities.[2] Human beings have the capacity to perceptually detect the layout of objects in their immediate environment, in part because previous instances of that capacity successfully hooked up with the layout of objects in the ancestral environment. It is, of course, only because such capacities went on to prompt fitness-enhancing behaviors that they were selected for, but, importantly in this case, the fact that the capacities tracked some independent truths is part of the explanation of why the behaviors they prompted were fitness-enhancing. Nothing comparable seems to apply in the moral case.[3]

It is noteworthy that neither the capacity hypothesis nor the tendency hypothesis (directly) explains individual moral judgments, what might be called token judgments (i.e., instances of a general type). It is one thing to explain why humans have the capacity to make moral judgments, or why we find "widespread human acceptance" of certain moral judgments, but quite another to explain a token moral judgment. One feature that is often overlooked in the evolutionary debunking literature is that if the explananda are token judgments, then it is far from obvious that evolutionary accounts can perform the explanatory work required. Sober (1984), for example, argues that evolution is suited to explain why a population consists of one set of individuals with certain traits rather than another set with different traits, but not to explain why a particular individual has the traits it does, rather than other traits.[4] If Sober is right, evolution is likewise not suited to explain why a particular individual makes a certain moral judgment rather than another. It might be tempting to insist that an evolutionary account will be part of the explanation of token moral judgments. Whether this temptation is to be resisted depends on one's views on what complete explanation consists in.[5] For present purposes we needn't resolve this complex issue. Whether or not an evolutionary account will partake in the explanation of a token moral judgment, it is clear that other ingredients will be required.[6] Indeed, the striking diversity in actual moral judgments seems to count against the view that all such judgments can be fully explained by evolution (Kahane 2011, 118; Fraser 2014, 469). So even if debunkers show that the evolutionary component of the explanation of token judgments is not truth-tracking in the relevant sense, there is still work to be done to establish that the best explanation of token judgments need not invoke the truth of the judgments in question.[7]

In the remainder of this chapter we focus on how the absence of an explicitly moral tracking element in the evolutionary explanation of morality is supposed to undermine the justification for moral judgments.

A Generic Debunking Argument

How might the foregoing considerations feature in a debunking argument? Consider the following:

(1) The best explanation of moral practice is an evolutionary explanation.
(2) An evolutionary explanation of moral practice does not rely on moral practice tracking moral truths.

Hence,

(C) all token moral judgments are unjustified.

This argument is not formally valid: there is a logical gap between the premises and conclusion – this is the "missing link" of our title. We think it is elusive.

Consider (2). What does "does not rely on moral practice tracking moral truths" mean? Synonyms include: "an off-track process" (Kahane 2011), "tracking failure" (Lillehammer 2010), and "a truth-mooting genealogy" (Mason 2010). Precisely what this tracking failure amounts to is a key question, and different answers generate different debunking arguments. But there seems to be considerable agreement on the following, minimal account: evolutionary explanations do not invoke or assume the truth of moral judgments.[8] As Ruse (1986, 254) notes: "You would believe what you do about right or wrong, irrespective of whether a 'true' right or wrong existed." This claim is supported by the fact that evolutionary explanations of moral traits are consistent with supposing that all moral judgments are false – the so-called error theory of morality (Joyce 2013).

Consider next (1). As previously noted, neither the capacity nor tendency hypotheses (completely) explain token moral judgments. Hence, if we are careful about what the "explanation of moral practice" in (1) means, we can see that (C) does not follow from (1) and (2). Toner, for example, raises the possibility that while the moral capacity might have originally been selected "for the broad purpose of increasing genetic fitness by enabling reciprocity," it later came to be "pressed into the service of other purposes … which do require sensitivity to evidence and truth" (2010, 533). On this view, the evolutionary explanation of the capacity to make moral judgments is not truth tracking, but the manifestation of this capacity is sensitive to moral truths. We do not have the space to delve into the details of this view here. Instead we note that debunkers usually overcome this hurdle by making the following assumption: that whatever non-evolutionary factors are involved in the explanation of token moral judgments, these are also truth-mooting with respect to moral truth (Joyce 2006a, 181; Mason 2010, 774). To incorporate this assumption, we can define an "evolutionary+" explanation of a moral judgment as one that involves the capacity or tendency hypothesis together with these additional morally truth-mooting factors. This generates the following argument:

(1*)   The best explanation of each token moral judgment is an evolutionary+ explanation.

(2*)   An evolutionary+ explanation of a token moral judgment does not rely on this judgment tracking moral truths.

(3)   If the best explanation of a token judgment does not rely on this judgment tracking the truth of its content, then that judgment is unjustified.

Hence,

(C) all token moral judgments are unjustified.

This is a valid argument. And granting both (1*) and (2*), the question is whether (3) is true or well supported.

Specific Debunking Arguments

One surprising feature of EDAs of this kind is the minimal role played by evolution.[9] That is, it may be possible to explain token moral judgments in morally truth-mooting yet non-evolutionary ways. Thus one might explain token judgments psychologically in terms of the "moral sensibility" of the judger (Harman 1977, 7), psychoanalytically (e.g., Freud), sociologically (e.g., Durkheim, Nietzsche), historically (e.g., Marx), or microphysically (when science is sufficiently advanced). If such explanations do, in fact, explain token moral judgments without relying on these judgment tracking moral truths, then each can be used to generate a debunking argument. If "the best explanation" in (1*) is a uniquely referring expression, then only one of these arguments will be sound. But note that it is far from obvious that an evolutionary explanation is necessarily superior to, say, a microphysical explanation. And most physicalists will probably think that in principle there should be a microphysical explanation of each token moral judgment. Whether the availability of an explanation of one kind excludes the viability of other kinds of explanation of the same phenomenon is a question we must put aside here. Nevertheless, it is worth noting that on this construal of EDAs, these arguments don't seem to advance the dialectic over and above, say, non-evolutionary arguments such as Harman's famous challenge, which urges moral skepticism on the basis that "we can have evidence for a hypothesis of a certain sort only if such hypotheses sometimes help explain why we observe what we observe" (1977, 13).

Despite its plausibility, the previous argument faces significant challenges. The notion of an explanation of a judgment that does not rely on it tracking relevant truth is central. One way to understand this notion is as the claim that a judgment can be explained without invoking its truth (this is the minimal sense of "tracking failure" noted earlier). On this understanding (3) claims the following:

(3*) If the best explanation of a token judgment does not invoke the truth of its content, then that judgment is unjustified.

But (3*) is controversial. First, there is the possibility of reduction. Suppose that property A is reducible to property B. Suppose further that a particular judgment concerning property A can be explained by citing property B. Then it seems that the judgment can be explained without invoking its truth, but it is far from obvious that the judgment is, therefore, unjustified. In particular, it may well be justified because the judgment is explained by invoking the property that is the reductive basis of the property that features in the judgment. This point is quite general. For example, physicists sometimes suggest that all other properties can be reduced to the properties quantified over by fundamental physics. If so, it seems that every judgment that we make (that isn't about physics) can be explained without explicitly invoking its truth. But, of course, this doesn't threaten the justification of all such judgments. So long as Bob's judgment about the location of his cat, say, is responsive to, and explained by, the reductive basis of catness, it will be justified (Quinn 1986, 537–549).

There are at least two possible replies to the reduction problem. The first is to deny that a reduction is available. This is the option preferred by Joyce in the moral case (2006a, 190–209). But once we realize that reduction threatens (3*), we can see that weaker metaphysical relations may do so as well. For example, suppose the property of being black locally supervenes on (without being reducible to) a particular surface spectral reflectance property "SSR." On one common understanding of the supervenience relation, this is to say that the blackness of an object cannot change without its SSR-ness changing. Now suppose that Bob's judgment that his cat is black is explained by his cat's possession of SSR. This seems grounds for thinking that Bob's judgment is justified. Again the point is general: it seems that a judgment can be justified even if it can be explained without invoking or assuming its truth, in particular when it can be explained by the relevant supervenience base.[10]

The second possible reply to the problem of reduction accepts that a reduction (or supervenience claim) may be available, but insists that, as a matter of fact, the relevant judgments are explained neither by the properties referred to be the judgment itself, nor by any plausible reductive or supervenience basis for those properties. But though this response delivers a more plausible version of (3*), it does so only by rendering the similarly modified versions of (1*) and (2*) less plausible. As mentioned previously, neither the capacity nor the tendency hypotheses show that the best explanation of each token moral judgment needs not invoke the truth of this judgment or the truth of any plausible relevant reductive or supervenient base. Indeed, it is consistent with both hypotheses that some moral judgments are explained in part by invoking a plausible reductive or supervenient base for the properties to which they refer (an example follows).

The second general problem with (3*) is that there are plausible counterexamples. Bob judges that the sun will rise tomorrow and that all men are mortal, both on the basis of previously observed instances, and both judgments are justified. But in neither case does the relevant truth seem to play a role in explaining Bob's judgment (see Dancy 1985, 34; White 2010, 583).

Further Arguments

What other ways might there be to capture the notion of an explanation of a judgment that does not rely on this judgment tracking moral truths? This section considers several recently discussed options.

(Lack of) Sensitivity

The first option for understanding the notion of an explanation of a judgment that does not rely on this judgment tracking moral truths is that the best explanation of token moral judgments reveals that these judgments are insensitive. A moral judgment is insensitive just in case we would make the same judgment that we make now even if the presumptive moral truths were actually false. More formally, the following arguments are generated:

(1*) The best explanation of each token moral judgment is an evolutionary+ explanation.

(2**) An evolutionary+ explanation of a token moral judgment reveals that this judgment is insensitive.

(3**) If the best explanation of a token judgment reveals that this judgment is insensitive, then this judgment is unjustified.

Hence,

(C) all token moral judgments are unjustified.[11]

Sensitivity is sometimes represented formally as "$\neg p \; \square \!\!\rightarrow \neg Bap$," where "p" is the content and "Bap" is a's judgment (or belief) that p and $\square \!\!\rightarrow$ stands for the subjunctive conditional.[12] Premise (3**) claims that insensitivity undermines justification.

Sensitivity is not all it's cracked up to be. The second premise is questionable. Consider Bob's judgment that Jones's torturing of Smith is wrong, which Bob possesses partly because Jones's torturing causes unwanted agony. Grant that we can explain this judgment without invoking or assuming its truth – Bob makes this judgment partly because evolution has endowed him with a moral capacity and partly because this capacity is sparked into life by the more proximal fact that Jones's torturing causes unwanted agony. Still, Bob's judgment can be sensitive to moral truth so long as the wrongness of Jones's act supervenes on (or is reducible to) its causing unwanted agony. Were Jones's act not wrong, it wouldn't be an act of causing unwanted agony, and hence Bob would not have judged it wrong.[13] In this case, Bob's judgment is sensitive to moral truth even though it can be explained in evolutionary(+) terms.[14]

In reply, the debunker might say: "You've only managed to show that some moral judgments are sensitive by assuming a substantive moral truth –

namely that the wrongness of Jones's action supervenes on it causing unwanted agony; but appeal to such claims is illegitimate, since it is precisely their justification that is at issue." To which the reply runs: "I may appeal to substantive moral claims to make a case for thinking that some moral judgments are sensitive, but by the same token you have to appeal to (or assume) substantive moral claims to make the case that moral judgments are insensitive to moral truth. To consider what would happen were a given moral 'truth' to turn out false, one needs to make assumptions about what that truth supervenes on." This point highlights a more general worry with many debunking arguments: there is clearly something problematic in trying to debunk all moral judgments on the basis of an argument that makes moral assumptions (Lillehammer 2010, 374–375).

A second set of problems for this argument is that sensitivity does not seem necessary for justification. If, as many believe, false judgments can be justified, and since false judgments are necessarily insensitive, we have many counterexamples to (3**).[15] Further, insofar as it is expressed in terms of counterfactual conditionals, sensitivity inherits well-known problems about the assessment of such claims. According to one popular view, for instance, the counterfactual, "if p were true, then q would be true," is true in the actual world just in case in all close possible worlds where p is true, then so is q. However, explicating a notion of "closeness" here can prove tricky (Lewis 1973).

(Lack of) Safety

A second option for understanding the notion of an explanation of a judgment that does not rely on this judgment tracking moral truths is that the best explanation of token moral judgments reveals that these judgments are unsafe. Here, to say that a judgment that p (as formed by method M) is safe is to say that in all nearby possible worlds where a subject (utilizing M) comes to judge that p, then p (i.e., Bap $\Box\!\!\rightarrow$ p). The thought is that justification requires that where the exercise of our cognitive capacities generates a particular judgment, then, even granting that this judgment is true, it better not be true by mere luck – that is, it better not be that the very same judgment, in a nearby scenario, would have been false. More formally, the arguments are as follows:

(1*) The best explanation of each token moral judgment is an evolutionary+ explanation.

(2***) An evolutionary+ explanation of a token moral judgment reveals that this judgment is unsafe.

(3***) If the best explanation of a token judgment reveals that this judgment is unsafe, then this judgment is unjustified.

Hence

(C) all token moral judgments are unjustified.

Here we can be fairly quick as the difficulties mirror those of the preceding argument. The same example (Bob's belief about Jones) shows that (2***) is questionable. Further, (3***) seems false because safety is too strong as a necessary condition for justification: all false judgments are unsafe, but some false judgments are justified.[16] And again, insofar as safety involves appeal to a vague notion of "nearby" worlds, claims of safety remain difficult to evaluate (Williamson 2009).

Luck

One way of understanding the sensitivity argument is relying on the thought that evolutionary explanations demonstrate the inflexibility of our moral judgments. In particular, that argument suggests that moral judgments are inflexible in the face of changes in moral truth. But there is a distinct sense in which moral judgments seem too flexible – namely, they seem to be beholden to morally irrelevant and highly contingent evolutionary factors (White 2010, 586–587). This suggests another way to understand the notion of an explanation of a judgment that does not rely on this judgment tracking moral truths: that the best explanation of token moral judgments reveals that these judgments are lucky. To say that a's judgment that p (as formed by method M) is lucky is to say that it is not the case that: were p to be true, then a would (employing M) judge that p. Again employing the possible worlds analysis, this is to say that: it is not the case in all nearby worlds were p, a (employing M) judges that p. Formally, $\neg(p \; \square\!\!\rightarrow Bap)$. So we get the following arguments:

(1*) The best explanation of each token moral judgment is an evolutionary+ explanation.

(2****) An evolutionary+ explanation of a token moral judgment reveals that this judgment is lucky.

(3****) If the best explanation of a token judgment reveals that this judgment is lucky, then this judgment is unjustified.

Hence

(C) all token moral judgments are unjustified.

Premise (3****) claims that luck undermines justification. Suppose – the argument can be put – that some moral claim, p, is true. Suppose also that evolution(+) has endowed us with the judgment that p. If it is the case that the same processes could have easily have endowed us with the judgment that not-p (or no judgment concerning p), then our initial judgment cannot be justified. This is to say, even granting the initial judgment is true, it better not be that it is held by luck – that is, it better not be that the same moral truth, in a nearby scenario, would not have been believed.

Unfortunately for the debunker, this argument faces familiar problems. Where a moral judgment is partly explained by a supervenient basis for the relevant moral property, that judgment will not be lucky. Here it is the case that in all nearby worlds where the moral truth obtains, the judgment occurs.[17] Furthermore it is doubtful that luck undermines justification. Suppose, as a schoolboy, Jones was given a copy of *Principia Ethica* as a prize. Subsequently, he came to form many true beliefs about what Moore says in that book. But in some nearby worlds the prize was *Principia Mathematica* and in these worlds no judgments about Moore resulted. Jones's judgments about the contents of *Principia Ethica* are lucky but still (it seems) justified.[18] Finally, insofar as luck is defined in the same counterfactual-conditional terms as sensitivity and safety, this argument faces the same problems of evaluating claims about distance in possible-world-space.

Diagnosis

Once is happenstance; twice is coincidence; a third time is somewhat dull. These last three arguments fail for similar reasons. Each claim that moral judgments exhibit a certain type of "tracking failure" (lack of sensitivity, lack of safety, luck) and that this type of tracking is necessary for justification. But it is not clear that these types of tracking failure are established by evolutionary explanations of morality, and there seem to be plausible examples of moral judgments that might track moral truths in the relevant ways that are consistent with evolutionary accounts. In any case, the proposed necessary conditions on justification are questionable. Perhaps other debunking arguments fail for similar reasons, but rehearsing these moves now seems redundant.[19] Might there be a different approach that will bring a great debunking dividend?

(Lack of) Method

One common response to the previous dialectic is as follows: Sure, it might be that some moral judgments are sensitive, safe, or what-have-you, and it might be that one can provide particular examples of moral judgments that satisfy these conditions. But this is not to provide any general account of the mechanisms by which humans form moral judgments such that those judgments might end up being sensitive, safe, or otherwise justified. In particular, it is not to provide any general account of how the mechanisms for forming moral judgments that we actually employ (or some suitably refined version of them) might constitute ways of getting into contact with moral truth. According to this line of thought, evolutionary explanations of our moral judgments debunk their justification insofar as they emphasize the absence of a positive epistemological story for those judgments (Fraser 2014, 470).

This line of thought has been expressed in several ways. Blackburn puts it in terms of whether we are able to explain why the "correspondence conditionals" given by sensitivity and safety obtain (1993, 161, 167). Joyce, citing Harman, puts the point thus: "we do *not* have a believable account of how moral facts could help explain the mechanisms and forces that give rise to moral judgements" (2016, 135). More formally, the argument might be as follows:

(4) We do not have a plausible general account of how moral truths could help explain token moral judgments.

(5) If we do not have a plausible general account of how certain truths in a domain could help explain token judgments in this domain, then those judgments are unjustified.

Hence,

(C) all token moral judgments are unjustified.

But this argument too is questionable. Premise (5), as stated, might lead to skepticism outside of morality in a way unpalatable to debunkers. For example, it is doubtful that we have a plausible general account of how truths about other minds could help explain token judgments about other minds, but few participants in this debate would like to endorse skepticism about other minds.[20] Likewise, it is doubtful that, say, Aquinas and his contemporaries had a plausible general account of how truths about perceptual objects could help explain token perceptual judgments, but few would be comfortable endorsing the view that all of Aquinas's perceptual judgments were unjustified.

Moreover, premise (4) is highly contested. Many philosophers of a realist persuasion, for instance, have put forward "moral epistemologies," that is, accounts of how our moral judgments hook up with moral truth. According to the moral coherentist, for example, moral judgments are justified insofar as they are part of a coherent system of beliefs, including particular and general moral beliefs as well as second-order beliefs about the nature of moral practice (Brink 1989, 100–143). According to foundationalists, on the other hand, certain moral judgments are justified non-inferentially insofar as they are self-evident (Audi 1993) or intuited (Huemer 2005). Such epistemologies are not obviously refuted by evolutionary considerations. Hence Joyce notes: "No one… thinks that genealogical empirical data alone can secure a sceptical victory; at most it battles alongside sceptical arguments of an a priori metaethical nature" (2016, 136).

But for our purposes the most salient feature of this argument is that it does not mention evolution at all – evolutionary considerations play no role in supporting its conclusion. So what, if anything, do evolutionary considerations bring to the table?

Our proposal is that the mistake in this argument, as in the preceding arguments, is a misidentification of the appropriate target of EDAs. All of the arguments considered here, like much of those discussed in the literature, aim to establish a first-order epistemological result: that all token moral judgments are unjustified. For the reasons previously canvassed it seems doubtful that evolutionary considerations can partake in successful arguments for this conclusion. Instead, the true force of evolutionary considerations for our understanding of moral practice may be located elsewhere. In particular, such considerations highlight the potential success of a naturalist, anti-realist, skeptical research program. This is the program that takes the central task in understanding moral practice to be locating that practice within a wider scientific view of human beings, which rejects the postulation of any judgment-independent realm of moral facts and which considers all moral judgments to be unjustified. Instead of trying to locate evolutionary explanations in a premise for an argument for this latter claim, we propose to view evolutionary explanations as aiming to show that if one adopts a naturalist, anti-realist approach, one has a better overall account of morality.

On this view, evolutionary considerations are held to be consistent with the existence of objective moral truths and justified moral judgments. But, the argument runs, a realist, non-naturalist, optimist research program is not as successful as the competing naturalist, anti-realist, skeptical one. The latter offers a clean, coherent, and satisfactory explanation of phenomena that the former struggles to address. On this view the debunker need not claim that all moral epistemologies are hopeless or that evolutionary considerations can establish that moral knowledge is impossible. Instead her goal is to show that the research program she favors is progressive, whereas the alternative is degenerative (for this terminology, see Lakatos 1970). The force of EDAs, we propose, is at the meta-level and not in first-order epistemology.

To see this in more detail, consider two competing theories of the nature of moral practice, initially held before any evolutionary account of that practice is known or believed. According to the former – a caricature of realism – human moral judgments are more-or-less reliable cognitions of an independently existing moral reality. According to the latter – a caricature of anti-realism – moral judgments answer to no independent reality but have a primary role in guiding action and behavior (error theory is one version of this anti-realism, but there may be others: see Blackburn 1993). Now suppose we add to this situation an evolutionary account of moral practice roughly along the lines Joyce suggests: where this practice involves traits selected for their downstream motivational effects. Which theory does this evolutionary appendix help the most? Arguably, it leads to an increased degree of coherence for the anti-realist theory – for it provides important detail as to the purpose of moral practice, if it is not to track an independently existing moral reality. Conversely, while the evolutionary story is consistent with the realist theory, it fails to add to its coherence. One may, thus, reasonably view developments in evolutionary psychology as an indication that the anti-realist program is progressive and, consequently, that one has good (pro tanto) reason to favor the anti-realist program to its nonprogressive realist competitor.

How damaging is an evolutionary account of morality to those pursuing the realist research program? The answer is not obvious. First, even granting that evolutionary considerations demonstrate that the anti-realist program is progressive whereas its realist competitor is degenerate, it does not follow that it is irrational to pursue the realist program. There are ample examples from the history of science in which major breakthroughs were made by scientists pursuing degenerative programs (see Matheson and Dallmann 2015). Moreover, the extent to which evolutionary considerations reveal that the anti-realist program is progressive whereas the realist one is degenerate is not obvious. Proponents of EDAs will claim that evolutionary considerations increase the coherence of their overall account of morality while leaving their realist opponents with unanswered questions about the mechanism for tracking moral truth. But a committed moral intuitionist, for example, who believes that intuitionism provides a satisfactory answer to the mechanism question, might not feel the bite of this argument. At best she will grant that recent evolutionary findings do more explanatory work for anti-realists than they do for her, but she will resist the claim that such considerations show that there are legitimate questions that her own research program leaves unanswered. Finally, the move to research programs suggests that we need to consider the successes and failures of such programs in all domains. For example, if, as some realists claim, there are successful indispensability arguments for realism – that is, arguments that purport to identify phenomena for which the best explanation must appeal to moral facts – then EDAs can hardly be claimed to have established that the anti-realist program is progressive. Expanding the scope in this way may also require us to consider the successes and failures of the competing research programs outside of morality, for example, in mathematics and religion.[21] If so, there is still much work to do before proponents of EDAs can celebrate the victory of the anti-realist agenda.

Conclusion

If the preceding reflections are correct, then evolutionary explanations of moral practice do add something distinctive to debates about the nature of morality (over and above other morally truth-mooting explanations). But it is not the knock-down argument for skepticism that some debunkers seem to think. The availability of evolutionary accounts of moral practice that nowhere invoke moral truth seems to increase the coherence of anti-realist positions in metaethics, whereas the coherence of realist positions at best remains unchanged and possibly decreases. But even on the most favorable interpretation EDAs can hardly be claimed to have established that the realist program should be abandoned.

Notes

1 For doubts, see Sinclair 2012 and White 2010, 588.
2 See Joyce 2006a, 182–184; Fraser 2014, 463; and Kitcher 2005, 176.
3 Another comparison is with the capacity to make mathematical judgments. Although it seems a stretch to say that this capacity was selected for so as to track mathematical truth, it does seem that any evolutionary explanation of this capacity must at least assume that mathematical truths obtain – again nothing similar seems to apply in the moral case (Joyce 2006a, 182).
4 For criticisms, see Neander 1988 and Nanay 2005.
5 See Leibowitz 2011.
6 For debunkers, these additional elements will likely be causal-psychological-sociological factors. The question then arises: what, if anything, does evolution add to the mix? We partially address this question in what follows.
7 Perhaps "third-factor" explanations can be understood as exploiting this possibility, although they are not typically presented in this way. See Enoch 2010, Wielenberg 2010, and Brosnan 2011.
8 See Joyce 2006a, 183–184, 211; Street 2006, 155; Mason 2010, 773; and Fraser 2014, 459.
9 See Lillehammer 2003, 570; Mason 2010, 770–771; and Brosnan 2011, 53–54.
10 Joyce seems inclined to reject supervenience claims along with reductionism, at least for moral properties (2006a: 209). Yet the basis for this view is an application of Ockham's Razor. In that case the epistemic debunking argument is not independent of a theory debunking argument. Further, one might argue that there is, after all, something that moral properties are required to explain – namely the intelligibility of our moral practices. See McDowell 1985; Shafer-Landau 2003, 13–52; and Enoch 2011.
11 For discussion, see Lillehammer 2010, 365; White 2010, 581; Wielenberg 2010; Clarke-Doane 2012; and Joyce 2016.
12 Strictly speaking, sensitivity needs to be relativized to a judgment as formed by a particular method. See Nozick 1981, 179.
13 See Sturgeon 1985.

14 Another point made by Sturgeon is that, if moral truths are necessary then if one's moral judgment that p is true, it is trivially sensitive because there are no worlds in which the antecedent of the counterfactual ($\neg p \:\square\!\!\rightarrow \neg Bap$) is true.

15 See White 2010, 580; and Dancy 1985, 39. It may be more plausible to suppose that sensitivity is necessary for knowledge; see Nozick 1981.

16 Again, safety may be more plausible as a necessary condition for knowledge; see Sosa 1999.

17 Clarke-Doane (2016) launches a bolder defense of lack of luck. See also Wielenberg (2016) and Joyce (2016) for doubts.

18 See Ichikawa and Steup 2014, sec. 8.

19 See, for example, the (lack of) reliability argument discussed in Wielenberg 2010. See White 2010, 584–585, for a diagnosis of why such arguments might seem attractive.

20 See Leibowitz 2014.

21 For discussions of indispensability arguments, their relation to EDAs, and their application to mathematics and other domains, see Leibowitz and Sinclair 2016.

16 Better Than Our Nature? Evolution and Moral Realism, Justification and Progress

Michael Vlerick

Introduction

The fact of evolution undeniably has consequences for moral philosophy. First of all, it raises immediate questions for the metaethical position of moral realism, because the origin of our moral faculties and dispositions in a contingent evolutionary process is, on the face of it, incompatible with the view that our moral beliefs are objective and track independent moral truths. Second, this metaethical worry seems to undermine the normative justification of our moral norms and beliefs. If we don't have any grounds to believe that the source of our moral beliefs has any ontological authority, how can our moral judgments be justified in an objective way?

Moral justification does indeed appear to be threatened to its very core by the fact of evolution. A contemporary of Darwin, Frances Cobbe, voices in horror that the evolutionary genealogy of our moral sense deals "a deadly blow at ethics, by affirming that, not only has our moral sense come to us from a source commanding no special respect, but that it answers to no external or durable, not to say universal or eternal, reality, and is merely tentative and provisional, the provincial prejudice, as we may describe it, of this little world and its temporary inhabitants, which would be looked upon with a smile of derision by better-informed people now residing on Mars" (Cobbe 1872, 10–11). This fear has all but disappeared. In recent work, Cottingham (2005, 54) points at the "unacceptably relativistic conclusion that rightness or wrongness depend on the contingences of species development."

If our moral beliefs and inclinations are the product of evolved faculties and dispositions, blindly shaping them for their instrumental value in the struggle for survival, what are we to make of those moral beliefs? Does it still make sense to speak of right and wrong, given that human morality is nothing but an "illusion foisted upon us by our [selfish] genes?" (Ruse 1986, 253).

In this chapter, I take these debunking arguments seriously, arguing that traditional moral realism is untenable given the evolutionary origin of our "moral sense." Nevertheless, in contrast to the recent influential account of evolutionary anti-realism proposed by Joyce (2006a), I do not hand the same fate to normative justification. In fact, I argue that it is precisely in their biological roots that we find the key to justifying our moral norms. This account, I will argue, has – in addition to averting Cobbe's fear that the fact of evolution takes away all normative justification – the virtue of allowing the naturalist or evolutionist to entertain the notion of moral progress.

I construct my argument by looking at the relation between our actual moral norms and beliefs and the rationale behind the evolution of our moral sense. This relationship is typically blackboxed by the evolutionary ethicist who often ignores actual moral norms and beliefs and derives her conclusions from the mere premise that our moral beliefs and evaluations are rooted in faculties and dispositions that have been shaped by an evolutionary process. I will argue that much in the same way that modern science has radically transcended the evolutionary intended output of our cognitive faculties, our moral behavior and norms have transcended the evolutionary intended output of our moral dispositions and faculties from which they emanate. The best explanation for such a surprising fact, I argue, is our possession of non-modular intuitive moral guidelines, which I call an innate "moral compass," in conjunction with our ability to reason. This I will argue opens the way toward an internalist justification of morality and enables the evolutionist to account for moral progress. Regarding the latter, I attempt to show both why we can rationally entertain the notion of progress in the domain of morality and how such progress comes about given the particular "moral sense" we possess.

Transcending Our Moral Nature

Morality is not a single adaptation, nor did it evolve to deal with a single set of problems. As Joyce (2006a, 140) puts it, "morality is a complicated and nebulous affair at the best of times, and moral judgments no doubt implicate many different psychological and neural mechanisms." The moral disgust with which we regard incest, for instance, evolved in the interest of our genetic heritage, whereas the moral disapprobation of in-group aggression evolved to maintain social cohesion. On the whole, however, the general consensus is that the central "function" of our moral sense is to promote cooperation within groups (Darwin 1871; Alexander 1985; Joyce 2006a; Krebs 2011; Kitcher 2011; Greene 2014). Tomasello (2013, 231) squarely states that "from an evolutionary perspective, morality is a form of cooperation."

When we look at our actual moral behavior, however, it appears that much of it is very hard to explain from an evolutionary perspective. So much in fact that I will argue that our moral beliefs and norms go radically beyond and even against the evolutionary raison d'être of the faculties and dispositions from which they emanate. On the one hand, most cultures have extended the scope of beneficiaries of our moral consideration and altruism from the small tribe for which it was intended when it was selected in Pleistocene hunter-gatherer societies, to the whole of humanity and even beyond (consider animal rights). On the other hand, we take on much stronger obligations vis-a-vis one another than the simple obligations necessary to maintain high levels of cooperation and enlightened reciprocity for which our moral sense was intended. To witness this we mustn't necessarily turn to the Mother Teresas and other moral heroes of our era – the average moral Joe will do. From helping old strangers cross the street, to donating blood to benefit a stranger anonymously, to something as banal as taking a quick survey on the Internet in the interest of science or an unknown company, we exhibit moral behavior that transcends its intended biological function (in-group reciprocity and cooperation) on a daily basis.

In fact, our morality has ventured so far from the intended behavior for which it was selected that in many cases we find ourselves in the paradoxical situation that – strictly speaking – our moral imperatives work against our biological fitness. Donating money to hungry children in Ethiopia deviates resources from oneself and one's close kin with virtually no chance of reciprocation in the future. Including animals in our moral circle has led a growing number of people to forsake precious nutrients from animal sources. Moreover, we spend important resources on animal conservation and – irony of all ironies – spend a small fortune on preserving some of our former predators.

Of course, the critical reader may object, we only do so because we can afford it. Vegetarians are very rare to come by in traditional hunter-gatherer tribes where the calories and nutrients from animal meat are vital to their survival, and people struggling to make ends meet won't typically donate large chunks of their meager earnings to charity (although you may be surprised). But this misses the point. Much as any other adaptation, our moral sense evolved to create a particular set of fitness-enhancing behaviors. What we witness however is that our moral sense has led us to a whole range of behaviors that it definitely did not evolve to produce. Note that we don't even need to invoke actual moral actions to make this point. The mere fact that we recognize these transcending behaviors as morally laudable is proof that our moral sense has come a long way from its evolutionary origin. This requires explanation.

An Evolved Moral Compass Powered by Reason

Faced with the fact that our moral beliefs and behavior often go radically beyond and against the behavioral patterns for which they evolved, three possible explanations come to mind. The first explanation is that the evolved dispositions making up our moral sense often "miss their mark" in terms of behavior because of imperfect design. Natural selection is a tinkering process and often yields less than perfect design (think about the vertebral column, which, although perfectly suited for our four-legged ancestors, leaves quite a few bipedal hominids with sore backs). With regards to our moral sense, however, this explanation is unlikely from the outset. We're not talking about small behavioral deviations from the intended purpose, but widespread, systematic, and radical departures – even outright opposite behavior to what its evolutionary origin would predict.

A second possible explanation is that the surprising behavioral output of our moral sense is the result of the radically different environmental conditions in which it is cast nowadays compared to the environment in which it evolved (the so-called environment of evolutionary adaptedness or EEA). There could be, in other words, a mismatch between both environments leading to different results than the ones "intended" by natural selection in our modern environment. A good example of such a mismatch is our craving for fat and sugar. This craving was vital for our survival for the most part of our evolutionary history when calories were hard to come by. However, it definitely doesn't serve our biological interests well in an environment filled to the brim with cheap fast-food options. Could this be the case for our moral sense? Perhaps donating money to Ethiopian children is just a kind of "misfiring" of our moral sense, which originally intended these altruistic gestures for in-group members only (people who could reciprocate and enhance my personal fitness by entering with me in cooperative undertakings).

There may be some truth to this last explanation; after all television and other media have brought the suffering African children into our living room and this "proximity" might trigger our pro-social instincts that, in the EEA, could only have been triggered by group members – the only individuals we would be exposed to on a daily basis (see Small et al. 2007). However, at closer inspection, this line of argument is not satisfactory. The moral imperatives we follow are often in tension with our intuitive social instincts. We try hard not to be racially biased (harder than you might think if you look at the results of "implicit association tests" measuring cognitive effort to withstand intuitive negative associations; Greenwald and Banaji 1995, Gladwell 2005). In-group/out-group differentiation has all but disappeared in our modern environment. For the most part, it is not our modern environment that confuses our evolved dispositions into "thinking" members of the out-group belong to the in-group and consequently extending our moral circle, it is our conscious moral reasoning that rejects our evolved bias. Moreover, the mismatch hypothesis doesn't explain the stronger moral obligations that we take on.

This leads me to a third explanation, which I favor. In shaping moral norms and beliefs we do more than blindly applying a set of emotional dispositions and intuitions that evolved to enhance our biological fitness. More particularly, moral behavior and evaluation is to an important extent carried by reason. This is not mere conjecture. Empirical research has brought to light that, in addition to emotions and intuitions, moral evaluation can and does involve reasoning processes. In a landmark experimental study subjecting participants to brain scans while presenting them with moral dilemmas, Greene and colleagues (2001) put forth a convincing case that next to an emotional cognitive subsystem, we also employ a reason-based cognitive subsystem in moral evaluation. Whereas the emotional system often floods our moral thinking automatically and subconsciously, the rational system can, in some cases, override its output and generally takes over when presented with moral problems for which we have no ready-made, automatic, emotion-based response (see also Greene 2013).

However, reason alone does not get morality off the ground. In fact, if reason ruled supreme in the absence of moral or pro-social dispositions (therefore by default serving only our self-interest), we would get notoriously immoral behavior – from the ruthless homo-economicus, who is happy to exploit people as long as there are personal gains to be made, to the outright psychopath, who has no qualms about murder as long as the projected benefits (perhaps the victim was encumbering some of his plans) outweigh the projected costs (the probability of getting caught and being punished). For morality to take a flight, as it has, we need moral intuitions guiding our reasoning processes – intuitions that provide our content-free reasoning processes with the necessary orientation to steer morality in a certain direction.

Importantly, the moral intuitions making up this "moral compass" cannot be modular. They cannot yield a fixed and automatic output reaction when fed a particular input (such as the automatic emotional reaction of moral disgust when confronted with incest [see Haidt 2001] or the automatic negative intuitive bias when confronted with members of the out-group as the previously mentioned "implicit association tests" reveal). They need to provide guidance to conscious reasoning rather than trigger automatic responses. Whereas reason by itself would be blind and would not lead to moral behavior, these intuitive principles would get us nowhere over and above the behavior they were intended to produce without reason. But when you combine both, you get a potent mix, pushing morality to a point where it radically transcends the purpose of its biological origin.

The basic ingredient in our moral compass, I believe, is what Tomasello et al. (2005) call "self-other equivalence." According to Tomasello, the unprecedented level of cooperation in human groups is the result of joint intentionality. This is established by having both joint goals and joint attention to establish how this goal is going to be realized. For joint intentions, all parties need to know that the others know the goal; they need to know their part in achieving the goal as well as the part of the others; and they need to trust that all other parties involved will do their part. In other words, we need to attribute thoughts and intentions to others – we need to "read their minds."

Consequently, according to Tomasello, we stop viewing others as mere elements of the social environment, instead viewing them as other "selves." As a result, "self-other equivalence" – as Tomasello puts it – emerges.

This self-other equivalence branches out into two important moral intuitions at the core of our moral compass. The first is empathy. When I believe that other individuals are the same as I am (with similar capacities for happiness and sorrow), I relate differently to their well-being. I can associate with their pleasure and pain. This ultimately leads us to the ubiquitous moral imperative we know as the Golden Rule: "don't do to others what you don't want others to do to you" (negative version) or "treat others the way you would want to be treated" (positive version). Interestingly, this rule has been formulated in very different cultural contexts ranging from Ancient Greece, to Ancient China and India, to Judeo-Christianity, Islam, and Buddhism, pointing at the universal and therefore innate nature of human empathy, arising out of the seeds of self-other equivalence. It incites us to refrain from causing harm to others (negative rule) and promote human flourishing where we can (positive rule), and it makes intuitive sense to everybody.

A second moral intuition rooted in self-other equivalence is fairness. According to Binmore (2005), human beings are endowed with a universal deep structure for promoting fairness. In analogy with Chomsky's (1959) influential view on human language, claiming that underlying all human languages is an innate deep structure or "universal grammar" constraining natural languages and supporting language acquisition, Binmore claims that a similar kind of innate framework underlies our thinking about fairness. This structure he argues is best captured by Rawls's (1971) original position (which explains its intuitive appeal).

Cross-cultural commonalities in our sense of fairness point at the innate nature of these moral intuitions (Binmore 2005). This is corroborated by Tomasello and colleagues, who studied young children and argue that we come equipped with a number of intuitions about distributive fairness we couldn't merely have absorbed through cultural learning. Young children share spoils equally with peers after having collaborated to obtain the goods even if they could easily monopolize the goods (Warneken et al. 2011), understand and defend the entitlement of others (Schmidt et al. 2013), understand fair as equal (Wittig et al. 2013), and give less to free riders than to collaborators (Melis et al. 2013). While Tomasello's research points out that these important fairness intuitions are uniquely human given that our closest evolutionary cousins the chimpanzees do not abide by them, some fairness intuitions may nevertheless have evolved in our primate lineage long before the human split off. Notably, Brosnan and de Waal (2003) found that capuchin monkeys refuse to cooperate with the experimenter if they receive an unfair treatment (less reward than their conspecifics for the same effort).

The fact that there is an innate basis for fairness does not mean that there is no cultural variation in fairness norms. Morality in general and fairness in particular, as Binmore (2005) points out, are both a product of biology and culture. Our innate sense of fairness provides the framework on which social contracts are built. It functions as a moral compass, not an automatic trigger of certain behavioral patterns, and precisely this aspect of our moral sense provides an opening for reason-based moral cognition and pushes our moral behavior way beyond its modest origins.

Presumably, this self-other equivalence was reserved for the in-group. According to an influential strand of research, an important factor underlying the evolution of adaptations allowing for a higher level of cooperation has been in-between group competition (Bowles and Gintis 2011; see also Vlerick 2016). We evolved the ability to read other minds and started seeing others as other selves because groups with these mind-reading, empathic individuals cooperated better and had an evolutionary edge over groups composed of individuals that didn't or did so to a lesser degree. Empathy, in other words, evolved in groups because it allowed those groups to out-compete other groups. It did not evolve to be extended beyond the in-group (which explains the coevolution of a pervasive negative out-group bias). In fact, it could only evolve if it wasn't extended to the out-group. Nevertheless, once you unleash reason on these moral guidelines (empathy, fairness), the circle is expanded, because rational thinking reveals that there is no principled difference between a member of the in-group and a member of the out-group.

At the core of this transcending moral behavior are those innate moral guidelines (rather than instinctive responses) that make up our moral compass. It is in this feature of our moral mind, I will argue, that we find the common ground necessary for the (internalist) justification of our moral norms and beliefs and a theoretical framework to assess moral progress and explain how it is realized. This is the subject of the following sections.

An Internal Road to Normative Justification

Evolutionary debunking arguments undermine moral realism (Ruse and Wilson 1986; Joyce 2006a). This, in turn, is typically considered to undermine normative justification, leading to relativism (Cobbe 1872; Cottingham 2005). In contrast, I will argue that relativism does not follow from the evolutionary debunking of moral realism. The fact of evolution, I maintain, does not undermine normative justification.

Moral realists generally claim that (1) there are moral facts that exist mind-independently and (2) these facts are therefore objective and universal. Our moral claims, traditional realists insist, can be true or false in the light of these facts, and we can be confident that at least some of our claims are true

(Sayre-McCord 2015). At the heart of moral realism are the claims that moral beliefs refer to mind-independent moral facts (1) and that moral truths are objective and universal (2). These two claims need to be kept distinct. They are not identical. If moral facts are mind-independent, it follows that they are objective, but this entailment relationship cannot be reversed. Moral truths could very well be objective or universal but not mind-independent. Take Kantian ethics, for example, in which moral truths are objective and universal, but not mind-independent. Indeed, they depend on reasoning, since Kant asks us to universalize the maxim of our actions and look for contradictions.

Evolutionary arguments aimed at undermining moral realism can target the mind-independent existence of moral facts and/or the objective or universal nature of our moral beliefs. The first line of arguments – targeting the independence criterion – claims that the genealogical story of moral beliefs makes the notion of independent moral truths superfluous (Joyce 2006a) or redundant (Ruse and Wilson 1986): we don't need objective independent moral facts to explain our moral experience (see also Street 2006). The second line of arguments – targeting the objective or universal nature of our moral beliefs – moves from the premise that different evolutionary conditions could have resulted in different moral beliefs to the conclusion that those beliefs are not objective, let alone universal norms as the realist hopes, but idiosyncratic species-specific norms (Ruse and Wilson 1986).

The first line of argument, challenging the independence criterion, undermines the epistemic justification of our moral beliefs. If there are no mind-independent moral truths, we are not epistemically justified to hold those beliefs, in the same way that we would not be epistemically justified to believe in the existence of God, if God does in fact not exist. Nevertheless, even in the absence of independent moral truths and the correlated epistemic justification of our moral beliefs, our moral beliefs could still be normatively justified insofar as they are objective. Kantian deontology as well as utilitarianism provides normative justification without reference to mind-independent moral facts.

You don't need the theory of evolution to undermine the independent existence of moral facts. Assuming that moral truths exist independently of moral beings means taking on quite a heavy metaphysical load. As Street (2006, 126) points out, it is more parsimonious and therefore a better explanation not to posit external moral facts. Would there be moral facts even if there were no conscious creatures around? Much as one might lean toward mathematical anti-realism – arguing that mathematical entities do not refer to an external realm of immaterial entities but are projected onto the world – one might take a similar point of view regarding moral truths. No need for evolution.

The real evolutionary challenge to normative justification, I think, resides in the line of arguments targeted at undermining the objectivity or universality of our moral beliefs. Darwin himself foresaw this disruptive consequence of his great theory when he asks us to imagine that humankind had evolved under the same environmental circumstances as hive bees:

[T]here can hardly be a doubt that our unmarried females would, like the worker-bees, think it a sacred duty to kill their brothers, and mothers would strive to kill their fertile daughters; and no one would think of interfering. Nevertheless the bee, or any other social animal, would in our supposed case gain, as it appears to me, some feeling of right and wrong, or a conscience.... In this case an inward monitor would tell the animal that it would have been better to have followed one impulse rather than the other. The one course ought to have been followed: the one would have been right and the other wrong. (Darwin 1871, 73)

In other words, moral right and wrong are fundamentally no different than other preferences. In the same way that dung flies are attracted to mammal feces and humans are disgusted by them (because those respective preferences served the biological fitness of both species), some actions – like killing a brother – could be perceived as morally right by one species and absolutely wrong by another depending on the environmental circumstances to which their moral sense is adapted. This genealogical threat seems to radically undermine the normative justification of our moral beliefs. How can we objectively state that certain actions are morally right and others are morally wrong if this reflects nothing but a contingent species-specific set of dispositions with no ontological authority whatsoever? Any attempt to justify our moral norms and beliefs appears misguided and hopelessly anthropocentric.

Nevertheless, the idiosyncratic nature of our moral sense doesn't force us to reject all forms of normative justification and capitulate to an "anything-goes" relativism, as was Cobbe's fear. In fact, the key to rescuing normative justification – providing an objective benchmark from which to distinguish moral right from wrong – from the challenge posed by the objectivity problem resides in the independence problem. Indeed, if we abandon the postulation of the existence of mind-independent moral facts, which given the "costly" metaphysical demand this imposes on us is not an attractive position for the naturalist to begin with, justification should not be sought in the outward correspondence with external moral truths. Instead, we should look inwards and search for common ground. In other words, viewing morality as a projection by moral beings rather than a mind-external quality magically drenched into the structures of the world entails that we cannot set up normative justification as correspondence with external moral facts or even as correspondence with a set of supposed norms on which all possible moral beings would converge (given that it is a species-specific projection). Rather, what makes moral beliefs objective for us and therefore normatively justified is the common ground we can find in our moral sense.

Such an internalist foundation would allow us to maintain that saying that it is wrong to kill a child for fun is not merely a matter of contingent taste or cultural convention but is objectively wrong in the context of human (inter) action. Why? Not because some feature of the world forces us to acknowledge this, but because our very nature forces us to project this moral judgment onto the world. It is not because some hypothetical being might see things differently, that it is any less true for us.

Compare this with color vision. Because of the particular way our eyes work, we see some wavelengths of light as the color red. There is arguably nothing intrinsic in these wavelengths that make them red rather than some other color. It is just the way our eyes perceive it. Another species with different eyes might very well perceive something different when those particular wavelengths hit its eyes. Nevertheless, I am no less justified in the context of human interaction to state that "this apple is red." The fact that there is nothing intrinsic in those wavelengths bouncing off the surface of the apple that makes it red and that other species might perceive the apple differently does not make it less justified in a human context.

Given our possession of what I have called an innate "moral compass," we have such a foundation that allows for objective moral judgments in the human context. On the one hand, the non-modular nature of the intuitions making up our moral compass enables us to make moral judgments and evaluations (rather than forcing reactions of aversion or attraction on us). On the other hand, the innate nature of these intuitions provides us with an objective human benchmark from which we can rationally state that harming an innocent person is wrong, while helping her is right.

Contrast this with the position of strong evolutionary anti-realism, most recently and most extensively defended by Joyce (2006a). Joyce argues that, while it is not practically irrational for us to act in accordance to our intuitive moral norms (he invokes the psychological discomfort we might experience by acting "immorally," and we can imagine that committing outright immoral acts is not the best strategy for self-preservation in most societies), holding moral beliefs is strictly speaking irrational (Joyce 2006a, 226–227). I disagree. It is no more irrational to state that child molesting is wrong than it is to call a red apple red. We don't need esoteric correspondence with external and universal moral truths (Platonic ideas perhaps) to make this claim. All we need is a proper context in which such a statement is rational. *In casu*, the context of human interaction.

I call this position internal realism – in reference to the like-named influential position defended by Putnam (1981) in epistemology. But I could equally well call it justificatory or non-relativistic anti-realism. It is indeed a halfway house between strong realism claiming that there are universal moral truths that can be known and strong anti-realism claiming that, since there are no moral truths, there can be no normative justification.

The point of internal realist accounts is to refrain from strong and ultimately untenable metaphysical commitments, all the while averting strong relativistic threats. In the theory of knowledge, Putnam's (1981) internal realism attempts to overcome both a problematic strong version of realism and a strong version of subjectivism. Against the realist he argues that we view and understand the world through conceptual schemes, but he doesn't follow the subjectivist in her anything-goes relativism. He argues that while our theories cannot be expected to correspond in an absolute sense to the world (they arise through our conceptual schemes), they can nevertheless be objective for us. For example, the claim that $\varpi = 3.14$ is not a matter of cultural convention or subjective opinion, but neither is it independent of conceptual schemes such as mathematics and geometry (see Vlerick 2014). Similarly, moral right and wrong depend on what the moral subject projects onto the world (compare with conceptual schemes), but this does not mean that our moral norms and beliefs lose all objectivity. It merely means that this objectivity is anchored in our biology and consequently that normative justification must be contextualized within the realm of human agency.

My analysis may not answer the skeptic who demands absolute foundations in Cartesian fashion. In epistemology, as Russell (1912) points out, there is no rebutting the skeptic who refuses to enter "the circle of knowledge" because she demands an absolute correspondence between representation and the world, which can never be provided. Of course, this doesn't mean all forms of justification go overboard. We can justify our beliefs to varying degrees. The proposition that the earth is round has stronger justification than the proposition that it is flat; the theory of human evolution has more evidence – that is, stronger justification – than the belief that we are the products of divine creation some 7,000 years ago. Not all theories are on the same footing, and if we want to make progress in the domain of knowledge, we can't be discouraged with the skeptic's impossible demands.

Similarly, in the domain of morality, the skeptic will dismiss such an internalist approach to justification. However, much as we can ignore the skeptic in epistemology and "enter the circle of knowledge," bootstrapping our way to better theories, we can do something similar in morality. Indeed, a moral compass from which we can rationally and objectively (for us) differentiate moral right from wrong, not only grants us (internalist) normative justification but also provides us with a framework enabling us to reason our way to better moral norms and beliefs. This is the subject of the next section.

Moral Progress

Apart from averting Cobbe's nightmare, the key virtue of the internalist normative framework I propose is that it allows the evolutionist to account for moral progress. If there is no normative framework to argue that some actions are better than others, there can logically be no assessment of moral progress. We just lack a foundation to state rationally that twenty-first century Europe with its liberal democracy is a morally superior regime than its sixteenth-century autocratic (and theocratic) counterpart when alleged witches were being burnt at the stake and slavery was an institutional reality. The meta-ethical worry that an evolutionary take on morality might undermine such a normative framework does not seem, however, to have deterred a number of naturalists (and evolutionists) such as Singer (2011), Pinker (2012), and Schermer (2015) to talk about moral progress. And it shouldn't. Given that within the human context there is an objective framework by which certain human actions are morally better than others, we can account for moral progress.

Providing a benchmark from which to assess moral progress is one thing, another one is to explain it. How is moral progress possible? An analogy with knowledge is in order here. How did we get from flat-earth geocentrism to round-earth heliocentrism or from Aristotelian physics to Einsteinian? While a full answer to this question requires a chapter or even a book of its own (see Vlerick 2012 for an attempt), the short answer is that we use our reasoning powers combined with a set of epistemic values (an epistemic compass if you will) guiding us in our search for better theories. Such values include the need for a theory to be coherent (both internally and with other accepted theories), its predictive accuracy, explanatory scope, and so forth. Without these values, reasoning powers are blind and ultimately powerless, with no benchmark for assessing that one theory is better than another and no way of identifying a defective theory in the first place (because it is incoherent or fails in its predictions for instance).

The same process is at work in the domain of morality. Our moral compass provides us with the necessary orientation (fairness, equality, human flourishing) and our reasoning powers provide us with the necessary "horsepower." This leads to what Singer (1995, 226) has called the "escalator" effect of reason. It accounts for the surprising fact that our moral norms and beliefs often go beyond and against the rationale behind the evolution of our moral sense.

Our possession of a moral compass powered by reason explains why we're expanding our moral circle (Singer 2011). Reason reveals that there is no principled difference between members of the in-group and members of the out-group and to some extent between the suffering of certain animals and human suffering. It also enables us to detect and root out contradictions in our behavior. When our nepotistic biases clash with our sense of fairness, or our in-group preference clashes with our sense of equality, we take note and improve. Finally, reason enables us to rid ourselves of intuitive aversions such as aversions toward homosexuality by pointing out – as Bentham (1978) did – that such behavior causes nobody any harm and allowing it promotes human happiness.

So far I hope to have convinced the reader that, rather than undermining normative justification, the biological roots of our moral sense support it, that the non-modular moral intuitions that are the distinctive feature of our moral minds provide us with a rational basis from which to assess moral progress and even explain the process by which moral progress comes about. Before resting my case, briefly, I would like to make an even stronger claim. Not only is an evolutionary take on the genealogy of morality compatible with normative justification and moral progress, such an evolutionary analysis can actually enable us to make moral progress. It can be a valuable instrumental tool to realize moral progress. It can do so by showing that moral dispositions that evolved to solve certain problems in an ancestral context are ill designed for a modern context and are likely to lead to poor moral outcomes.

Greene (2013), for instance, points out that many of our moral instincts and emotions have been selected to solve the problem of me-versus-us (promoting group cooperation). Those instincts, however, are poorly suited to deal with problems of us-versus-them, the problems we typically encounter in the modern globalized world. Another example is our evolved aversion to cause physical harm to others. Cushman and colleagues (2012) found that even "pretend" harmful acts, such as pointing a fake gun at somebody, causes strong physical reactions of aversion (*in casu*, increased peripheral vasoconstriction). This gut-wrenching aversion of hurting somebody at close range, however, disappears when we are removed from the scene. Ironically, and in the context of modern war technology dauntingly, it is much "harder" to harm a person with bare fists than it is to hurl missiles at a distance or even an atomic bomb with a single flip of the switch. While it makes sense from an evolutionary perspective that our emotion-based harm aversion only kicks in when there is actual physical contact (the only possibility to cause harm in ancestral times), we should be conscious of the fact that in our modern environment this aversion is not triggered precisely in those circumstances where we can inflict the most harm.

As Singer (2005) and Appiah (2008) point out, we should not follow our moral intuitions blindly. They are heuristics that yield good results in most cases but can (dramatically) fail in some cases. We can and should use empirical research to learn when these cases arise and take the necessary precautionary measures. An evolutionary perspective on our moral wiring is a valuable tool in this process. It teaches us to regard the output of some of our evolved moral dispositions with a healthy dose of skepticism, given that the behavior they evolved to produce cannot always be expected to lead to good (moral) results (as determined by our reason-powered moral compass). This puts us one step ahead in our constant strive to become better than our nature.

Conclusion

The evolutionary origin of our moral sense is often taken to undermine both moral realism and normative justification. Darwin was aware of this disruptive consequence and so were some of his contemporaries (Cobbe 1872). In this chapter, I argued that while the fact of evolution undermines strong realism, it need not undermine normative justification. In fact, it is precisely in the biological foundation of our moral sense that I locate the roots of normative justification. In other words, rather than undermining justification, the biological nature of our moral sense provides us with an internalist road to justification, warding off an otherwise unstoppable anything-goes relativism in the absence of traditional moral realism.

The fact of evolution, as I hope to have convinced the reader, does not vindicate normative justification (as has been argued successfully against Spencerian ethics and other vindicative projects), nor does it undermine it (as I argue against evolutionary moral skeptics). In fact, normative justification does not relate to it at all. It relates to those aspects of moral reasoning in which we can find common ground and that form the basis of moral conscious reasoning (as opposed to modular automatic reactions) – a sense of fairness and equality, a willingness to promote human flourishing. In the bigger scheme of things, these values are an idiosyncratic evolutionary heritage, but that does not make them any less objective within the human context. No less, indeed, than the redness of apples and the blueness of skies, and ultimately – powered by reason – they can help us make our moral skies an even brighter shade of blue.

Part IV

Elaborations

17 Darwinian Ethics: Biological Individuality and Moral Relativism

Frédéric Bouchard

Introduction

Immediately after the publication of Darwin's *On the Origin of Species*, questions arose about how evolutionary theory would affect our understanding of human morality: if we are members of a species that has evolved by natural selection, and if natural selection acts on the most significant traits and behaviors – pruning those that reduce fitness and favoring those that increase it – shouldn't we expect behaviors that we classify as moral behaviors to be liable to be modified by evolution? Given the significance of behaviors and attitudes associated with morality, shouldn't we expect natural selection to latch on to those mechanisms generating those behaviors in ways that increase fitness? Such questions have served as the foundations for attempts to "Darwinize" parts of ethics since Darwin's own *Descent of Man*. For the purpose of this text, I will define Darwinian ethics as a project that aims to naturalize some or all ethical claims, based on evolutionary explanations of human traits and behaviors. We will examine how a better understanding of biological individuality, and of the gut microbiome and its role in defining human nature, should inform projects in Darwinian ethics and why it seems to warrant a new type of ethical relativism.

Simply put, most attempts to Darwinize ethics rest on two related working hypotheses:

1. If there is such a thing as a moral agent, it's a human individual agent. That human individual is an evolved animal. Therefore, the story of the evolution of Homo sapiens is a necessary (some would even say sufficient) condition for understanding moral agents. We'll call this the Animal Being Hypothesis (ABH).[1]
2. Some, many, or all behaviors that we would classify as moral have significant fitness costs and benefits at the individual or at the group level. Therefore, we would expect some, many, or all moral behaviors to be liable to be selected and evolve. If those behaviors cannot be selected for directly, then we would expect that some of the mechanisms generating those behaviors would be liable to be selected for and to evolve. We'll call this the Adapted Moral Being Hypothesis (AMBH).

While all Darwinians would embrace some form of ABH, they may not all embrace AMBH. For some (e.g., E. O. Wilson, Boyd, and Richerson), AMBH is the inexorable extension of Darwinian evolution that underpins ABH. But other Darwinians (e.g., S. J. Gould) have embraced at least part of ABH while thoroughly rejecting AMBH as being an unsubstantiated theoretical overreach based on a naive understanding of genetics and its explanatory power (e.g., Allen et al. 1975; Gould 1977, 1978; Caplan 1980). In other words, most discussions of Darwinian ethics take for granted some form of ABH and focus instead on the optimistic adoption or vehement rejection of AMBH (see Ruse 2012 for a survey of the relationship between those two hypotheses).

Here I will give some reasons to believe that, like most dichotomies, this opposition is overblown. The enthusiastic Darwinians are right to believe that (at least some) moral behaviors fall under the shadow of natural selection and the less enthusiastic are right in believing that the moral story is not an exclusive genetic one subject to natural selection. But what can we offer as a compromise naturalistic view? I will explain how both "extreme" positions have adopted an incomplete account of human identity (i.e., they are wrong about ABH) and therefore that most attempts to naturalize morality on a Darwinian foundation rest on an inexact account of our biological nature. ABH, intentionally or not, overemphasizes the role of genetic determinism (or the gene's eye view to adopt Dawkins' 1976 usage) and downplays the role of the environment in shaping who we are. In most accounts of the ABH we usually conceive of human beings as members of a higher primate species that is part of a long but relatively simple evolutionary history (e.g., Trivers 1971; Cavalli-Sforza and Feldman 1981; Cosmides and Tooby 1987; Boyd and Richerson, 1995). We will see how individual humans are in fact to be conceived as multispecies communities encompassing, yes, a big primate, but also a plethora of bacteria and other microorganisms. We'll call this the Multispecies Being Hypothesis (MSBH).

What is significant for our purposes is that some of these species (let's call them our symbionts to simplify) are acquired horizontally via interactions with our environment during each human being's lifetime. If this is correct, then Darwinians need to take more seriously the role of ecological interactions and how those interactions shape the elaboration of human individuals. Any future attempt to Darwinize ethics should take into consideration our multispecies identity and its ecological plasticity.

Looking at our microbiome also changes what kind of biological explanations could serve as the naturalistic foundation for morality. We will see how skepticism about the AMBH is warranted, not because of some misguided view of the human mind as a tabula rasa molded solely by culture and not biology, but because we may be tracking only one of the species in the relevant ecological community that constitutes each human individual.

This chapter is not intended to resolve all outstanding issues related to Darwinian ethics: I merely wish to enrich how we think about the evolved biological bearer of moral behaviors and the preferred locus of explanations in Darwinian ethics. To truly Darwinize ethics, we should move from an ABH to a MSBH. Unsurprisingly, this will force us to revise our understanding of the AMBH. The goal here is not to address metaethical questions (addressed in other chapters in this book) or fine-grained naturalized moral psychology or normative questions. I am concerned with understanding the complex biological nature of the moral agent and how that complex nature could lead (or not) to some moral behaviors as adaptations. I will argue that our complex nature gives some biological support to a new form of moral relativism.

We will begin in Section 2 with a short overview of how the link between the ABH and the AMBH was drawn and criticized. We will then (in Section 3) show why the ABH rests on a simplified understanding of human individuality. This will lead us into Section 4 to a reassessment of the AMBH in favor of an Ecological Moral Being Hypothesis (EMBH). As we will see in the conclusion, the unexpected upshot is that it warrants the consideration of a new type of ethical relativism not anchored solely in cultural diversity but also in diversity of ecological contexts.

A Very Brief Impressionistic History of the Relationship between Animal Beings Hypotheses and Adapted Moral Beings Hypotheses

I should point out that the focus of Darwinian ethics is often behaviors and not intentions for a simple reason: although intentions have causal efficacy on behaviors, natural selection cannot directly "see" intentions, rather it can only "see" the resulting behaviors. Take two primates, one with reliable social behaviors caused by specific intentional states and another with the same behaviors but without being caused by the aforementioned intentional states: they will both be selected in the same fashion if the selective environment is selecting for these behaviors. More importantly, if one primate group has reliable adaptive behaviors unencumbered with moral intentions, while another group displays less reliable adaptive behaviors accompanied with moral behaviors, the first group may be selected positively: a well-trained psychopath or a hypothetical zombie could in fact have more reliable behaviors than a well-meaning but fickle "normal" agent, and therefore the first group may have fitter behaviors in certain social environments. These are theoretical limit cases for those interested in human behaviors, and therefore we usually give an important role to intentions in our explanations of human behavior, but, because intentions have not always been necessary for social behaviors to emerge (e.g., eusocial insects), I would argue that an intentionless approach is more useful in an evolutionary context. For that reason, I would argue that, as far as Darwinian ethics is concerned, it's usually not the thought that counts when it pertains to adaptation. With this in mind and for simplicity's sake, let me offer a provisional definition of what I will count as a moral behavior in a Darwinian ethics framework:

Moral Behavior: Behavior with social behavioral input and outputs with the goal of regulating and improving social interactions in the future.

Note that this is not intended as an exhaustive definition of moral behavior, but as a provisional definition of moral behavior as it is used specifically in the Darwinian ethics project (i.e., one should not see this proposition as providing a foundation for a normative ethics project). In a Darwinian framework, it is safe to assume that behaviors that we would qualify as moral behaviors have significant impact on survival for social species such as ours, where complex collective behaviors are so prevalent. So it is reasonable to assume that this regulation and improvement could increase fitness (this was the guiding hypothesis of Richard Dawkins's *Selfish Gene*). In fact, for a Darwinian, it would be surprising if those behaviors weren't affected by selective pressures at all. This intuition was foreshadowed in Darwin's *Origin of Species* and put at the forefront of his inquiries in the *Descent of Man*. In Darwin's time, what was lacking, of course, was a compelling story of the heritability mechanisms that would allow for the transmission of moral attitudes or capacities for moral behaviors from one generation to the next. Evolution by natural selection is usually construed as working by the (slow) accumulation of intergenerational change. Such accumulation depends on the heritability of characters. No heritability, no evolution (see Lewontin in Levins and Lewontin 1985 for a description of the conditions for evolutionary explanations). Given the labile nature of behaviors and attitudes, the obvious cultural transmission of many said behaviors and attitudes, and the lack of a convincing story as to how these could be transmitted intergenerationally in a robust fashion, a Darwinian explanation of morality seemed, for some critics (e.g., Gould 1977, 1978), at best overly optimistic and at worst ideological overreach.

The emergence of molecular genetics in the 1950s and of mathematical models of kin selection in the 1960s changed this initial assessment of a Darwinian morality. Complex behaviors such as the seemingly anti-Darwinian behaviors associated with altruistic behavior could at least in theory be modeled in terms of response to selection in small groups of related individual organisms. The altruistic behavior between parent A and offspring B is understandable in part because, from the point of view of the genes, one bearer of the genes is helping another bearer of many of the same genes. This gene-based and kin-based account of the possibility of the emergence of altruistic behavior gave a scientific toehold on attempts to provide a Darwinian account of morality (e.g., Hamilton 1964; Wilson 1971, 1978; Dawkins 1976). Such a view was then enriched by E. O. Wilson's account of the emergence of social behaviors (beyond altruism) using revised or extended kin selection and then group models. We now have from both philosophers' and biologists' detailed arguments of how altruism and other collective-level traits could emerge via evolution by natural selection (see Sober and Wilson 1998 and Okasha 2006).

The pushback against such Darwinization of social or collective behaviors that seemed to have a moral component was extremely strong on both scientific grounds (e.g., it's not clear what would be the relevant units of selection) and on ideological grounds (e.g., many believe that complex animal behaviors have a degree of freedom unlinkable to gene complexes on which selection could act on). The wholesale tabooification of sociobiology in the 1970s and 1980s (thanks in large part to the campaign initiated by the Sociobiology Study Group of Science for the People, 1976) led to other related but "distinct enough" projects to Darwinized complex animal behaviors. Thus, evolutionary psychology arose in the aftermath of sociobiology; that discipline focused on the evolution of human individuals (as opposed to groups) and the capacities acquired and maintained since the Pleistocene. A common criticism of such projects is that, because they don't actually have access to fossilized brains or to ancestral social environments, they can only provide just-so stories for the evolution of behavioral traits (Buller 2005). These criticisms led to revisions of sociobiology and evolutionary psychology that have attempted to include a better integration of culture in the modeling of population change (e.g., Cavalli-Sforza and Feldman, 1981; Boyd and Richerson, 2004).

We see how our current understanding of the possibilities for Darwinian ethics is based on a broad acceptance of ABH and a polarized response to AMBH. The blunt and simplistic dilemma for Darwinians has been to have to choose between adopting a gene's eye about most significant human behaviors (those most liable to affect Darwinian fitness) *or* to completely reject the possibility that key behaviors could be biologically determined, as if humans were for some reason or another ripped from evolutionary constraints. But all protagonists would recognize that genetic determinism as it relates to behavior is at best promissory but also that a maximalist "culturalist" view is highly optimistic: evidence that complex human behaviors are genetically determined is absent but it's reasonable to hypothesize that at least some parts of complex human behaviors have a genetic component. Conversely, cultural relativism (that assumes that the human mind is a tabula rasa later furnished by culture) assumes that the higher primates – including human beings – have special capabilities that free them of any selective impediments. Like many who have tried to revise sociobiology and evolutionary psychology have argued, it seems obvious that Darwinians interested in morality have to chart a way between the Charybdis of genetic determinism and the Scylla of cultural relativism.

I want to propose such a way. I argue that strong belief in genetic determinism of (moral) behavior is based on a reductive and erroneous view of what constitutes the Darwinian agents that we are. It assumes in large part that the determined behaviors (or selected propensities for certain behaviors) belong to a coherent Homo sapiens genetic lineage. But is such a lineage biologically exhaustive of what humans are as biological individuals? As we will now see, it has been argued by others (e.g., Margulis 1993) and myself (Bouchard 2013) that to be a human being is to be a composite of many species, a holobiont. We won't discuss in detail holobiont theory, but we will show how this theory puts into question the adequacy of ABH as a foundation of Darwinian ethics.

From Animal Beings to Multispecies Beings

To understand why individual human beings are to be understood as multi-species communities, one must first understand general considerations about biological individuality. Elsewhere, I (Bouchard 2013) have built on other philosophers' works (e.g., Dupré, O'Malley, J. Wilson, R. A. Wilson, E. Sober) on defining what is a biological individual; this effort at definition goes beyond common intuitions about what counts as a biological organism. One must not conflate what is a biological individual with what is an individual organism. Assuming that biological individuality doesn't depend on possessing a soul, most accounts of biological individuality focus instead on how individuals are defined and maintained via the functional integration of their parts. Following this view, one must recognize that many multiorganismic biological systems function as individuals via the coordination of behaviors or division of labor of the many distinct organisms constituting the system, whether these organisms are of the same species or not: in other words, collectives of individual organisms can become emergent individuals with emergent traits. In many respects, the behavioral integration of eusocial insects (that inspired E. O. Wilson's work on sociobiology) exemplifies potential emergent individuals arising at the collective level (what we usually refer to as a superorganism in the case of an ant colony; see Haber 2013 for discussion of this case). Take a termite mound: it doesn't "belong" to any individual termite and is the result of the work of a multitude of termites, and yet it provides respiration and cooling at the colonial level, acting as a wind turbine expelling unbreathable air and reintroducing outside air within the mound to allow for the survival and growth of the colony (Turner 2004, 2013). In some sense, the mound is an artificial lung necessary for colony development. This mound system benefits the colony beyond benefiting individual termites, since each individual termite could just as well step out of the mound to breath on its own. At the colonial level, respiration is more than the aggregate respiration of individual termites, because the mound temperature, the amount of CO_2, and humidity result in part from in the internal aerodynamics, for some species the fungal gardens and so forth. Colony respiration is more than termite respiration and is handled differently as well. Such emergent individuals force us to reconsider how we wish to define individuality. Eusocial insect colonies are the better known examples of this emergent individuality, but one can find other complex examples in nature.

If we accept that distinct organisms can add up to become an emergent organism, why should we assume that the process would occur only with same-species organisms? Symbiotic associations between organisms of different species "working" as one are striking examples of multispecies integration leading to emergent individuals (Bouchard 2009, 2013). Some termites (e.g., *Macrotermes natalensis*) cannot digest their food source (cellulose) without having it previously composted by a fungus, *Termitomyces* (see Bouchard 2013 for discussion). In this case, digestion is made possible by a fungus symbiont at the individual termite level, but the fungus is grown at the colony level in fungus gardens within the mound. I say it is grown at the colony level and not at the individual termite level because the fungus garden is maintained by some and not all termites via division of labor. These associations between organisms are as crucial to the survival of the relevant individuals as my liver is essential to my own survival. Such associations are rampant in nature and force us to step back from simplified single organism-centered notions of individuality: in the biological world there are multi-level concurrent overlapping emergent individuals.

Elsewhere, I (Bouchard 2013) discuss other biological examples and the various philosophical notions of individuality that can help us better understand biological individuality for multispecies associations. I argue that the best definition of biological individuality was provided by Wilson and Sober (1989) in their attempt to rehabilitate the concept of superorganism in the case of insect colonies. To paraphrase their proposal:

A biological individual is a functionally integrated entity whose integration is linked to the common fate of the system when faced with selective pressures of the environment.

Adopting Sober and Wilson's definition of individuality brings to the fore the functioning of a system as a whole. But some process needs to provide and maintain functional integration. This role is played by natural selection and adaptation, which generate the overall common fate of the system. In many biological cases, organism A plays a vital function for organism B that in the framework discussed here translates in the construction of a higher level individual constituted of A + B.

While many may be willing to accept that individuals can emerge from collectives (be it a collective of same-species or different-species organisms), it's another matter to show that human beings are such collectives (the argument proposed here). And yet, this is exactly what a better understanding of our gut microbiome indicates. If we accept that the most general notion of individuality hinges on functional integration, it is fair to ask whether the microbes associated with Homo sapiens are functionally integrated into an emergent whole (see Hutter et al. 2015). To make this claim, one would have to show that our gut microbiome – the focus of my argument here but I could have taken other bacteria (e.g., cutaneous) with which we are also associated – provides a function that is helpful or necessary for the hosts' (Homo sapiens) survival and/or reproduction, i.e., that the gut microbiome is a fitness-enhancing part of a higher-order system. Again, this is what current research on the gut microbiome seems to indicate.

Studies of the human gut microbiome (e.g., Turnbaugh et al. 2009; Arumugam et al. 2011; and a plethora of other recent studies) describe the various essential functional roles played by the microbes living in our gut: unsurprisingly many of these functions concern digestion (e.g., Bacteroidetes and Firmicutes allow for the fermentation enabling bacteria to flourish in an oxygen-poor environment, which in turn generates fatty acids, i.e., energy; see Tremaroli and Bäckhed 2012 and Rosenberg and Zilber-Rosenberg 2011 for discussion and other examples), but there is also an indication that the gut microbiome plays an essential role in the "training" and the response of our immune response and other developmental processes (see Pradeu 2012 for philosophical discussion). Just like the termite, our regular functioning depends on other organisms from other species. In many respects, the role of these symbionts is more significant to my survival than the role played by my toes, regardless of the fact that the first are not Homo sapiens gene-based while the latter are.

Why is this significant to understand the nature of the moral agent that we are? Assuming that we are multispecies is one thing, but the actual issue is how some of these parts are acquired. How do we "get" this gut microbiome? The answer is that we are colonized by microbes living in our environment, from gestation onward, and the functioning of these bacteria is affected by our diet and other metabolic processes that may be triggered by environmental cues. Parts of us, the non-Homo sapiens parts, are acquired from the happenstance interactions we have with our environment and are not vertically transmitted via genes. Our digestive system is in part beyond Homo sapiens genome; and even though its functioning is dependent on capacities emanating from other species' genomes, we cannot reduce the acquisition of said species to the genetic properties of the symbiont. Rather, we must investigate what type of ecological interactions made it possible for individuals of different genomes to meet, to associate, and to form a novel emergent individual.

We get some of our fundamental biological subsystems from our environment. Therefore, to be a human individual is to be at least partly a biological multispecies individual that is ecologically constructed and maintained.

This should weaken the appeal, even for a fully committed orthodox Darwinian, of a gene's eye view of our human nature. The less orthodox should also be concerned about the limits of adopting a view of human evolutionary history that is based solely on the evolutionary history of our Homo sapiens parts.

The bad news for the "traditional" Darwinian ethicists (or another flavor of the sociobiologist 2.0) is that this shows how they were wrong about the unit of adaptation under investigation. The good news for the ethical naturalist is that it enriches the ways in which biological processes determine who we are. A Darwinian account of morality must not latch onto the evolutionary transformation of a monogenomic primate individual bearing moral attitudes and behaviors, but rather onto the evolutionary and ecology transformation of multispecies community being.

From Adapted Moral Beings to Ecological Moral Beings

A better understanding of the importance of our microbiome and its role in our constitution is essential for naturalistic accounts of morality for two reasons. First, although the microbiome system does not contradict attempts to anchor human nature to our biological evolved nature, it does weaken the hypothetical exhaustivity of genetic reductionism about human nature. We are biological entities, and yet we are way more than the "lumbering robots" of Homo sapiens genes (see Dawkins's vehicles or Hull's interactor concept). Second, we see how we are not monogenomic, but multigenomic. But more importantly, broad sections of our humanity are biological *and* context dependent. In relation to many traits, we are, yes, evolved interactors, but we are also multispecies ecologically constructed and maintained consortia.

These are two general points about how to reconceive of human nature. By themselves, they are sufficient to force us to revisit biological accounts of human nature beyond genetics and evolution and to construct a more ecological and developmental account of the multispecies individuals that we are. In short, the understanding of the role of bacteria in human individuality forces us to rethink our individuality beyond the ABH and move toward the MSBH. This move in turn brings the ecology of the interactions between our species parts to the forefront of naturalistic accounts of morality.

But where this complex view of human nature may transform Darwinian ethics is not only in the way it recasts the boundaries of the human animal or alters notions about the origins of its parts but also in the way we have to assess moral attitudes, which may issue from this ecologically structured creature.

It should be noted that, while the arguments presented in Section 3 in favor of MSBH were on firm theoretical and scientific footing, the arguments that will be presented here are based on recent scientific findings that are of a more tentative nature. The possibility briefly explored in this section is that some of our moral attitudes and behaviors are biological *and* dependent on our environment (the environment that is putting us in contact with these other species that then become part of us and our functioning).

Many active research projects focus on what is called the "gut-brain axis." The basic idea is that changes in gut functioning affect the central nervous system without being mediated by prior mental states. The most active research project is looking into anxiety and depressive states (see Carabotti et al. 2015 for review). For our purposes, depression is especially significant since it deeply affects social interactions. Agents with depressive states tend to reduce social interactions. Remember how moral behavior was defined for the purposes of this discussion: behavior having the goal of regulating and improving social interactions in the future. Gut-brain axis research hints that some deeply social behaviors such as anxiety and depression may be modulated by gut microbiota.

Some of those parts constituting the human are acquired from the environment during the individual's lifetime. This is a culturally mediated biological relativism. What we eat and how we live determines what microbes become part of us. More importantly for this discussion, these heterogeneous parts affect emotions that are key to a general understanding of moral psychology.

The gut-brain axis hypothesis is significant for it may explain dissociating the effects of gut microbiology and culture on some of our moral reasoning (I am indebted to Nick Byrd for raising this point and the following example to me). There are cases where cultural norms may hold that a behavior is bad or non-desirable in a top-down kind of way, while other cultural norms may indirectly promote the same behavior in a bottoms-up kind of way. In the Western world, many believe that overeating is "bad" in various ways (e.g., because it is hypothetically linked to higher health and social costs: note that my use of this example does not imply agreement with this view). So we have strong cultural norms that try to control how we eat. But other cultural norms about hygiene may actually promote a gut microbiome that increases the likelihood that we will attempt to overeat (e.g., Sclafani 2004; de Araujo et al. 2013). While our culture may "directly" affect how our brain generates some eating restrictions behaviors, other aspects of our culture may affect our gut microbiome that may affect our brains and how it generates contradictory eating behaviors. In other words, some cultural norms via top-down cognitive processes may guide us to adopt behavior A, but other cultural norms via bottom-up processes (via the gut-brain axis) may guide us to adopt behavior Not-A. Ecology and culture determine parts of our biological nature in many ways, sometimes conflicting ways. These conflicts may explain why we appear to have inconsistent behaviors.

To recap, we have treated the gut microbiome as an organ mediating various digestive processes. In this view, to be human is to be a multispecies construct. These species are acquired via our environment in every generation. This process raises the significance of horizontal transmission in determining human functioning and being. Some transmission channels for these symbionts are ecologically *and* culturally mediated. In this section we have looked at the possibility that some of these ecological interactions influence some behaviors and responses associated with morality, the moral behaviors defined in the first section of this chapter.

If this is correct, the biological picture of our moral behaviors is enriched. Our behaviors (including some moral behaviors) may have a genetic component, but we now know why, for biological reasons, this component is necessarily limited. This limitation was previously thought to result from the power and flexible aspect of human culture and its significance in a top-down cognitive process. But the understanding of the functioning of our gut microbiome hints that the limitation of many outstanding Darwinian ethics accounts may in fact result from the power and ecological aspect of our partial microbial nature and how it also affects bottom-up cognitive processes. The argument presented in this section is more tentative, but it seems sufficient to warrant that Darwinian ethicists and ethical naturalists in general should consider moving from an AMBH to an EMBH.

Conclusion

Getting back to our initial presentation of the project of Darwinian ethics, we now see how our understanding of the roles of the microbiome recasts the likely candidates of a Darwinian explanation. When thinking about ways in which ethics or morality could be naturalized via Darwinism, the view of human individuality presented here is significant for two reasons. First, most attempts to Darwinize ethics assume that our genetic "destiny" is based on the evolution of the Homo sapiens genome (what I called the Animal Being Hypothesis). Many have assumed that tracking the evolution of that single genome is relatively exhaustive of human nature (in its biological respects). It was assumed that it was the molding of this genetic "history" by natural selection that underpinned the revised moral framework suggested by Darwinian ethicists. But by recognizing that the biological substrate that is a human individual is way more than the traditional view of Homo sapiens, it becomes unlikely, even under the most optimistic scientific scenario, that a full and complete story of Homo sapiens genes could give us a solid and expansive foundation on which to ground ethics. If this result is correct, this would put stress on traditional accounts of sociobiology and evolutionary psychology. It does not necessarily condemn them to falsification, but it restricts their potential applicability and complexifies the explanations that they would have to provide. The first lesson to would-be ethical naturalists is that future projects attempting to Darwinize ethics have to rely on a MSBH with all its afferent community and ecology-based explanations.

Second, if there is some value to the gut-brain axis hypothesis, and if the constitution of our gut microbiome is in large part non-Homo sapiens and horizontally acquired, then a Darwinization of ethics will at least in part have to be community (i.e., multispecies assemblage) based at the behavior level as well. The AMBH was based on the evolution of monogenomic individuals, but the process of bacteria acquisition and their potential effects on some key behaviors raises the significance of ecological interactions over evolutionary interactions. Many Darwinian ethics projects have focused on group traits (e.g., the evolution of altruism in same-species kin or non-kin groups) and have anchored their argument in group-selection explanations. The picture drawn here hints that same-species interactions will not be exhaustive of how moral attitudes and behaviors can be mediated gut-brain interactions, and their hypothetical effect on some morally relevant behaviors raises the importance of providing a more detailed account of the community selection involved, and a better understanding of how our behaviors affect our environment that in turn affect the types of bacteria that will affect some of our behavior. The necessity of examining horizontally acquired species makes the current (or proximate) context as essential to the naturalistic story as studying selective pressures from the Pleistocene. Naturalism about ethical claims will have to include, yes, evolutionary considerations but ecological considerations as well.

The first point, Multispecies Being Hypothesis, is in many respects more fundamental, for it reduces the appeal of any genetic reductionism view (or gene's eye view) regardless of the trait under study: if digestion is not solely Homo sapiens-based, why should we assume a priori that any complex behavioral traits to be solely Homo sapiens genes-based? The second point is more controversial for it depends on some empirical facts about how those other species interact with some specific attitudes and behaviors. Although recent tentative results seem to give some justification for this approach, it remains an open empirical result as to whether and to what extent our gut microbiome affects behaviors that we would qualify as moral. Study of the microbiome should not and cannot completely displace studies of the Homo sapiens genome and its relationship to behaviors. But those studies are an important reminder that contemporary environments affect fitness-related behaviors in ways that are relatively independent from Homo sapiens genes but that may affect moral behaviors. The fact remains that, even if we reject the second point about the need to go from AMBH to EMBH, Darwinian ethics has to take seriously the fact that to be a human individual is to be multispecies and that those species are under separate but intersecting evolutionary and ecological trajectories.

What about the potential "ablation" of those microbial communities (via broad-spectrum antibiotics)? One can lose microbial communities and still have moral behaviors. If one can be considered a moral agent without being a multispecies individual, doesn't that imply that those microbial communities are not that interesting for moral accounts? (My thanks to Robert J. Richards for raising this question.) Yes and no. One can still be considered a responsible moral agent after ablation of certain parts of his brain, even if those parts were related to causing moral behaviors. The fact that one can lose a part and still function does not mean that that part did not play a significant role in certain contexts. I am not arguing that bacteria are necessary to cause many or most moral behaviors. I am arguing that because they can have a big effect and because they are introduced via ecological interactions, they should enrich our view of biological accounts of significant behaviors.

Notice that my account recasts how to think about some traditional accounts of the rejection of the AMBH. S. J. Gould's and other tabula rasa-like accounts suggest that while we are evolved animals, our mental states and cognition are relatively unconstrained by our evolutionary history. This general purpose processor view of the brain raises the purported significance of human cultures and how they shape our behaviors. This understanding of how behaviors are generated serves as the foundation for ethical relativism based on culture: different cultures produce different ethical norms and behaviors. MSBH and EMBH provide additional ammo for the relativist but not on a pure cultural footing. Who we are and how we act is in part a function of our biology, our microbiology. This microbiology is ecologically acquired via cultural and non-cultural ways. Some bacteria that may affect moral behaviors via the gut-brain axis are acquired and mediated via various cultural practices (diet, hygiene practices, and other culturally dependent practices), but they are also acquired and mediated via purely biological and environmental context (e.g., the gut microbiome of the mother and how part of it colonizes the infant at birth, or the type of dirt in the household during early developmental stages). Context matters in more ways than the "tabula rasa" camp usually argues for, and pure culturalist accounts will miss out on how microbes shape who we are, how we think, and how we act.

For Darwinian ethics proponents, this is both good news and bad news. Biology is even more significant than previously believed but not in the way that it is usually argued for. For the Darwinian ethicists, the way forward is to rethink how evolution works in general beyond change in gene frequencies. In the end though, it can be of some solace that Naturalized and Darwinized ethics remains a viable project if it can leave behind some of its Hamiltonian, Wilsonian, or Dawkinsian reductionism and embrace the importance of ecological context and culture. There is grandeur in this view of human nature.

Note

1 Most literature on biological individuality uses the term "individual" instead of "being" (I have favored the term "individual" as well in other writings, e.g., Bouchard and Huneman 2013). In the context of Darwinian ethics, however, I will use the term "being" to distinguish the discussion at hand from traditional discussions about biological ontology (that focus on "individuals" irrespective of the species of interest, e.g., Hull 1978, 1980; Dupré 1993; Wilson 1999; Wilson 2004) and the discussions about moral agents (that focus on the bearers of moral responsibility and/or the actors of moral behaviors, e.g., Jaworska and Tannebaum 2013). For our purposes, we will take "beings" to refer to biological individuals that are involved in behaviors of interest for moral theories.

18 Evolutionary Psychology, Feminist Critiques Thereof, and the Naturalistic Fallacy

Lynn Hankinson Nelson

Introduction

In this chapter we explore relationships between what is called "the naturalistic fallacy," evolutionary psychology, and feminist critiques of the hypotheses about gender proposed in that program. The British empiricist David Hume (1711–1776) is generally credited with first identifying the pattern of reasoning that has become known as "the naturalistic fallacy" (although Hume did not use that phrase) as being fallacious (i.e., flawed) reasoning. Others (notably G. E. Moore) have also given accounts of a naturalistic fallacy. The patterns of reasoning in question have been defended as well as criticized. We will use Hume's formulation of the naturalistic fallacy because it is the formulation assumed in the arguments on which we focus. We do not consider the question of whether Hume was correct that the reasoning in question is fallacious, although this is an important question, because it is not relevant to the arguments considered here. These arguments, I suggest, are important in their own right.

Hume drew a distinction between two kinds of statements: statements about "matters of fact" and statements concerned with what he called "relations of ideas." Matter of fact statements are what we would call empirical or factual statements. "It is below freezing today in Minnesota" is a statement of this sort. Whether it is true depends on what the weather in Minnesota is like today. Statements about relations of ideas, on the other hand, are not about what the world is like, but rather about concepts. Their truth or falsity depends on whether the concepts that they relate are or are not so related. They are generally taken to include what we call logical truths, for example, "A square has four sides," as well as logical falsehoods, such as "Some bachelors are married."

Hume reports having observed that authors frequently put forward one or more empirical claims (i.e., "matter of fact claims"), and then without comment proceed to advance a normative conclusion – that is, something ought to be the case.

In every system of morality, which I have hitherto met with, I have always remarked that the author proceeds for some time in the ordinary way of reasoning [noting or establishing is, or is not, statements] … when of a sudden I am surpriz'd to find that instead of the usual copulations of predicates is, and is not, I meet with no proposition that is not connected to an ought, or an ought not. (Hume 1739, 335)

By "I meet with no proposition that is not connected to an ought, or an ought not," Hume means that the propositions now before him are all "ought" or "ought not" propositions. "This change," Hume continues, "is imperceptible [not noted or remarked on by the author]; but is, however, of the last [i.e., greatest] consequence." Such reasoning, Hume argued, is fallacious: "ought" or "ought not" statements do not follow from "is" or "is not" statements. One subtlety should be mentioned. Some normative claims, for example, "inflicting gratuitous suffering is wrong," are put forward in the grammatical form of "is" claims. But Hume would rightly say of this example that it is not a report of a matter of fact to be confirmed or disconfirmed by observing instances of the infliction of gratuitous suffering. It is, instead, a judgment about the moral status of inflicting such suffering. The bottom line of Hume's argument is that no claim about what is or is not morally right can be established by appealing only to empirical claims.

We are now in a position to briefly anticipate the arguments we consider. The first, offered by evolutionary psychologists, is that they do not in fact commit the naturalistic fallacy. They posit specific features of human psychology and propose evolutionary explanations of them, and they contend that both are purely factual claims. Because they agree with Hume that to draw normative conclusions from them would be fallacious, they do not do so. In a second line of argument, evolutionary psychologists maintain that feminist scientists and science scholars who criticize their hypotheses about gender do engage in reasoning of the kind Hume maintained is fallacious. They take feminist scientists and science scholars to wrongly draw normative implications (i.e., social, political, and/or ethical implications) from evolutionary psychologists' hypotheses about gender, implications that in fact do not follow from them.

A third line of argument is offered by authors who are not evolutionary psychologists but strongly support the goals of evolutionary psychology (hereafter, EP) but maintain that those in the program misunderstand Hume's argument. This misunderstanding, they argue, has resulted in evolutionary psychologists refusing to engage important ethical issues that are raised by their hypotheses. In another argument, the authors maintain that EP neglects Darwin's arguments about the evolution, nature, and significance of "the ethical sense" – a trait he took to be the "most noble" outcome of evolution and unique to humans. So while they agree that evolutionary psychologists do not commit the naturalistic fallacy, they argue that EP's avoidance of ethical issues and of morality is unwarranted and misguided. The last argument we consider questions whether the hypotheses EP proposes about gender are purely factual or empirical. Feminists' critiques of these hypotheses, I argue, reveal that the hypotheses in question make use of terms that are "evaluatively thick" – that is, terms that have both empirical and normative content – and that the hypotheses in which they figure are evaluatively thick. If feminists are correct, a surprising conclusion follows: in terms of EP's hypotheses about gender, neither evolutionary psychologists nor feminist critiques of these hypotheses commit the naturalistic fallacy – that is, reason from purely factual claims to "ought" or "ought not" claims, because the claims about which they disagree have both empirical and normative content.

To engage the details of the arguments summarized, some background is in order.

Evolutionary Psychology

Evolutionary psychology emerged as a research program in the late 1980s and early 1990s. The overarching goal of the program, as its architects describe it, is to bring about "conceptual integration" between the human sciences, particularly psychology, and evolutionary biology (Cosmides et al. 1995). Its core tenets include that adaptive problems our ancestors faced during the later stages of Pleistocene (the period ending about 8,000 to 10,000 years ago) led to "selection for" specific psychological features (cognitive predispositions, mechanisms, and/or modules); that these are heritable; and that they are features of current human psychology. Other tenets of EP include that these features of human psychology were relevant to specific adaptive problems, and because our ancestors living in the Pleistocene were faced with a large number of adaptive problems, the resulting heritable predispositions, mechanisms, and/or modules are many. Together, according to evolutionary psychologists, they constitute "a universal human nature." EP also maintains that some psychological features we have inherited may now be "maladaptive."

EP understands its questions, methods, and hypotheses about human psychology as firmly rooted in the theories and theorizing of Darwin and Darwinism. They assume that natural and sexual selection result in adaptations conducive to survival and reproductive success, and that this is true of psychological traits. Like Darwin, they also propose that there are important differences between women and men, which EP construes as psychological differences. Evolutionary psychologist David M. Buss describes the assumption that there are psychological differences between men and women as one of several "meta-theoretical" assumptions guiding EP. In summarizing its rationale, Buss paraphrases an argument offered by Donald Symons in 1992:

> To an evolutionary psychologist, the likelihood that the sexes are psychologically identical in domains in which they have recurrently confronted different adaptive problems over the long expanse of human history is essentially zero. (Buss 1996, 302)

Evolutionary Psychologists on Gender

EP maintains that the differences in the adaptive problems associated with reproductive success that faced ancestral women and men are the result of gametic dimorphism – differences in the size and number of eggs and sperm. Eggs are "more expensive" than sperm, and gestation and lactation make women's "investment" in offspring much greater than men's. EP holds that these differences in parental investment lead to gender differences in mating and parenting strategies. Here they appeal to core tenets of the Parental Investment Theory, which was developed in the 1960s.

Among the gender-differentiated predispositions evolutionary psychologists propose and explain by appeal to the Parental Investment Theory are women's preference for men with resources (a preference, they maintain, which has no correlate on the list of the qualities men prefer in a prospective mate) and men's predisposition to sexual jealousy. The latter they take to be a function of men not wanting to invest in offspring who are not their own, which they, in turn, take to explain men's coercive behaviors, such as their predisposition to engage in domestic violence. They also posit that men have a predisposition to rape. This predisposition, they argue, is either an adaptation directly related to reproductive success or a by-product of an adaptation related to such success, such as the desire for multiple mates.

How do evolutionary psychologists arrive at these and related hypotheses? In the introduction to *The Adapted Mind*, the editors, who are among the architects of EP, describe the methods used in the program to predict or to explain features of human psychology (Cosmides et al. 1995, 3–18). Briefly put, Cosmides and colleagues maintain that knowledge of the adaptive problems facing our Pleistocene ancestors allows predictions about psychological features likely to have been selected for and that these predictions can then be tested through studies of contemporary humans. Explanations can be proposed of some known feature of current human psychology by identifying the ancestral adaptive problem that led to its selection.

We have so far introduced evolutionary psychology's core tenets, methods, and approach to gender in the terms of those engaged in the program use. We take a similar approach in describing feminist science scholarship.

Feminist Science Scholarship

Feminism is, of course, a political commitment and movement. But it is also a research tradition now well more than four decades old. Part of that tradition, the part that is relevant to the present discussion, has come to be called "feminist science scholarship." Many scientists, including Galileo, Boyle, and Einstein, and contemporary scientists (such as particle physicist Leon Lederman and geneticist Richard Lewontin), as well as philosophers and historians of science, have engaged in "science scholarship." Those so engaged ask, among other questions: What is science? What is the nature and strength of the evidence that supports scientific hypotheses and theories? How is science different, if it is, from other human endeavors, including other kinds of inquiry? How are science's practices and/or hypotheses related, if they are, to the historically specific contexts – social and scientific – in which science is undertaken?

To be sure, these and related questions are also asked by those interested in challenging specific scientific hypotheses (e.g., that evolution occurs or that climate change is real) as well as science's credibility. In what I will call "serious science scholarship," these questions are pursued in an effort to understand the nature and strength of the evidence central to scientific practice. Today, those undertaking such inquiries are likely to focus on a specific science or subfield, recognizing that the methods and priorities of the sciences often differ. So understood, science scholarship is an empirical endeavor. But it also has normative dimensions. Scientists and science scholars often ask whether the theories and models they study are empirically adequate, whether the methods employed in a field are effective and/or appropriate, and/or whether a historical or a contemporary science was or is partly influenced by its historically specific social context.

Those working in feminist science scholarship – a research tradition engaged in by laboratory and field scientists, philosophers of science, historians of science, and scholars in other disciplines – view their engagements as "serious science scholarship." They emphasize evidence, and engage both empirical and normative questions of the kind previously noted. What distinguishes feminist approaches to science from other traditions in science scholarship is attention to and use of the analytic category, gender, in efforts to understand science.

So there are parallels between EP and feminist science scholarship. Both programs emphasize gender. Moreover, both view questions about it (and there are many) as appropriately pursued using scientific methods and evidence and both take themselves to be involved in scientific research. One might well say the parallels end there – while also recognizing, as we will see, that emphasis on scientific methods and evidence they share contributes to the serious disagreements that divide them.

Evolutionary Psychologists on the "Naturalistic Fallacy"

Books and textbooks devoted to EP often include a description of the naturalistic fallacy that is based on Hume's account and maintain that the hypotheses proposed or discussed in the volume do not commit that fallacy – i.e., do not entail normative conclusions. For example, Steven Gaulin and Donald McBurney, authors of the textbook, *Psychology: An Evolutionary Approach*, state:

Evolutionary psychology explains behavior; it does not justify it. Imagining that it offers a justification is known as the naturalistic fallacy (Buss 1990). In a nutshell, the naturalistic fallacy confuses "is" with "ought". It confuses the situation that exists in the world with our ethical judgement about that situation. (Gaulin and McBurney 2001, 2)

According to Gaulin and McBurney, evolutionary explanations of behavior are just like scientific explanations of any natural phenomena. They do not take the behaviors so explained to be "good" because they are "natural" any more than scientific explanations of earthquakes take them to be good because they are natural. Rather, like scientific explanations of earthquakes, evolutionary explanations of behavior seek to understand those behaviors so that we will "be better able to eradicate them or alleviate their effects." "We hold," they continue, "that studying the possible evolutionary origins of child abuse … is a good way to understand and therefore address [the problem]" (Gaulin and McBurney 2001, 16). We should note that although EP views itself as identifying and explaining features of human psychology, like the authors of this textbook, evolutionary psychologists often focus on behaviors. This, as we later explore, raises questions.

In *A Natural History of Rape: Biological Bases of Sexual Coercion*, biologist Randy Thornhill and anthropologist Craig T. Palmer argue that rape is either an adaptation (i.e., the product of natural or sexual selection) or a by-product of other adaptations (for example, aggressiveness or sexual desire), and they insist that these explanations do not entail that it is morally justified. In an argument similar to that about child abuse offered by Gaulin and McBurney, Thornhill and Palmer maintain that knowing the biological bases of rape can lead to preventative measures, for example, to education programs for teenage boys devoted to rape prevention (Thornhill and Palmer 2001, 179).

Evolutionary psychologists' arguments that a cognitive predisposition or behavior that was once adaptive may now be "maladaptive" are understood to demonstrate that evolutionary psychologists do not draw normative conclusions from the factual hypotheses they propose. But what does it mean for a trait, including a psychological trait, to be maladaptive? To term an adaptation as maladaptive might seem to imply that the "adaptation" now works against, rather than for, survival or reproductive success. But the EP literature makes it clear that this is not always (or even generally) what evolutionary psychologists mean when they describe an adaptation as now maladaptive. Consider, for example, Steven Pinker's argument that our fear of snakes, adaptive in ancestral environments, is now maladaptive given that today most humans live in urban environments. In this and similar cases (e.g., in discussions of our fear of heights), "maladaptive" can only mean "irrational" or "unwarranted." There is no reason to think, nor does Pinker argue, that such fears work against our survival or reproductive success (Pinker 2009).

But then why do evolutionary psychologists use the term "maladaptive"? "Mal" derives from the Latin *mal*, meaning "bad." So is describing an adaptation as "maladaptive" saying that the adaptation produces ethically wrong or bad behavior? Do arguments that describe child abuse or rape as now "maladaptive" – taken to demonstrate that EP does not conflate "is claims" with "ought claims" – assume that the behavior in question is unethical? The claim that EP's explanations of now maladaptive adaptations allows us to intervene, through education or counseling, and thus prevent such behavior, certainly suggests that evolutionary psychologists consider such adaptations to result in behavior that harms others.

There is a way to avoid this conclusion. Within the framework of Darwinism, a predisposition toward a behavior or the behavior that results is maladaptive if and only if it acts against the survival or reproductive success of an individual so predisposed. Given that individuals who engage in child abuse or rape are subject to imprisonment or even death, such behavior is not conductive to their survival or reproductive success, and thus it is maladaptive. So any preventive methods made possible by identifying the cause of the behavior are accurately understood as benefiting those at risk of engaging in them, not as preventing harm to others. This interpretation is certainly in keeping with how "maladaptive" is understood in evolutionary theory. But, as far as I know, no evolutionary psychologist has advocated it. Nor does it seem plausible that this is how evolutionary psychologists understand the arguments in question.

Alternatively, evolutionary psychologists might argue that although they describe the predispositions or behaviors in question as currently "maladaptive" because they cause harm to others, in such contexts they are not speaking as scientists but rather as members of societies that deem the behaviors in question to be unethical, judgments with which they agree. But if this is the case, then it is a mistake to describe the dispositions or behaviors in question as "maladaptive" – the claims in which the term appears are not evolutionary claims, but ethical claims.

In addition, such claims (construed as normative) raise questions about describing such predispositions or behaviors as adaptive in ancestral populations. Even if we grant EP's arguments that the coercion of our female ancestors by their male counterparts contributed to men's reproductive success – for example, that sexual jealousy that explains domestic violence contributed to some assurance that a man wouldn't invest in offspring that were not his own, or that rape was selected for because it was conducive to reproductive success – did they not harm the female ancestors subjected to them? If time travel were possible, would evolutionary psychologists claiming to know the biological bases for them intervene – travel back to the Pleistocene and use the kinds of measures they say their explanations might now lead to – to prevent or limit instances of such behaviors? And when they describe the predispositions or behaviors as once adaptive, why do they refrain from acknowledging that they caused harm? (Still another way of understanding the use of "maladaptive" by evolutionary psychologists is that, in some cases, they just "blurt it out.")

Evolutionary Psychologists on Feminist Critiques and the Naturalistic Fallacy

Evolutionary psychologists often charge feminists with committing the naturalistic fallacy by mistakenly attributing normative implications (i.e., political, social, and/or ethical implications) to the programs' empirical hypotheses about gender. For example, in their account of the history of feminists' engagements with EP, David M. Buss and David P. Schmidt argue:

Some feminists saw evolutionary approaches as antithetical to political goals, such as achieving gender equality. Some expressed concern that if gender differences exist and are evolved, then some might claim that gender differences "ought" to exist, and these theories might therefore be used to oppress women and interfere with achieving gender equality. (Buss and Schmidt 2011, 770)

Their use of the past tense does not indicate that they no longer attribute such views to feminists. It reflects that they no longer attribute them to all feminists because there are feminist evolutionary psychologists.

A second line of argument taken to demonstrate that feminist critiques of EP commit the naturalistic fallacy relies on the assumption that the goals of evolutionary psychology and of feminists are inherently at odds. EP, it is argued, is like any science in that it seeks to discover "how things are." In contrast, feminism is a political commitment focused on "how things should be" and, as such, is not science. In the first edition of the textbook, *Evolutionary Psychology: The New Science of the Mind*, David M. Buss explicitly made this argument. And although he now describes feminism in ways that recognize its scientific dimensions, he and Schmitt continue to see the interests in and approaches to gender that characterize the two as "in some ways incommensurable" (Buss and Schmitt 2011, 770). Citing surveys that indicate evolutionary psychologists vary in their political orientations and beliefs, Buss and Schmitt argue that, in contrast, feminism is partly a scholarly enterprise, but also often contains explicitly political agendas. They present evolutionary psychology as a scientific meta-theoretical paradigm designed to understand human nature and argue that it has no political agenda (Buss and Schmitt 2011, 770). So, although the two disciplines' interests in gender overlap, in these senses, they conclude, the two are in some ways incommensurable.

Several issues addressed in these lines of argument are important. As earlier mentioned, I will suggest that a useful way of understanding feminists' critiques of evolutionary psychology's hypotheses about gender is that feminists do not view many such hypotheses as purely factual, but as incorporating normative as well as empirical content. Here we consider a different interpretation of feminists' arguments that hypotheses about gender advanced in EP carry normative (social, political, and/or ethical) implications.

For many feminists (myself included), it is difficult to view arguments that such predispositions are core features of "a universal human nature" that were conducive to survival and reproductive success, as not carrying implications for social or political policies and/or beliefs. It is true that, unlike human sociobiology, EP does not assert direct causal relationships between genes and behavior, but rather proposes that psychology establishes an intermediate level constituted by evolved cognitive predispositions and modules, and it is also true that it maintains that these may or may not be expressed behaviorally. This is because there are so many predispositions and modules that they contribute to significant flexibility and/or because their expression in behavior require relevant "triggering conditions" (e.g., Buss and Schmitt 2011).

But many feminists view evolutionary psychologists' claims to have discovered universal features of human psychology as deterministic in and of themselves, even if they are not always expressed behaviorally. Genes selected for during the Pleistocene function in EP as the "ultimate cause" of the features that constitute human psychology. Moreover, as noted previously, evolutionary psychologists often claim to explain actual behaviors.

The foregoing focuses on whether feminists' critiques of EP commit the naturalistic fallacy. What has not yet been addressed is that many such critiques focus on the evidential warrant and empirical adequacy of the hypotheses about gender EP proposes, the methods it employs to generate them, and the evidence for the reconstructions of Pleistocene life on which EP relies. These things challenge the argument positing an incommensurability between feminism and science, as does the fact that many feminists critical of EP are scientists.

Buss and Schmitt also claim that, as a science, evolutionary psychology is apolitical or "value-free." But this view of science has been and remains a matter of considerable debate, as does the view that motivates it – namely, that social or political values necessarily compromise science.

Does EP "Misunderstand" Hume's Account of the Naturalistic Fallacy? Does It Neglect Darwin's Arguments about Morality, and If So, Is This Misguided?

In this section, we focus on an article written by evolutionary biologist David Sloan Wilson, philosopher Eric Dietrich, and anthropologist Anne B. Clarke (Wilson et al., 2003). In it, they argue that evolutionary psychologists misunderstand Hume's account of the naturalistic fallacy and thus make "inappropriate" use of it. Wilson and colleagues correctly point out that EP's understanding of Hume is incomplete. They note that although Hume did hold that no "ought claim" follows from one or more claims that are exclusively "is claims," he did not hold that an "ought claim" can never be the conclusion of a valid argument. That is, Hume does not deny that an ethical claim can follow from a combination of premises, at least one of which is itself an ethical claim. To illustrate the difference, Wilson and colleagues contrast the following arguments, noting that Hume would regard the first as invalid and the second as valid. An argument is valid if and only if it is truth preserving – that is, if and only if it is not possible for the premises to be true and the conclusion false. An argument is invalid if and only if it is not truth preserving – that is, if and only if it is possible for the premises to be true and the conclusion false.

Torturing people for fun causes great suffering (factual premise).

Torturing people for fun is wrong (ethical conclusion).

Torturing people for fun causes great suffering (factual premise).
It is wrong to cause great suffering (ethical premise).

Torturing people for fun is wrong (ethical conclusion).

Wilson and colleagues use the fact that one can derive normative conclusions from a set of premises that includes a normative claim to argue that the ethical issues raised by EP's hypotheses should be acknowledged by evolutionary psychologists and engaged.

After noting that "human societies around the world are governed by moral systems that classify behaviors into 'right' and 'wrong' "Wilson and colleagues use EP's hypothesis that rape is an evolutionary adaptation as an example of research whose ethical "implications need to be addressed by evolutionary psychologists" (Wilson et al 2003, 677). Suppose, they argue, that we accept Thornhill's and Palmer's arguments that rape is an adaptation (arguments Wilson and colleagues confess to finding problematic), the notion of common welfare calls for moral evaluation of that adaptation.

[Rape] clearly qualifies as immoral because it benefits the actor at great expense to others, not just the rape victim but society at large. The fact that the actor benefits [in the sense of promoting his reproductive success] does nothing to change its moral status since morality is defined in terms of common welfare. (Wilson et al. 2003, 677)

One might or not might not find Wilson and colleagues' appeal to "the common welfare" convincing, but agree that rape is a behavior that raises serious ethical issues. And Wilson and colleagues are correct that Hume allows that one can validly infer an ethical conclusion from an ethical premise. But does it follow, as they suggest, that evolutionary psychologists should be the ones to grapple with the ethical implications of their research? They are, after all, not normally trained in ethical reasoning or knowledgeable of various ethical theories. This question, although it warrants serious consideration, lies beyond the scope of the present discussion.

In a second line of criticism, Wilson and colleagues argue that, although EP presents itself as Darwinian, it does not take account Darwin's arguments about the evolution in humans of what he called "the ethical sense." Wilson and colleagues take this "failure" to contribute to EP's unwillingness to identify and engage the ethical issues their hypotheses raise. Finally, they note that in contrast to many of the psychological traits EP proposes, Darwin's examples of the traits and behaviors that he attributes to "the ethical sense" are normatively positive – that is, ethically good. In the *Descent of Man*, for example, Darwin states:

I fully subscribe to the judgment of those writers who maintain that of all the differences between man and the lower animals, the ethical sense or conscience is by far the most important. This sense, as Mackintosh remarks, "has a rightful supremacy over every other principle of human action." (Darwin 1871, 66)

Indeed, Darwin adds, "it is summed up in that short but imperious word *ought*, so full of high significance" (Darwin 1871). And he describes the ethical sense as the most noble result of evolution, "leading [man] without a moment's hesitation to risk his life for that of a fellow creature" to sacrifice himself by "the deep feeling of right or duty" (Darwin 1871).

Given the significance Darwin attributed to morality, and that most if not all human societies characterize behaviors as "good" or "bad," Wilson and colleagues ask:

How is it possible for a subject so central to human nature and so clearly recognized by Darwin himself to be neglected by the discipline that claims to explain human nature from a Darwinian perspective? (Wilson et al. 2003, 679).

Wilson and colleagues propose that what they describe as "the curious" failure of human evolutionary psychologists to address the ethical implications of evolved cognitive predispositions, or indeed "the very subject of morality," is explained by EP's adherence to a model of natural selection that assumes it does not operate at the level of groups. Although most of Darwin's arguments focus on individual organisms and their traits, his explanation of the evolution of the ethical sense, as Wilson and colleagues note, assumes group selection: specifically, that groups who exhibited an ethical sense would be more likely to thrive than those that did not.

We think that the answer to [the question we ask of why EP fails to address the subject of morality] can be traced to the fact that Darwin explained the ethical sense as a product of group selection, which most evolutionary psychologists do not accept as an important evolutionary force. (Sober and Wilson, 1998; Wilson et al. 2003, 679)

The status of group selection remains controversial for reasons we can only briefly describe. As Wilson and colleagues note, many evolutionary theorists were convinced by arguments offered in the 1960s that apparent instances of group selection could be better explained on the basis of selection at the level of genes. Although some evolutionary theorists and philosophers of biology are beginning to reevaluate group selection, many share evolutionary psychologists' views that selection only takes place at the level of genes. Wilson and colleagues cite an argument advanced by Thornhill and Palmer that is representative:

One cannot grasp the power of natural selection to "design" adaptations until one abandons the notion that natural selection favors traits that are ethically good and the notion that it favors traits that function for the good of the group. (Thornhill and Palmer 2001, 6)

Wilson and colleagues note that, if one takes "individual self-interest" at the level of genes or individual organisms as "the only explanatory principle," it becomes impossible to give an evolutionary explanation of morality. One must assume that unselfish and/or altruistic behavior are only "apparently so" and that what they actually promote are the self-interests of genes or individual organisms. From this perspective, the behaviors in question (or, in the case of EP, the psychological features in question) are not "good" or "bad" in an ethical sense.

To be fair to evolutionary psychologists, many scientists and philosophers who identify themselves as Darwinists share the assumptions and approaches of EP that Wilson and colleagues criticize. So the obstacles to changing minds about the issues in question are formidable. What the arguments in this section demonstrate is that one can take the goals of EP, and presumably many hypotheses offered in the program, as appropriate and significant, yet criticize its understanding of Hume and the results this has had in limiting EP's scope. And one can accept EP's goals and also criticize its avoidance not only of specific ethical issues raised by its hypotheses but also of morality itself, as unfortunate and unwarranted.

Evaluatively Thick Concepts, Evolutionary Psychology, and the Naturalistic Fallacy

We earlier cited several hypotheses about gender that evolutionary psychologists propose. In this section we consider whether, as evolutionary psychologists maintain, their hypotheses about gender are purely factual – or whether they are, instead, evaluatively thick – that is, they include both empirical and normative content. The idea that there are concepts that include both kinds of content did not originate in feminist scholarship; nor do the critiques feminists offer of hypotheses in EP use the term "evaluatively thick." But I suggest that feminist critiques of many of the concepts and hypotheses of EP research can be understood as arguments that these are evaluatively thick, not purely factual.

To illustrate the idea that there are concepts that are evaluatively thick, we consider some relatively straightforward examples. Consider John Locke's claim in the *Second Treatise of Government* that "all men are equal" (Locke was using "men" to refer to males). He took pains to make it clear that he was not claiming that men are equally intelligent or equally industrious (indeed, he would go on to deny that they are). Rather, he was using the term to claim that no man has a natural or rightful authority over any other. Similarly, when feminists state that women and men are equal, "equal" has both empirical and normative content: there are no differences that render one gender "superior" to the other.

How are evaluatively thick concepts relevant to evolutionary psychology? We begin with the presence, function, and consequences of gender stereotypes in the research program. To describe a term or phrase as a "stereotype" is to call its warrant into question: one might argue that using it to describe a group is unwarranted because not all of its members have the characteristic or trait built into the stereotype, or because it is not an accurate description of any members of the group to which it is attributed. For example, when evolutionary psychologists describe men as less interested than women in caring for their children, they are assuming gender stereotypes. Not all men are less interested than women in caring for their children, and there are women who are uninterested or less interested than some men are in caring for children. Similarly, when women are described as seeking mates able to provide for them and their offspring, a gender stereotype is invoked; not all women seek or need such support.

It is of course true that generalizations to which there are exceptions can be useful in scientific inquiry, provided the exceptions are relatively few in number and/or explainable. But the generalizations EP advances are not of this sort. The exceptions to them are numerous and important, and those exceptions cannot be explained as aberrations from a norm (as white crows can be explained as aberrations from the norm, "All crows are black"). More specifically, we know that variations in parenting arrangements are tied to cultural, generational, socioeconomic, and other variables. These connections pose serious challenges to EP's claim that the "universals" it cites – universals with many exceptions – result from gametic dimorphism.

The examples so far considered show that there is insufficient evidence to warrant the stereotypes we have considered. But gender stereotypes also carry normative content. EP's appeal to gametic dimorphism is far from harmless – as the attribution of a predisposition to parenting to women has long been used to limit their access to education and full participation in all aspects of the public sphere. In addition, EP's stereotypes tend to rule out exceptions to its generalizations, exceptions that suggest that evolutionary explanations of some of the phenomenon EP identifies might be uncalled for. These also significantly shape the reconstructions of the adaptive problems our ancestors faced in the Pleistocene – reconstructions that are necessary to EP's hypotheses about gender-specific predispositions. Arguably, these reconstructions are not warranted for two reasons. First, lacking any direct evidence for how social life was then organized, EP relies on accounts provided by anthropologists in the mid-twentieth century of *contemporary* hunter/gatherer groups. The problem is not just that these groups are not our ancestors and have been affected by cross-cultural influences, including missionaries. Many anthropologists today also recognize the early accounts of contemporary groups on which EP relies as androcentric and ethnocentric. Secondly, recent research into gender dynamics in ancient civilizations, undertaken in archaeology by non-feminists as well as feminists, indicates that divisions of labor by gender are neither universal nor stable throughout human history. So, once again, the gender stereotypes imposed on the Pleistocene by evolutionary psychologists can be viewed to be empirically unwarranted as well as evaluatively thick.

Feminists have also pointed to the presence and role of "gendered metaphors" in evolutionary psychology. A lot of work has been undertaken to define what metaphors are. For our purposes, it is sufficient to understand metaphors as words or phrases that attribute a characteristic or quality appropriately attributed to one kind of object to a different kind of object to which it is not literally applicable – for example, we speak metaphorically when we say that genes are "selfish" or that someone "is drowning in paperwork." Metaphors are not only common in science, but many take them to facilitate scientific theorizing. They are viewed as unproblematic provided that they are recognized as metaphors rather than literal and empirically appropriate descriptions.

Among the gendered metaphors feminists identify in evolutionary psychologists' reasonings, methods, and hypotheses are some we encountered earlier, including "expensive" eggs and "relatively cheap" sperm. The notion of "parental investment," and that of different levels of expense in relation to eggs and sperm, apply economic concepts and properties to women's and men's biological contributions to conception and birth. Men's contribution (metaphorically described as their "investment") is argued to be small given that they have many sperm, only one of which is needed for conception, and that they are able to reproduce many offspring by engaging in sex with a large number of women. From the outset, women's contributions (again metaphorically described as their "investment") are described as substantially larger. The average adult woman has about 400 eggs and these are much larger in size (and, it has been argued, require more energy to sustain) than sperm. In the case of humans and other mammals, females' biological contributions to offspring, unlike males', do not end at conception but include a long period of gestation and lactation. Moreover, when they are pregnant, women cannot further their reproductive success by engaging in heterosexual sex. As earlier noted, EP argues that these differences explain gender differences in terms of predispositions to childcare. Women's greater investment, evolutionary psychologists maintain, explains their interest in taking care of offspring; men's smaller investment, and their ability to impregnate many women, explains their being less interested in providing such care. The metaphors that motivate hypotheses about differences in women's and men's interests in offspring are not only not literal, empirical descriptions; they also include normative content. "Expensive" and "cheap" have normative content, as do "lesser" and "greater" investment in offspring. This normative content carries over to evolutionary psychologists' explanations of gender-differentiated mating and parenting strategies because they fail to consider alternative, sociological explanations of whatever gender differences there are in parental investment.

Finally, we consider the hypothesis that domestic violence is the result of men's predisposition to sexual jealousy and that such jealousy is the result of their uncertainty about paternity. Is it purely factual or is it evaluatively thick? Feminists argue that the concept of "domestic violence" assumed in this hypotheses is inappropriately limited. For example, Heather Douglas notes that this conception incorrectly assumes that it involves heterosexual couples that are focused on having children (Douglas 2011). But she and other feminists point out that domestic violence occurs in same-sex relationships and in relationships in which partners are not interested in or are unable to have children. I suggest that hypotheses that domestic violence is the result of predispositions that are adaptations carry normative as well as empirical content. The normative content here is that heterosexual couples that want and are able to have children are the preferred model for relationships in which domestic violence occurs. In fact, they are not the only individuals among whom domestic violence occurs so there are likely other causes of such violence that evolutionary psychologists fail to explore.

If feminists are correct that hypotheses in EP assume gender stereotypes and gendered metaphors, if it is appropriate to describe these things as evaluatively thick, and if some of the hypotheses themselves are evaluatively thick, then feminists' critiques do not commit the naturalistic fallacy. They do not do so because they argue from premises that they take to have ethical content, and ethical conclusions can be drawn from such premises. Nor, it follows, do evolutionary psychologists who advance hypotheses about gender commit the fallacy – albeit, for a different reason. It is not because they do not draw normative conclusions from purely factual hypotheses. It is because the hypotheses in question are not purely factual.

Conclusion

The foregoing discussion makes it clear that the issues involved in clarifying the relationships between the naturalistic fallacy, evolutionary psychologists' hypotheses about gender, and feminist critiques thereof are complex. Readers may view some of the arguments considered to be unsuccessful. But if the arguments encourage others to engage one or more of the issues involved, I will be satisfied. Evolutionary psychology is taught at many colleges and universities, publishes books addressed to the lay public, and has been largely dismissive of feminist critiques of it. So the issues here considered warrant attention.

19 A Theological Evaluation of Evolutionary Ethics

Michael L. Peterson

Evolutionary theory provides important insights into the nature, content, and function of morality. Here I assess Evolutionary Debunking Arguments (EDAs) which cite evolutionary theory in their attack on metaethical realism and support of meta-ethical anti-realism. I show that the force of any anti-realist argument – whether epistemological or metaphysical – is ultimately driven by worldview commitments regarding the nature of reality, knowledge, and the human person rather than by the evolutionary facts. I argue that philosophical naturalism is typically the worldview behind anti-realist metaethical positions but that it has weaknesses which have always made morality problematic and thus predictably make morality problematic in connection to evolution. I argue an alternative thesis: that a theistic worldview which accepts evolution can support a version of metaethical realism that provides a more adequate explanation of morality.

Naturalism, Theism, and Science

Conflicts in metaethics between anti-realism and realism should not be categorized too strictly as either epistemological (pertaining to whether moral knowledge is justified) or metaphysical (pertaining to whether there are moral facts) because questions about moral epistemology inevitably raise questions about moral ontology. Most evolutionary anti-realist approaches are rooted, implicitly or explicitly, in philosophical naturalism. However, the inadequacies of naturalism have always made morality problematic and predictably make morality problematic in connection to evolution.

Naturalism rests on several broad assumptions. Central to naturalism is the metaphysical commitment that only physical nature is real. As Ernest Nagel says, naturalism is belief in "the executive and causal primacy of organized matter in the executive order of nature" (Nagel 1954, 9). Naturalism denies the existence of the supernatural or divine and thus entails atheism. Naturalists are typically materialists about human persons, which means that versions of physicalism prove attractive. For several decades, attempts have been made to "naturalize" key areas of philosophy, making them continuous with the natural sciences, emulating scientific method, referring only to scientifically respectable objects, and the like – such as Quine's call for a "naturalized epistemology" (Quine 1969).

Naturalists are convinced that an adequate account of morality can be given within the framework of philosophical naturalism. Simon Blackburn explains that the naturalist about ethics "asks no more than this: a natural world, and patterns of reaction to it" (Blackburn 1984, 182). Science, then, supplies the needed details. Since E. O. Wilson declared that ethics should be "removed temporarily from the hands of the philosophers and biologicized" (Wilson 1975, 562), the field of evolutionary ethics has undergone significant development. Let us call the conjunction of naturalism with evolutionary theory "evolutionary naturalism."

Theism is an important philosophical alternative to naturalism. Standard theism is the belief that there is a supreme personal being, God, who is omnipotent, omniscient, and perfectly good, and who created, sustains, and interacts with the universe and all it contains. The physical universe is a creation of this infinite personal-rational-moral being; this universe is rich and complex, but not ultimate. Theism is generally positive toward science as an avenue of inquiry into nature, including evolutionary science and its factual findings. Let us call this view "evolutionary theism" or "theistic evolutionism."

Evolutionary Epistemology and Anti-Realism

Most evolutionary debunking arguments (EDAs) are epistemological, advancing an evolutionary explanation to challenge the justification or epistemic standing of our moral beliefs. According to E. O. Wilson and Michael Ruse,

[w]hat Darwinian evolutionary theory shows is that this sense of "right" and the corresponding sense of "wrong," feelings we take to be above individual desire and in some fashion outside biology, are in fact brought about by ultimate biological processes. (Ruse and Wilson 1986, 179)

Ruse explains that the pressures of natural selection have exerted enormous influence on human psychology, including the hardwiring of epigenetic rules, giving us widely distributed propensities to believe and behave in certain ways that have "adaptive value" (Ruse 1998, 223). Likewise, Richard Joyce argues that it was "useful for our ancestors to form beliefs concerning rightness and wrongness independently of the existence of rightness and wrongness" (Joyce 2006a, 183). The epistemic disconnect between our moral beliefs and any would-be truth-makers for those beliefs results in moral skepticism. This is the morality – naturalized by being biologicized – that Wilson would return to the philosophers.

Although various commentators have analyzed the structure of different EDAs in detail (e.g., Kahane 2011), a simple model will serve present purposes. Every epistemological EDA contains at least three types of premises:

1. A causal premise stating that evolutionary forces shaped our moral capacities and/or beliefs.
2. An epistemic premise asserting that evolutionary forces are concerned with adaptive fitness, not with moral truth.
3. A probability premise maintaining that it is highly unlikely that our moral beliefs are in touch with anything that might be called "moral truth."

Then,

4. A conclusion holding that our moral beliefs do not constitute knowledge because they are unjustified.

The causal premise reflects only one of many causal stories that have been advanced in the history of philosophy to debunk the traditional understanding of morality, as well as to debunk theories of rationality and free will. The premise is not straight science but rather an extrapolation from evolutionary theory – a "how possibly" story about how the ethical project "might" have proceeded, as Philip Kitcher says (Kitcher 2011, 239). The implicit thought is that morality is a purely natural phenomenon – a naturalist commitment.

The epistemic premise does not state that evolutionary forces make us incapable of true moral beliefs but claims that there is no reason to expect that evolutionary forces on their own would make us capable of grasping objective moral truth. The underlying assumption is again naturalistic: that blind evolutionary forces are the only causal factors relevant to whether we can know moral truth. Hence, the common theme echoing through EDAs is that an evolutionary explanation provides a sufficient and total account of moral beliefs.

Generated from premises 1 and 2, the probability premise estimates the improbability that evolutionary forces per se would be concerned with anything like moral truth. Sharon Street states: "[T]he realist must hold that an astonishing coincidence took place – claiming that as a matter of *sheer luck*, evolutionary pressures affected our evaluative attitudes in such a way that they *just happened* to land on or near the true normative views among all the conceptually possible ones" (Street 2008a, 208–209; emphasis mine). Clearly, this probability estimate is made within a naturalistic framework that does not envision any other relevant factors.

The conclusion is that our moral sense is not trustworthy to deliver truth and, thus, that our moral beliefs are unjustified. But the denial of moral knowledge by Street and other epistemological debunkers is filtered at every step through a naturalist metaphysical perspective.

Contemporary Epistemology and Evolutionary Debunking

Work in epistemology over the past few decades has clear bearing on epistemological EDAs. EDAs are said to provide an epistemic defeater for our moral beliefs by supplying a reason to question the reliability of our grounds for holding those beliefs (see Pollock and Cruz 1991, 196–197). Such an argument is an undercutting defeater – a reason why the relevant belief-forming capacity or process is unreliable in producing true judgments. On the other hand, a rebutting defeater attacks the truth of a belief, providing a reason for holding the belief's content to be false.

Two aspects of defeaters modulate our assessment of EDAs. First, all defeaters are person-relative: each person has a total set of beliefs, making up his or her noetic structure, which factor into whether he or she feels that a proffered defeater has probative force. If a given person feels the force of a defeater for the belief that p, then that person needs to come up with a "defeater defeater" to continue being warranted in believing p. Contrary to common claims for EDAs, there really is no such thing as a defeater simpliciter.

Second, the notion of background beliefs applies to the debunker's premise that, given the evolutionary story, the probability is low that our moral faculties are reliable for truth. The premise invokes antecedent probability, a notion of what we would reasonably expect given certain assumptions. So, where Pr is probability, M is the claim that our moral capacity is reliable in delivering moral truth, and E is the relevant evolutionary story, the debunker is claiming that

$$\Pr(M/E) \text{ is low.}$$

Read: "the probability of M given E is low" or "the probability of M on E is low." But this claim is not at all obvious. At best, the debunker could be taken as saying, "all else we know held equal, the probability of M on E is low." But why think this? M on E doesn't just have a probability value. Wittingly or unwittingly, the debunker brings his total set of beliefs to these evaluations. Besides, ideally, we want to arrive at our considered judgments based on "total evidence."

The debunker's assessment of the low probability of M on E is very much a function of his background beliefs, which include naturalist commitments – that science provides the only acceptable type of explanation, that the world contains only natural facts, and so forth. Thus, it is not the putative scientific facts regarding the evolutionary shaping of morality but the philosophical interpretation of them that drives the debunking argument. Debunkers such as Street are mistaken in insisting that the burden of proof falls on the realist (Street 2006). Realists bear no special burden of proof, particularly if their noetic structures contain the belief that science does not offer the total explanation of reality or that morality has an objective quality or that God has guided evolution to condition us to know moral truth.

Puzzling Darwinian Counterfactuals

Tamler Sommers and Alex Rosenberg explain why moral skepticism results from epistemological EDAs: "[I]f our best theory of why people believe P does not require that P is true, then there are no grounds to believe P; therefore there are no grounds to believe P is true" (Sommers and Rosenberg 2003, 667). The "problem with ethics," as Gilbert Harman says, is that moral facts are explanatorily irrelevant in a way that natural facts are not. Harman indicates that, to explain moral beliefs, "You need only make assumptions about the psychology or moral sensibility of the person making the moral observation" (Harman 1977, 6).

Charles Darwin thought that the specifics of our evolutionary history determine our moral beliefs. Mark Linville finds this idea expressed in what he calls the "Darwinian Counterfactual" (Linville 2012, 403):

If … men were reared under precisely the same conditions as hive-bees, there can hardly be a doubt that our unmarried females would, like the worker-bees, think it a sacred duty to kill their brothers, and mothers would strive to kill their fertile daughters, and no one would think of interfering. (Darwin, 1871, 73)

Further discussion pertains to whether moral judgments are best interpreted as hard-wired or as modified by reflective equilibrium and cultural advance. But it is not surprising that evolutionary debunkers speak of our moral beliefs as being insensitive to objective moral facts.

My present interest is what I dub the Darwinian Epistemological Principle:

> (D) If human morality is a result of natural selection, operating according to reproductive advantage, without regard to truth, then there is no moral knowledge.

Even if there is objective moral truth, our moral beliefs are instead aimed at biological fitness. Hence, debunkers endorse what we may call the Independence Thesis:

> (I) The process of forming moral beliefs is independent of any would-be truth makers for the beliefs formed.

However, this overreaching thesis is much harder to establish than one might imagine.

Elliott Sober has pointed out what is required to establish independence. The debunker would first have to identify (a) all of the processes of moral belief formation and (b) the would-be truth-makers for moral beliefs, and then show that (a) and (b) are independent (Sober 1994, 93–113). This daunting project in metaethics is not guaranteed by any "how possibly" story and depends heavily on naturalist background beliefs to arrive at the slim odds that adaptiveness and truth would coalesce. All of this sets the stage for begging the question against the moral realist.

We have to be careful with the logic here. It is one thing to suggest that there are reasons for suspecting epistemic independence. But it is quite another thing to assert that there are no reasons to think that a relevant dependence relation obtains between the causes of our moral beliefs and whatever would make them true. Nicholas Sturgeon notes that thoughtful people commonly think that moral facts are indeed good explanations of their moral beliefs and that "many of these explanations look plausible enough on the surface to be worth taking seriously" (Sturgeon 1988, 239).

Metaethical epistemological anti-realism generates another odd and troubling Darwinian counterfactual regarding our belief about Joseph Stalin's extermination and suppression of millions of people:

(S1) If Stalin had done just what he did but was not morally depraved, we still would have believed that he was depraved.

The metaethical evolutionary anti-realist explains this and all our moral beliefs by reference to natural facts about the adaptive pressures shaping them rather than to objective moral facts and our ability to access them.

Neither does the best metaethical anti-realist explanation of Stalin's actions involve his actual depravity, leading to another puzzling counterfactual:

(S2) Even if Stalin were not morally depraved, he would have done just what he did.

Stalin's natural properties – including abuse by his father, depression, paranoia, alcoholism, Marxist views, and aim to form a perfect communist state – are taken as sufficient to explain his actions, without reference to any moral properties, such as depravity.

Yet, we would ordinarily explain Stalin's depravity – provided we do not suspect or know of mitigating factors bearing on our moral evaluation – as involving his particular volitional response in the context of the relevant natural facts, while acknowledging that someone else might have responded in a more morally approvable way in the same natural context. Here we see that the epistemological debunking strategy relies on the underlying metaphysical assumption that there simply are no moral properties. But this metaphysical view cannot simply be assumed in support of epistemological anti-realism since it is a key point at issue. We must now face metaphysical matters directly.

Metaphysical Anti-Realism and Error Theory

As Hastings Rashdall observed over a century ago:

So long as he is content to assume the reality and authority of the moral consciousness, the Moral Philosopher can ignore Metaphysic; but if the reality of Morals or the validity of ethical truth be once brought into question, the attack can only be met by a thorough-going enquiry into *the nature of Knowledge and of Reality*. (Rashdall 1907, 192; emphasis mine)

Not only do epistemological EDAs inevitably rely on metaphysical assumptions, but also there are EDAs that are officially metaphysical.

In 1977, J. L. Mackie famously originated the modern error theory, "There exist no objective values." His explanation is straightforward:

[T]hat values are not objective, are not part of the fabric of the world, is meant to include not only moral goodness, which might be most naturally equated with moral value, but also other things that could be more loosely called moral values or disvalues – rightness and wrongness, duty, obligation, an action's being rotten and contemptible, and so on. (Mackie 1977, 15)

If there were objective values, Mackie continues, "they would be entities or qualities or relations of a very strange sort, utterly different from anything else in the universe" (Mackie 1977, 38). Objective values would be "metaphysically queer" – and they could be known only by some "special faculty of moral perception or intuition" that is itself weird and unlike any other mode of knowledge.

Joyce, Street, and Ruse incorporate evolution into their renditions of error theory (Joyce 2006a; Street 2006; Ruse 1998). Although their EDAs clearly have an epistemological dimension, involving evolutionary explanations of why we have our moral beliefs, their fuller views include the denial of objective moral truth to which our beliefs can be connected. This metaphysical moral anti-realism continues the tradition of Mackie and Harman.

Michael Ruse's work on error theory, spanning several decades, is illustrative. Ruse and Wilson declare that "ethics as we understand it is an illusion fobbed off on us by our genes in order to get us to cooperate" (Ruse and Wilson 1989, 51). According to Ruse,

[t]he Darwinian argues that morality simply does not work (from a biological perspective), unless we *believe* that it is *objective*. Darwinian theory shows that, *in fact*, morality is a function of *(subjective) feelings*; but it shows also that we have (and must have) the *illusion* of objectivity. (Ruse 1998, 253; emphasis mine)

Ruse defends the epistemological independence of moral beliefs from objective moral facts because evolution has produced in us certain useful subjective responses that are the actual ground of morality. Yet Ruse's explanation of why we have the moral beliefs we do is connected to his rejection of ontologically objective moral facts.

Even if the weaker epistemological version of the Independence Thesis could be established, its truth would not entail that moral beliefs are "in fact" false. At best, an evolutionary explanation of the process of moral belief formation might raise caution that we should treat moral beliefs "as if" they are false in the absence of any reason to think they have an epistemic relation of dependence on their truth conditions. Remembering Sober's point, it's highly unlikely that the metaethical project of establishing epistemological independence can be accomplished anyway.

Ruse's stronger ontological interpretation of independence is that there simply are no truth-makers for our moral beliefs. As he says, there is no "*independent, objective, moral code* – a code which ultimately is unchanging and not dependent on the contingencies of human nature" (Ruse 1989, 269; emphasis mine). The absence of moral properties or facts means that there are no objective moral truths to be epistemically acquired. Metaphysical EDAs maintain that the content of our moral judgments, construed literally in realist terms as being independent in important ways from our attitudes or stance, are false in toto.

Ruse's "evolutionary naturalism" – as well as Street's "value naturalism" – puts metaphysical naturalism on full display. The world as generally understood by the Darwinian naturalist produces or contains only natural properties or facts, which entails that there are no moral properties or facts. As Joyce says of the evolutionary story: "It was no background assumption of that explanation that any actual moral rightness or wrongness existed in the ancestral environment" (Joyce 2006a, 183). Such assertions about the ontology of the evolutionary world generate yet another Darwinian counterfactual:

(S3) If no moral properties exist, such as depravity, then Stalin was not actually depraved, but we would still believe that he was.

Although there are no moral properties, assuming that the evolution of the human moral capacity remains the same, we would react morally in the same way to Stalin's natural properties.

Forms of moral realism have been offered to counter the typical anti-realist orientation of naturalist metaethics. Some naturalists have proposed realist metaethical theories, maintaining that there are objective moral properties or facts but that they are identical to or reducible to natural properties or facts (e.g., Jackson 1998). Other realist metaethical theories have been developed that are nonnaturalist, positing moral properties or facts that are sui generis, irreducible to natural properties or facts (e.g., Sturgeon 2002, 190). Some moral nonnaturalists are philosophically naturalistic or nontheistic (e.g., Wielenberg 2005), while others are theistic (e.g., Wainwright 2010).

Both naturalist realisms and nontheistic nonnaturalist realisms encounter serious difficulties. Problems for moral naturalisms include their inability to explain the intrinsic worth of human beings and the normative force of moral judgments. Moral nonnaturalist realisms face difficulty in explaining what nonnatural "brute" properties and facts are as well as their supposedly "supervenient" relation to natural facts. Some form of robust metaethical realism that can make full sense of morality must be sought elsewhere.

Metaethics in Light of Meta-Philosophy

Considering the unacceptable metaethical consequences of anti-realist EDAs, as well as the metaethical difficulties of both naturalist and nonnaturalist nontheistic realisms, we might ask whether there are any constraints on metaethical theorizing. A prior question is, are there are any constraints or guidelines on the doing of philosophy generally? It seems unexceptionable that any philosophical theory or piece of philosophical analysis should strive to make sense of (a) the relevant scientific facts, including evolutionary facts, and (b) our ordinary beliefs that can be deemed basic and common to the human race, such as belief in the existence of an external world and other minds. This is a fair-minded and auspicious philosophy of how to do philosophy in general – indeed it is a meta-philosophy. A philosophical theory or piece of philosophical analysis that satisfies both (a) and (b) is rationally preferable to a philosophical theory or piece of philosophical analysis that does not.

To put a finer point on this meta-philosophical approach, let us state just three criteria it entails. A philosophical theory or piece of philosophical theorizing

1. must not contradict the truth values of ordinary propositions;
2. must not contravene the meaning of ordinary propositions discernible to a nonphilosopher; and
3. should explicate the implicit ontology of ordinary propositions, which when articulated is competitive with alternative ontologies.

The concept of ordinary proposition here refers to a commonsense proposition that is virtually universal – that is, a proposition that is either evident or self-evident to all normal people who consider it. But the concept would not include, for example, propositions that reflect group prejudices or revisable factual descriptions. Thomas Reid, for example, followed this meta-philosophy in defending perceptual judgments against Berkleyan subjective idealism and perceptual and moral judgments against Humean skepticism. G. E. Moore largely followed a commonsense realist approach in supporting basic perceptual beliefs against Bradleyan absolute idealism. For present purposes, this realist meta-philosophy provides guidance as we pursue a philosophical explanation of our moral beliefs and convictions that is more adequate than all anti-realist approaches as well as the various realist approaches that in aggregate change the truth value, meaning, or implicit ontology of our moral beliefs.

The meta-philosophy sketched previously takes with utter seriousness the common beliefs of the human race – such as the existence of an external world, the past, and so forth – as well as the general reliability of the rational powers forming these beliefs, such as perception, memory, and so forth. Humans are implicitly realist about such beliefs – a fundamental theme in Aristotle, Aquinas, Thomas Reid, and, to some extent, G. E. Moore. A key idea here is that our native belief-forming capacities produce beliefs that are so fundamental epistemically that no philosophical interpretation of them that violates their truth value, meaning, or implicit ontology can be more rationally acceptable than they are. Indeed, such violation is a signal that something is wrong with the philosophical interpretation – a point we will pursue in regard to EDAs and the metaethical debate they initiate. The contemporary renaissance of realism in many areas of philosophy, such as metaphysics, epistemology, philosophy of science, and philosophy of religion, displays the philosophical power of this insight.

The renewal of realism in ethics, including in virtue ethics, natural law ethics, and metaethics, has been evident since the late 1970s (e.g., Sayre-McCord 1988; Foote 2001; Budziszewski 2011). In discussions of evolutionary ethics, then, it is reasonable to expect that metaethical theorizing should seek to fulfill the previously stated twin goals of taking into account scientific facts about morality and our ordinary understanding of the nature of our moral beliefs. Metaethical anti-realists explain away ordinary moral beliefs as either unjustified or false, leaving a morality unrecognizable to the common person. Ruse's error-theoretic position, for example, contradicts the truth value of our most basic ethical propositions and installs an alien naturalistic ontology that recognizes no moral properties or facts. Interestingly, although Ruse locates the epistemic root of morality in a subjective response shaped by evolution, another part of his metaethical view is cognitivist (rather than non-cognitivist) regarding the common meaning of moral language. Moral statements are capable of being true, but they aren't true.

Ruse's naturalistic worldview inclines his anti-realist metaethics toward science as the only mode of knowledge and the objects in its domain as exhausting what there is. Yet Ruse's position involves not just evolutionary science but an interpretation of the science based on naturalistic assumptions, which results in violations of meta-philosophical criteria that all good philosophizing should meet. When its objections to realism are defanged and its own naturalist assumptions revealed, evolutionary metaethical anti-realism is hardly a devastating threat to metaethical realism. Following the meta-philosophical approach recommended here in metaethics, it is clear that our common moral experience provides good reason to be realists. Put simply, metaethical moral realism, which embraces the relevant evolutionary facts and elucidates ordinary morality, is rationally preferable to anti-realist theories that do not. Even naturalist and nonnaturalist nontheistic realisms that seek to preserve the truth value of ordinary moral judgments still violate in their own ways the implicit ontology or meaning of moral judgments.

A Theistic Genealogy of Morals

According to Darwinian naturalists who offer EDAs, a basic genealogy of morality goes as follows. Given the contingencies of the evolutionary landscape, it is not simply certain organic structures that are adaptive but certain behaviors as well. Insofar as adaptive behaviors are genetic, they are heritable. Through genetic variation, some early hominids developed something approximating a moral sense leading to certain behavior. This slight distinction enabled them to survive and reproduce at a rate higher than the rate of hominids without this feature. The human moral sense – what Darwin called the "social instincts" – includes the ability to feel moral sentiments and make moral judgments. Since the natural world is our only frame of reference, human morality as we know it must be seen as ultimately produced by natural selection rather than based on objective norms and the ability to discern them.

Different evolutionary naturalists expand or modify this schema in navigating such issues as whether morality is a direct or indirect product of biological evolution or whether rational reflection and cultural evolution change the evolutionary formula. Yet the metaethical anti-realism remains: (a) morality is completely a result of natural selection, and (b) certain construals of this claim about morality raise epistemological and metaphysical objections to realist metaethical views. Unfortunately, Darwinian evolutionary theory conjoined with philosophical naturalism fails to provide a worldview framework within which an adequate metaethical view can be developed because it makes moral reality and our ability to know it problematic. Of course, a large number of important phenomena of life and the world are also made problematic on naturalism, which projects a universe that is fundamentally nonrational, nonpersonal, and nonmoral. No wonder perspectives that combine naturalism and evolution have to stretch Darwinian principles to try to account adequately for rationality and personhood in addition to trying to account for morality.

Theism combined with evolutionary theory implies its own genealogy of morals. God somehow guided natural processes within creation to achieve various higher purposes, including the development of reliable moral capacities in human creatures. Contrary to Street's low probability assessment that truth and adaptiveness coalesce (Street 2006, 13), the odds in a theistic evolutionary world are exactly 100 percent. Robert Adams states:

> If we suppose that God directly or indirectly causes human beings to regard as excellent approximately those things that are Godlike in the relevant way, it follows that there is a causal and explanatory connection between facts of excellence and beliefs that we may regard as justified about excellence, and hence it is in general no accident that such beliefs are correct when they are. (Adams 1999, 70)

Theism essentially identifies a "tracking relation": we have the basic moral beliefs we do because they are true, and our moral faculty is reliably truth aimed.

On evolutionary theism, moral beliefs possess reproductive fitness because they are true; fitness is not the only reason we have them. This is the core of a theistic answer to Street's question about whether the realist can identify a relationship between evolutionary forces that guided our evaluative views and the values that exist independently of those views (Street 2006). There is no necessary dichotomy between fitness and truth. The adaptive value of a human father caring for his children is unquestionable on pragmatic grounds, but it is also objectively true that fathers ought to care for their children regardless of the adaptive value of doing so.

The theistic tradition contains ample conceptual resources for explaining how God could act to allow fitness-related factors to shape reliable moral capacities. For example, the Thomistic view of God as the primary cause working with creaturely secondary causes allows for scientific and theological explanations to be complementary. I have argued elsewhere that, under these categories, evolutionary science studies secondary causes and gives mechanical explanations, while theology puts these matters in larger perspective in terms of God's purposes (Peterson and Ruse 2017).

A Worldview Home for Morality

Street advances her view as a "scientific hypothesis" or "scientific explanation," which is the preferred explanation because it is supposedly the most parsimonious in not invoking additional entities or qualities (Street 2006). A quick refresher in philosophy of science, however, reminds us that there are various criteria for theory choice and that parsimony is recommended only if all other relevant explanatory factors are equal. In discussions of the bearing of evolution on our understanding of morality, all other explanatory factors are not equal because morality purports to involve realities beyond the purview of science, such as values, rights, and duties, and the capacity to form true beliefs about them. There is no nonquestion-begging way for naturalism – including its totalizing claims for science – to exclude these items as having no explanatory relevance.

A complete explanation of morality is not scientific but philosophical. In a manner of speaking, morality is an important human phenomenon looking for a suitable worldview home. In secular Western culture, naturalism and theism continue as rival worldview competitors. Now I ask: In what kind of universe does morality make best sense? In a universe described by a naturalist worldview? Or in a universe described by a theistic worldview? Naturalism combined with evolution envisions a universe in which the common understanding of morality is epistemologically and metaphysically problematic. For those who argue that the physical is the fundamental reality, morality must ultimately be explained by the nonmoral – as must rationality be explained by the nonrational and personhood by the nonpersonal.

Evolutionary theism, by contrast, entails that rationality, personhood, and morality are integral parts of a total, coherent, and meaningful universe. God created a physical and biological system that eventually produced creatures complex enough to embody personal-rational-moral life. Christianity and Judaism, for example, teach that humans are created "in God's image" and are intimately connected to the material world. What has happened in the evolution of the living world – the emergence of life, and then conscious life, and then self-conscious rational and moral life – reveals an unfolding teleology that makes far better sense on theism than it does on naturalism.

Theism affirms that personhood, with all of its powers and properties, is ontologically fundamental in the order of reality, not derivative. God is the supreme person, the ultimate rational-moral-relational being. God and humans share an overlap of kind membership in personhood itself. Human dignity is found precisely in membership in that kind: personhood just is the kind of reality that is intrinsically and unconditionally valuable. "On theism," explains Mark Linville, "human persons have been fashioned, in one morally relevant respect, after the most ultimate and sacred feature of reality and thus participate in that sacredness" (Linville 2012, 445). Interestingly, if humans were not biological beings, some moral considerations would not apply to us – say, regarding sexual behavior or gratuitously inflicting physical pain. But even our biology belongs to a certain kind of personal being – a rational moral animal of exquisite worth. The personal and all it entails find a fitting metaphysical home in a theistic universe.

A theistic universe is also an epistemological home for rational moral beings. From the perspective of contemporary epistemology, all of the elements of warrant are present in a theistic universe – generally reliable cognitive capacities, such as reasoning and perception and memory; a suitable environment for the operation of these capacities; a sufficient degree of confidence in our beliefs; and so forth (Kvanvig 1992). In regard to moral beliefs specifically, it is not simply, as Darwin indicated, that rationality adjudicates among instincts rooted in our animal ancestry but that our rationality elevates human morality far above even the highest instincts of our animal ancestors. We have warranted moral beliefs about the truths of moral reality. There is indeed a dependence relation between our moral beliefs and their truth-makers such that genuine moral knowledge is possible.

Properly interpreted meta-philosophically, our moral experience provides good reason to be realist. But not just any version of realism makes equally good sense of morality. Moral objectivity cannot be based solely on natural facts about humankind and its environment or on an independent realm of abstract moral properties or facts. Moral anti-realism cannot be effectively countered by realist theories emanating from naturalist or nonnaturalist nonthestic visions of reality because neither can affirm the fundamental ontological status of personhood. Theistic moral realism – which grounds the objectivity of moral norms in the personal and relational nature of reality itself rather than in, say, decrees of the divine will – is an effective and superior alternative.

Interestingly, the realist account of morality I'm advocating clearly falls under the rubric of naturalized ethics – utilizing relevant findings from the natural sciences to supplement and refine our common understanding of morality. Ironically, naturalized ethics (just like naturalized epistemology) flourishes much better within the context of supernatural theism than in the context of metaphysical naturalism (Plantinga 1993, 217–237). The theistic worldview – which is both scientifically informed and profoundly realist – underwrites moral life.

Theistic realism philosophically explains morality better than anti-realist naturalism does. "To explain" here may be defined by the previous notion of antecedent probability. Antecedently, morality as we know it from evolutionary science and common experience is more likely on the assumption that theism is true than on the assumption that naturalism is true. Where Pr is probability, M is the proposition that morality is reliable for moral truth as commonly understood, T is theism, N is naturalism, and E is the evolutionary facts, my contention is that

$$Pr(M/T\&E) > Pr(M/N\&E).$$

Theism also explains morality better than naturalist and nonnaturalist nontheistic versions of realism. Theistic moral realism follows the most reasonable meta-philosophical approach in projecting a total, coherent view of reality in which morality and all its features find a fitting home.

Morality occurs within a very particular kind of world that contains rationality, personhood, and many other important phenomena. Naturalism, whether anti-realist or realist, cannot adequately explain these phenomena. Take rationality. Where R is the claim that rationality is reliable for truth,

$$Pr(R/T\&E) > Pr(R/N\&E).$$

See Alvin Plantinga's devastating argument against Daniel Dennett's attempt to explain rationality in terms of evolutionary naturalism (Plantinga 2011, 16–21).

Consider personhood. Where P is the claim that personhood is fundamental and intrinsically valuable,

$$Pr(P/T\&E) > Pr(P/N\&E).$$

We could reframe the comparative antecedent probabilities of naturalism and theism in terms of "epistemic surprise." Thus, it is more epistemically surprising (because less probable) that the world would contain the fundamental reality of personhood given naturalism than it is given theism. On naturalism, morality and rationality are very surprising as well.

Since the debate over the impact of evolution on metaethics prominently involves science, we must remember that science can only exist in a certain kind of world. Science must assume, but it cannot by its own methods establish, that there is a lawlike, intelligible world and that rational inquirers can investigate and know it with increasing accuracy. Where S is science,

$$Pr(S/T) > Pr(S/N).$$

Even science itself must rest on philosophical foundations – foundations that are more obviously entailed by a theistic worldview than by a naturalist worldview (Peterson 2012, 301; Peterson and Ruse 2017, ch. 2).

As long as approaches to evolution and ethics assume a naturalistic outlook, whether anti-realist or realist, no adequate account of morality is possible. For naturalism, the physical universe is the ultimate brute fact, an inexplicable surd, that serves as the basis for a philosophical account of everything – not just of morality but also of rationality, personhood, science, and other important phenomena. Even ethical nonnaturalism, which attempts to be compatible with either naturalism or theism, posits a realm of abstract objects that are themselves brute or surd, with no coherent account of their existence, let alone of the existence of personhood and its significance. Consider Wielenberg's hopeless nonnaturalist thesis: "From valuelessness, value sometimes comes" (Wielenberg 2009, 40 n68).

Philosophical explanations of basic features of life and the world inevitably entail worldview commitments about the character of reality. A given view of ultimate reality, in turn, grounds the explanation of everything else. In the comparison between theism and naturalism, God as ultimate provides an explanatory framework that is much more enlightening than nature as ultimate. The intellectual poverty of naturalism in accounting for the whole of reality – of which morality is a part – is obvious. To borrow familiar language from the debate over DBAs, for a naturalistic universe to contain such things as morality, rationality, and personhood, and even such a thing as science, is an "astonishing coincidence" indeed!

Bibliography

Abbot, Patrick, Jun Abe, John Alcock, Samuel Alizon, Joao A. C. Alpedrinha, Malte Andersson, Jean-Baptiste Andre, et al. 2011. "Inclusive Fitness Theory and Eusociality." *Nature* 471 (March 24): E1–E4.

Abend, G. 2010. "What's New and What's Old about the New Sociology of Morality?" In *Handbook of the Sociology of Morality,* eds. S. Hitlin and S. Vaisey, 561–584. New York: Springer.

Adams, Robert M. 1999. *Finite and Infinite Goods.* Oxford: Oxford University Press.

Adams, R. S., et al. 1995. "Calculating Drinking Water Intake for Lactating Cows." *Dairy Reference Manual* (NRAES-63). Ithaca, NY: Northeast Regional Agricultural Engineering Service.

Addams, Jane. 1893. "The Subjective Necessity for Social Settlements." In *Philanthropy and Social Progress: Seven Essays*, ed. Henry C. Adams, 1–26. New York: Thomas Y. Crowell.

    1902. *Democracy and Social Ethics.* New York: Macmillan.

    1907. *Newer Ideals of Peace.* New York: Macmillan.

Alcock, John. 2003. *The Triumph of Sociobiology.* New York: Oxford University Press.

Alexander, R. 1985. "A Biological Interpretation of Moral Systems." *Zygon* 20 (1): 3–20.

Alexander, Samuel. 1889. *Moral Order and Progress: An Analysis of Ethical Conceptions.* London: Trübner.

    1920. *Space, Time and Deity.* London: MacMillan and Co.

Allee, Warder Clyde. 1927. "Animal Aggregations." *Quarterly Review of Biology* 2: 367–398.

    1938. *The Social Life of Animals.* Norton: New York

    1940. "Concerning the Origin of Sociality in Animals." *Scientia* LXVII: 154–160.

Allen, Elizabeth, Barbara Beckwith, Jon Beckwith, Steven Chorover, David Culver, Margaret Duncan, Steven Gould, et al. 1975. "Against Sociobiology." *New York Review of Books* 22: 43–44.

Allhoff, Fritz. 2003. "Evolutionary Ethics from Darwin to Moore." *History and Philosophy of the Life Sciences* 25: 51–79.

Anderson, Elizabeth. 1991. "John Stuart Mill and Experiments in Living." *Ethics* 102: 4–26.

    1998. "Pragmatism, Science, and Moral Inquiry." In *In Face of the Facts: Moral Inquiry in American Scholarship*, eds. Richard Wightman Fox and Robert B. Westbrook, 10–39. Washington, DC: Woodrow Wilson Center.

    2014a. "Dewey's Moral Philosophy." Stanford Encyclopedia of Philosophy. http://plato.stanford.edu/archives/spr2014/entries/dewey-moral/.

2014b. "Social Movements, Experiments in Living, and Moral Progress: Case Studies from Britain's Abolition of Slavery." Lindley Lectures. University of Kansas. http://hdl.handle.net/1808/14787.

Anderson, Melissa E. 2004. "Jane Addams' Democracy and Social Ethics: Defending Care Ethics." *Macalester Journal of Philosophy* 13(1): Article 2.

Annas, J., and J. Barnes. 2000. *Sextus Empiricus: Outlines of Scepticism*. Cambridge: Cambridge University Press.

Appiah, K. 2008. *Experiments in Ethics*. Cambridge, MA: Harvard University Press.

Aquinas, Thomas. Summa Theologica (1265–1274 – 3 vols.).

Arumugam, Manimozhiyan, Jeroen Raes, Eric Pelletier, Denis Le Paslier, Takuji Yamada, Daniel R. Mende, Gabriel R. Fernandes, et al. 2011. "Enterotypes of the Human Gut Microbiome." *Nature* 473 (7346): 174–180. doi:10.1038/nature09944.

Atran, S., and J. Henrich. 2010. "The Evolution of Religion: How Cognitive By-Products, Adaptive Learning Heuristics, Ritual Displays, and Group Competition Generate Deep Commitments to Prosocial Religions." *Biological Theory* 5 (1): 18–30.

Audi, R. 1993. "Ethical Reflectionism." *The Monist* 76: 295–315.

1996. *The Good in the Right*. Oxford: Oxford University Press.

2004. *The Good in the Right: A Theory of Intuition and Intrinsic Value*. Princeton: Princeton University Press.

Axelrod, R. 1984. *The Evolution of Cooperation*. New York: Basic Books.

Ayala, F. 2010. "The Difference of Being Human: Morality." *Proceedings of the National Academy of Sciences of the USA* 107: 9015–9022.

Baillargeon, R. 2008. "Innate Ideas Revisited for a Principle of Persistence in Infants' Physical Reasoning." *Perspectives on Psychological Science* 3: 2–13.

Bandura, A. 1999. "Moral Disengagement in the Perpetration of Inhumanities." *Personality and Social Psychology Review* 3: 193–209.

Barkow, J. H., L. Cosmides, and J. Tooby. 1992. *The Adapted Mind: Evolutionary Psychology and the Generation of Culture*. New York: Oxford University Press.

Baron, S. 2007. *Thinking and Deciding*, 4th ed. Cambridge: Cambridge University Press.

Bartels, D., C. Bauman, F. Cushman, D. Pizarro and A. P. McGraw. 2016. "Moral Judgment and Decision Making." In *The Wiley Blackwell Handbook of Judgment and Decision Making*, Volume 1, eds. 479–491. Chichester, UK: Wiley.

Beck, N. 2012. "Be Fruitful and Multiply: Growth, Reason and Cultural Group Selection in Hayek & Darwin. *Biological Theory* 6(4): 413–423.

2013. "Social Darwinism." In *The Cambridge Encyclopedia of Darwin and Evolutionary Thought*, ed. M. Ruse, 195–201. Cambridge, UK: Cambridge University Press.

Bedke, Matt. "Intuitive Non-Naturalism Meets Cosmic Coincidence." *Pacific Philosophical Quarterly* 90: 188–209.

Bentham, J. 1978. "Offences Against One's Self." *Journal of Homosexuality* 4 (3): 389–405.

Bentley, Michael. 2011. *The Life and Thought of Herbert Butterfield: History, Science and God*. Cambridge: Cambridge University Press.

Bergman, M., and P. Kain, eds. 2014. *Challenges to Moral and Religious Belief: Disagreement and Evolution.* Oxford: Oxford University Press.

Bergson, Henri. 1911. *Creative Evolution.* New York: Henry Holt and Co.

Berker, S., 2009. "The Normative Insignificance of Neuroscience." *Philosophy and Public Affairs* 37: 293–329.

Berker, Selim. 2014. "Does Evolutionary Psychology Show that Normativity Is Mind-Dependent?" In *Moral Psychology & Human Agency. Philosophical Essays on the Science of Ethics*, eds. Justin D'Arms and Daniel Jacobson, 215–252. Oxford: Oxford University Press.

Bernhardi, F. V. 1912. *Germany and the Next War (Deutschland und der Nächster Krieg)*, trans. A. H. Powells. Stuttgart und Berlin: J. G. Cotta.

Besaude-Vincent, Bernadette. 2014. "Concluding Remarks: A View of the Past through the Lens of the Present." *Osiris* 29: 298–309.

Binmore, K. 2005. *Natural Justice.* Oxford: Oxford University Press.

Blackburn, Simon. 1984. *Spreading the Word.* Oxford: Clarendon Press.

1985. "Supervenience Revisited." In *Exercises in Analysis*, ed. Ian Hacking, 47–67. Cambridge: Cambridge University Press.

1993. *Essays in Quasi-Realism.* Oxford: Oxford University Press.

Boehm, C. 2012. *Moral Origins: The Evolution of Virtue, Altruism and Shame.* New York: Basic Books.

Borrello, Mark. 2004. "'Mutual Aid' and 'Animal Dispersion': An Historical Analysis of Alternatives to Darwin." *Perspectives in Biology and Medicine* 47: 15–31.

2005. "The Rise, Fall, and Resurrection of Group Selection." *Endeavour* 29: 43–47.

Bouchard, Frédéric. 2009. "Understanding colonial traits using symbiosis research and ecosystem ecology." *Biological Theory* 4 (3): 240-246.

2013. "What Is a Symbiotic Superindividual and How Do You Measure Its Fitness." In *From Groups to Individuals Evolution and Emerging Individuality*, Vienna Series in Theoretical Biology, eds. Frédéric Bouchard and Philippe Huneman, 243–264. Cambridge, MA: MIT Press.

Bouchard, Frédéric, and Philippe Huneman, eds. 2013. *From Groups to Individuals Evolution and Emerging Individuality*, Vienna Series in Theoretical Biology. Cambridge, MA: MIT Press.

Bowles, S., and H. Gintis. 2006. "The Evolutionary Basis of Collective Action." In *The Oxford Handbook of Political Economy*, eds. Barry G. Weingast and Donald A. Wittman, 951–70. Oxford: Oxford University Press.

2011. *A Cooperative Species: Human Reciprocity and Its Evolution.* Princeton: Princeton University Press.

Boyd, Robert, and Peter J. Richerson. 1995. "Why Culture Is Common, but Cultural Evolution Is Rare." *Proceedings of the British Academy* 88: 77–93.

2004. *The Origin and Evolution of Cultures (Evolution and Cognition).* New York: Oxford University Press

Boyer, P. 2002. *Religion Explained: The Evolutionary Origins of Religious Thought.* New York: Basic Books.

Braddock, Matthew, Walter Sinnott-Armstrong, and Andreas Mogensen. 2012. "Comments on Justin Clarke-Doane's 'Morality and Mathematics: The

Evolutionary Challenge." Ethics at PEA Soup. http://peasoup.typepad.com/peasoup/2012/03/ethics-discussions-at-pea-soup-justin-clarke-doanes-morality-and-mathematics-the-evolutionary-challe-1.html.

Brink, David O. 1989. *Moral Realism and the Foundations of Ethics.* Cambridge: Cambridge University Press.

Broad, C. D. 1925. *The Mind and Its Place in Nature.* New York: Harcourt, Brace and Co.

1944. "Critical Notice of Julian Huxley's Evolutionary Ethics." *Mind* 53: 156–187.

1944. "Review of Evolution & Ethics by Julian S. Huxley." *Mind* 53: 344–367.

Brobjer, Thomas. 2008. *Nietzsche's Philosophical Context.* Champaign: University of Illinois Press.

Brosnan, K. 2011. "Do the Evolutionary Origins of Our Moral Beliefs Undermine Moral Knowledge?" *Biology and Philosophy* 26: 51–64.

Brosnan, S., and F. de Waal. 2003. "Monkeys Reject Unequal Pay." *Nature* 425: 297–299.

Browne, Janet. 2010. "Asa Gray and Charles Darwin: Corresponding Naturalists." *Harvard Papers in Botany* 15: 209–220.

Budziszewski, J. 2011. *What We Can't Not Know: A Guide.* Rev. ed. San Francisco: Ignatius.

Bulbulia, J. 2004. "Religious Costs as Adaptations that Signal Altruistic Intention." *Evolution and Cognition* 10 (1): 19–38.

Buller, David J. 2005. *Adapting Minds: Evolutionary Psychology and the Persistent Quest for Human Nature.* Cambridge, MA: MIT Press.

Burgess, J. 2007. "Against Ethics." *Ethical Theory and Moral Practice* 10 (5): 427–439.

Buss, D. M. 1996. "Evolutionary Insights into Feminism and the 'Battle of the Sexes.'" In *Sex, Power, Conflict: Evolutionary and Feminist Perspectives*, eds. D. M. Buss and N. Malamuth. Oxford: Oxford University Press.

1998. *Evolutionary Psychology: The New Science of the Mind.* Boston, MA: Allyn & Bacon.

2005.*The Handbook of Evolutionary Psychology.* Hoboken, NJ: Wiley.

Buss, D. M., and D. P. Schmitt. 2011. "Evolutionary Psychology and Feminism." *Sex Roles* 64: 768–787.

Butterfield, Herbert. 1931. *The Whig Interpretation of History.* London: Bell and Sons.

Butterworth, Brian. 1999. *What Counts? How Every Brain Is Hardwired for Math.* New York: Free Press.

Callebaut, Werner, and Diego Rasskin-Gutman, eds. 2005. *Modularity: Understanding the Development and Evolution of Natural Complex Systems,* Vienna Series in Theoretical Biology. Cambridge, MA: MIT Press.

Campbell, R. 1996. "Can Biology Make Ethics Objective?" *Biology and Philosophy* 11: 21–31.

Cannon, Walter B. 1932. *The Wisdom of the Body.* New York: Norton and Co.

Caplan, Arthur L. 1980. *Sociobiology Debate: Readings on Ethical and Scientific Issues.* New ed. New York: Joanna Cotler Books.

Carabotti, Marilia, Annunziata Scirocco, Maria Antonietta Maselli, and Carola Severi. 2015. "The Gut-Brain Axis: Interactions between Enteric Microbiota,

Central and Enteric Nervous Systems." *Annals of Gastroenterology: Quarterly Publication of the Hellenic Society of Gastroenterology* 28 (2): 203–209.

Carr-Saunders, A.M. 1922. *The Population Problem; A Study in Human Evolution.* Oxford: Clarendon Press.

Casebeer, W. D. 2003. *Natural Ethical Facts: Evolution, Connectionism, and Moral Cognition.* Cambridge, MA: MIT Press.

Cavalli-Sforza, Luigi Luca, and Marcus W. Feldman. 1981. *Cultural Transmission and Evolution: A Quantitative Approach.* Princeton, NJ: Princeton University Press.

Chomsky, N. 1959. "Review of Skinner's Verbal Behavior." *Language* 35: 26–58.

Chudek, M., W. Zhao, and J. Henrich. 2013. "Gene-Culture Coevolution, Large-Scale Cooperation, and the Shaping of Human Social Psychology." In *Cooperation and Its Evolution*, eds. K. Sterelny, R. Joyce, B. Calcott, and B. Fraser, 425–458. Cambridge, MA: MIT Press.

Churchland, P. 2011. *Braintrust: What Neuroscience Tells Us About Morality.* Princeton, NJ: Princeton University Press.

Clark, Maudemarie, and David Dudrick. 2012. *The Soul of Nietzsche's Beyond Good & Evil.* Cambridge: Cambridge University Press.

Clarke-Doane, Justin. 2012. "Morality and Mathematics: The Evolutionary Challenge." *Ethics* 122: 313–340.

    2015. "Justification and Explanation in Mathematics and Morality." In *Oxford Studies in Metaethics*, Volume 10, ed. Russ Shafer-Landau, 80–104. New York: Oxford University Press.

    2016. "Debunking and Dispensability." In *Explanation in Ethics and Mathematics*, eds. Neil Sinclair and Uri Leibowitz. Oxford: Oxford University Press.

    2017. "What is the Benacerraf Problem?" In *New Perspectives on the Philosophy of Paul Benacerraf: Truth Objects Infinity*, ed. Fabrice Pataut, 17–43. Dordrecht: Springer.

Clements, Frederic. 1905. *Research Methods in Ecology.* Nebraska: University Pub. Co.

    1916. *Plant Succession.* Washington DC: Carnegie Institution.

    1949. "Competition in Plant Societies (1933)." In *Dynamics of Vegetation*, ed. B. W. Allred. New York: H. W. Wilson Company.

Cobbe, F. P. 1872. *Darwinism in Morals and Other Essays.* London: Williams and Norgate.

Coe, Sophie D. 1994. *America's First Cuisines.* Austin: University of Texas Press.

Cohen, G. E. 2003. "Facts and Principles." *Philosophy and Public Affairs* 31: 211–245.

Cohen, T., R. Montoya, and C. Insko. 2006. "Group Morality and Intergroup Relations: Cross-cultural and Experimental Evidence." *Personality and Social Psychology Bulletin* 32 (11): 1559–1572.

Cook, L. 2003. "The Rise and Fall of the Carbonaria Form of the Peppered Moth." *Quarterly Review of Biology* 78 (4): 399–417.

Copp, David. (2008). "Darwinian Skepticism about Moral Realism." *Philosophical Issues* 18: 186–206.

Coren, S., L. M. Ward, and J. T. Enns. 2004. *Sensation and Perception*, 6th ed. Hobokan, NJ: John Wiley & Sons Inc.

Corning, Peter. 2005. *Holistic Darwinism: Synergy, Cybernetics, and the Bioeconomics of Evolution.* Chicago, IL: Chicago University Press.

Cosmides, Leda, and John Tooby. 1987. "From Evolution to Behavior: Evolutionary Psychology as the Missing Link." In *The Latest on the Best: Essays on Evolution and Optimality*, 1st ed., ed. John Dupré, 277–306. Cambridge, MA: MIT Press.

Cosmides, L., J. Tooby, and J. H. Barkow. 1995. "Introduction: Evolutionary Psychology and Conceptual Integration." In *The Adapted Mind: Evolutionary Psychology and the Generation of Culture*, eds. J. H. Barkow, L. Cosmides and J. Tooby, 163–228. Oxford University Press: Oxford.

Cottingham, J. 2005. *The Spiritual Dimension*. New York: Cambridge University Press.

Cracraft, James. 2012. *Two Shining Souls: Jane Addams, Leo Tolstoy, and the Quest for Global Peace*. Lanham, MD: Lexington.

Crippen, Matthew. 2010. "William James on Belief: Turning Darwinism against Empiricist Skepticism." *Transactions of the Charles S. Pierce Society* 46: 477–502.

Crisp, Roger. 2006. *Reasons and the Good*. Oxford: Clarendon.

Cross, Stephen J., and William R. Albury. 1987. "Walter B. Cannon, L. J. Henderson, and the Organic Analogy." *Osiris* 3: 165–192.

Cuneo, Terence, and Russ Shafer-Landau. 2014. "The Moral Fixed Points: New Directions for Moral Nonnaturalism." *Philosophical Studies*. doi:10.1007/s11098-013-0277-5.

Cunningham, S. 1996. *Philosophy and the Darwinian Legacy*. Rochester, NY: University of Rochester Press.

Cushman, F., K. Gray, A. Gaffey, and W. Mendes. 2012. "Simulating Murder: The Aversion to Harmful Action." *Emotion* 12: 1–7.

Dahaene, Stanislas. 1997. *The Number Sense: How the Mind Creates Mathematics*. Oxford: Oxford University Press.

Daly, C., and D. Liggins. 2010. "In Defence of Error Theory." *Philosophical Studies* 149: 209–230.

Dancy, J. 1985. *An Introduction to Contemporary Epistemology*. Oxford: Blackwell.

Daniels, Norman. 1979. "Wide Reflective Equilibrium and Theory Acceptance in Ethics." *Journal of Philosophy* 76: 256–282.

Darwall, S., A. Gibbard, and P. Railton. 1992. "Toward Fin de siècle Ethics: Some Trends." *Philosophical Review*, 101(1): 115–189.

Darwin, C. 1859. *On the Origin of Species by Means of Natural Selection, or the Preservation of Favoured Races in the Struggle for Life*. London: John Murray.

 1871. *The Descent of Man and Selection in Relation to Sex*. London: Penguin.

 1872. *The Origin of Species*, 6th ed. London: John Murray.

 (1879) 2004. *The Descent of Man*. London: Penguin Classics.

 1881. Letter 13230. Darwin Correspondence Database. http://www.darwinproject.ac.uk/entry-13230.

 1882. *The Descent of Man and Selection in Relation to Sex*, 2nd ed. New York: D. Appleton & Company.

Daston, Lorraine, and Robert J. Richards, eds. 2016. "Introduction." In *Structure of Scientific Revolutions at Fifty*, 1–11. Chicago, IL: University of Chicago Press.

Dawkins, Richard. 1976. *The Selfish Gene*. New York: Oxford University Press.

 2006. *The God Delusion*. Boston: Houghton Mifflin.

2012. "The Descent of Edward Wilson." *Prospect Magazine* (May 24). http://www.prospectmagazine.co.uk/magazine/edward-wilson-social-conquest-earth-evolutionary-errors-origin-species/.

de Araujo, Ivan E., Tammy Lin, Maria G. Veldhuizen, and Dana M. Small. 2013. "Metabolic Regulation of Brain Response to Food Cues." *Current Biology* 23 (10): 878–883. doi:10.1016/j.cub.2013.04.001.

de Lazari-Radek, K., and P. Singer. 2012. "The Objectivity of Ethics and the Unity of Practical Reason." *Ethics* 123: 9–31.

DeMarco, J. P. 1996. *Moral Theory, A Contemporary Overview*. Boston, MA: Jones and Bartlett Publishers.

DePaul, Michael. 1993. *Balance and Refinement: Beyond Coherence Methods of Moral Inquiry*. London: Routledge.

De Waal, Frans. 1997. *Good Natured: The Origins of Right and Wrong in Humans and Other Animals*. Cambridge: Harvard University Press.

2006. *Primates and Philosophers: How Morality Evolved*. Princeton, NJ: Princeton University Press.

Deegan, Mary Jo. 1988. *Jane Addams and the Men of the Chicago School, 1892–1918*. New Brunswick: Transaction.

Dennett, D. 1995. *Darwin's Dangerous Idea: Evolution and the Meanings of Life*. New York: Simon and Shuster.

2003. *Freedom Evolves*. New York: Penguin.

2006. *Breaking the Spell: Religion as a Natural Phenomenon*. New York: Penguin.

Dewey, John. 1891. *Outlines of a Critical Theory of Ethics*. Ann Arbor, MI: Inland Press.

1892. "Green's Theory of the Moral Motive." *Philosophical Review* 1: 593–612.

1894. "Moral Philosophy." In *Johnson's Universal Cyclopaedia: A New Edition*, Volume 5, ed. Charles Kendall Adams, 880–885. New York: A. J. Johnson.

1898. "Evolution and Ethics." *The Monist* 8: 321–341.

1902. "The Evolutionary Method as Applied to Morality." *Philosophical Review* 11: 107–124, 353–371.

1910. *The Influence of Darwin on Philosophy, and Other Essays in Contemporary Thought*. New York: Henry Holt.

2010. "Sociology of Ethics (1902–1903)." In *The Class Lectures of John Dewey*, Vol. 1, ed. Donald F. Koch, 2246–2650. Charlottesville: Intelex.

Dewey, John, and James H. Tufts. 1908. *Ethics*. New York: Henry Holt.

Donnellan, Brendan. 1982. "Friedrich Nietzsche and Paul Rée: Cooperation and Conflict." *Journal of the History of Ideas*, 43: 595–612.

Douglas, H. 2011. "Domestic Violence Research: Valuing Stories." In *Qualitative Criminology*, 129–139.

Driesch, H. 1908. *The Science and Philosophy of the Organism*. Aberdeen: University of Aberdeen Press.

Dugatkin, Lee Alan. 1997. *Cooperation among Animals: An Evolutionary Perspective*. New York: Oxford University Press.

Dupré, John. 1993. *The Disorder of Things: Metaphysical Foundations of the Disunity of Science*, New ed. Harvard, MA: Harvard University Press.

Dupree, A. Hunter. 1959. "The First Darwinian Debate in America: Gray versus Agassiz." *Daedalus* 88: 560–569.

Durkheim, É. 1973. "The Division of Labour in Society." Extracts in *On Morality and Society*, ed. R. N. Bellah. Chicago, IL: Chicago University Press.

1982. *The Rules of Sociological Method*, ed. S. Lukes. New York: Macmillan.

1993. *Ethics and the Sociology of Morals*, ed. R. T. Hall. New York: Prometheus Books.

Dwyer, S. 2006. "How Good Is the Linguistic Analogy?" In *The Innate Mind*, Volume 2: Culture and Cognition, eds. P. Carruthers, S. Laurence, and S. Stich, 237–255. Oxford: Oxford University Press.

Eddy, Beth. 2010. "Struggle or Mutual Aid: Jane Addams, Petr Kropotkin, and the Progressive Encounter with Social Darwinism." *The Pluralist* 5: 21–43.

2016. *Evolutionary Pragmatism and Ethics*. Lanham, MD: Lexington.

Eldakar, Omar Tonsi, and David Sloan Wilson. 2011. "Eight Criticisms Not to Make about Group Selection." *Evolution* 65: 1523–1526.

Emerson, Alfred E. 1939. "Social Coordination and the Superorganism." *American Midland Naturalist* 21: 182–209.

Enoch, David. 2010. "The Epistemological Challenge to Metanormative Realism." *Philosophical Studies* 148: 413–438.

2011. *Taking Morality Seriously*. Oxford: Oxford University Press.

Feffer, Andrew. 1993. *The Chicago Pragmatists and American Progressivism*. Ithaca, NY: Cornell University Press.

Fesmire, Steven. 2003. *John Dewey and the Moral Imagination: Pragmatism in Ethics*. Bloomington: Indiana University Press.

Field, Hartry. 1989. *Realism, Mathematics, and Modality*. Oxford: Blackwell.

2005. "Recent Debates about the A Priori." In *Oxford Studies in Epistemology*, Volume I, eds. Tamar Szabo Gendler and John Hawthorne, 69–88. Oxford: Oxford University Press.

Fischer, Marilyn. 2013. "Reading Addams's Democracy and Social Ethics as a Social Gospel, Evolutionary Idealist Text." *The Pluralist* 8: 17–31.

FitzPatrick, William. 2008. "Morality and Evolutionary Biology." In *The Stanford Encyclopedia of Philosophy* (Winter 2008 Edition), ed. Edward N. Zalta. http://plato.stanford.edu/archives/win2008/entries/morality-biology/.

2014. "Why There Is No Darwinian Dilemma for Ethical Realism." In *Challenges to Moral and Religious Belief: Disagreement and Evolution*, eds. M. Bergmann and P. Kain, 237–255. Oxford: Oxford University Press.

2015. "Debunking Evolutionary Debunking of Ethical Realism." *Philosophical Studies* 172: 883–904.

Foot, Philippa. 2001. *Natural Goodness*. Oxford: Clarendon Press.

Francis, M. 2007. *Herbert Spencer and the Invention of Modern Life*. Ithaca: Cornell University Press.

Francis, Mark. 2014. *Herbert Spencer and the Invention of Modern Life*. New York: Routledge.

Francis, Mark, and Michael Taylor. 2015. *Herbert Spencer: Legacies*. New York: Routledge.

Frank, R. H. 1988. *Passions within Reason: The Strategic Role of the Emotions*. New York: Norton.

Fraser, B. 2014. "Evolutionary Debunking Arguments and the Reliability of Moral Cognition." *Philosophical Studies* 168: 457–473.

Fry, Iris. 1996. "On the Biological Significance of the Properties of Matter: L. J. Henderson's Theory of the Fitness of the Environment." *Journal of the History of Biology* 29: 155–196.

Gardner, Andy. 2011. "Kin Selection Under Blending Inheritance." *Journal of Theoretical Biology* 284: 125–129.

Garner, R. 1994. *Beyond Morality*. Philadelphia, PA: Temple University Press.

　2007. "Abolishing Morality." *Ethical Theory and Moral Practice* 10(5): 499–513.

Gaulin, S. J. C., and D. McBurney. 2001. *Psychology: An Evolutionary Approach.* Upper Saddle River, NJ: Prentice Hall.

Gayon, Jean. 1999. Nietzsche and Darwin. In *Biology and the Foundation of Ethics*, eds. Jane Maienschein and Michael Ruse. Cambridge: Cambridge University Press.

Gert, B. 2012. "The Definition of Morality." In *The Stanford Encyclopedia of Philosophy* (Fall 2012 Edition), ed. Edward N. Zalta. http://plato.stanford.edu/archives/fall2012/entries/morality-definition/.

Gibbard, Alan. 2003. *Thinking How to Live*. Cambridge, MA: Harvard University Press.

Gintis, H., J. Henrich, S. Bowles, R. Boyd, and E. Fehr. 2008. "Strong Reciprocity and the Roots of Human Morality." *Social Justice Research* 21: 241–253.

Girón, Alvaro. 2003. "Kropotkin between Lamarck and Darwin: The Impossible Synthesis." *Asclepio* 55: 189–213.

Gissing, G. 1891. *New Grub Street*. Melbourne: E.A. Petherick.

Gladwell, M. 2005. *Blink: The Power of Thinking without Thinking*. New York: Back Bay Books.

Godfrey-Smith, Peter. 1996. *Complexity and the Function of Mind in Nature.* Cambridge: Cambridge University Press.

Goodwin, G., and J. Darley. 2008. "The Psychology of Metaethics: Exploring Objectivism." *Cognition* 106 (3): 1339–1366.

　2010. "The Perceived Objectivity of Ethical Beliefs: Psychological Findings and Implications for Public Policy." *Review of Philosophy and Psychology* 1 (2): 161–188.

　2012. "Why Are Some Moral Beliefs Seen as More Objective than Others?" *Journal of Experimental Social Psychology* 48 (1): 250–256.

Gouinlock, James. 1972. *John Dewey's Philosophy of Value*. New York: Humanities.

Gould, Stephen Jay. 1977. *Ever since Darwin Reflections in Natural History.* New York: Norton.

　1978. "Sociobiology: The Art of Storytelling." *New Scientist* 80: 530–533.

　1987. "The Panda's Thumb of Technology." *Natural History* 1 (Jan.): 68–74.

　1988. "On Replacing the Idea of Progress with an Operational Notion of Directionality." In *Evolutionary Progress*, ed. M. H. Nitecki, 319–338. Chicago, IL: University of Chicago Press.

　1989. *Wonderful Life: The Burgess Shale and the Nature of History*. New York: W. W. Norton Co.

Gould, S. J., and R. C. Lewontin. 1979. "The Spandrels of San Marco and the Panglossion Paradigm: A Critique of the Adaptationist Programme." *Proceedings of the Royal Society: Biological Sciences*, Series B 205: 581–598.

Grant, B. 1999. Fine Tuning the Peppered Moth Paradigm. *Evolution* 53 (3): 980–984.

Gray, Asa. 1877. *Darwiniana: Essays and Reviews Pertaining to Darwinism.* New York: Appleton.

Gray J. 1984. *Hayek on Liberty.* New York: Basil Blackwell Publishers.

Green, Judith M. 2010. "Social Democracy, Cosmopolitan Hospitality, and Intercivilizational Peace: Lessons from Jane Addams." In *Feminist Interpretations of Jane Addams,* ed. Maurice Hamington, 223–253. University Park: Pennsylvania State University Press.

   2013. *Moral Tribes. Emotion, Reason, and the Gap between Us and Them.* New York: Penguin Press.

Greene, Joshua D. 2008. "The Secret Joke of Kant's Soul." In *Moral Psychology, Vol. 3. The Neuroscience of Morality: Emotion, Brain Disorders, and Development,* ed. Walter Sinnott-Armstrong, 35–80. Cambridge, MA: MIT Press.

Greene, J. 2014. "Beyond Point-and-Shoot Morality: Why Cognitive (Neuro)science Matters for Ethics." *Ethics* 124: 694-726.

Greene, J., R. Sommerville, L. Nystrom, J. Darley, and J. Cohen. 2001. "An fMRI Investigation of Emotional Engagement in Moral Judgment." *Science* 293 (5537): 2105–2108.

Greenwald, A., and M. Banaji. 1995. "Implicit Social Cognition: Attitudes, Self-esteem, and Stereotypes." *Psychological Review* 102 (1): 4–27.

Guttenplan, Samuel. 1994. *A Companion to the Philosophy of Mind.* Oxford: Blackwell.

Haber, Matt. 2013. "Colonies Are Individuals: Revisiting the Superorganism Revival." In *From Groups to Individuals: Perspectives on Biological Associations and Emerging Individuality,* Vienna Series in Theoretical Biology, eds. Frédéric Bouchard and Philippe Huneman, 195–217. Cambridge, MA: MIT Press.

Haidt, J. 2001. 'The Emotional Dog and Its Rational Tail: A Social Intuitionist Approach to Moral Judgment." *Psychological Review* 108: 814–834.

   2012. *The Righteous Mind: Why Good People Are Divided by Politics and Religion.* New York: Pantheon.

Haidt, J. and J. Graham. 2007. "When Morality Opposes Justice: Conservatives Have Moral Intuitions that Liberals May Not Recognize." *Social Justice Research* 20 (1): 98–116.

Haidt, J. and C. Joseph. 2007. "The Moral Mind: How Five Sets of Innate Intuitions Guide the Development of Many Culture-Specific Virtues, and Perhaps Even Modules." *The Innate Mind* 3: 367–392.

Haines, Valerie. 1991. "Spencer, Darwin, and the Question of Reciprocal Influence." *Journal of the History of Biology* 24: 409–431.

Haldane. 1932. *The Causes of Evolution.* Ithaca, NY: Cornell University Press.

Hale, Piers J. 2014. *Political Descent: Malthus, Mutualism, and the Politics of Evolution in Victorian England.* Chicago, IL: University of Chicago Press.

Hamilton, W. D. 1964a. "The Genetical Evolution of Social Behaviour I." *Journal of Theoretical Biology* 7: 1–16.

   1964b. "The Genetical Evolution of Social Behaviour II." *Journal of Theoretical Biology* 7: 17–52.

   1975. "Innate Social Aptitudes of Man: An Approach from Evolutionary Genetics." In *ASA Studies 4: Biosocial Anthropology,* ed. R. Fox, 133–153. London: Malaby Press.

Hamington, Maurice. 2009. *The Social Philosophy of Jane Addams.* Urbana: University of Illinois Press.

Hardin, G. 1968. "The Tragedy of the Commons." *Science* 162 (3859): 1243–1248.

Harman, Gilbert. 1977. *The Nature of Morality: An Introduction to Ethics.* New York: Oxford.

Harman, Oren. 2010. *The Price of Altruism: George Price and the Search for the Origins of Kindness.* New York: W. W. Norton.

Harms, W. F. 2000. "Adaption and Moral Realism." *Biology and Philosophy* 15: 699–712.

Harrington. 1996. *Reenchanted Science.* Princeton: Princeton University Press.

Harris, S. 2004. *The End of Faith: Religion, Terror and the Future of Reason.* New York: Norton.

Hauser, Mark. 2006. *Moral Minds: How Nature Designed Our Universal Sense of Right and Wrong.* New York: Ecco.

Hayek F. A. 1958. "Freedom, Reason and Tradition." *Ethics* LXVIII(4): 229–245.

    (1960) 1971. *The Constitution of Liberty.* Chicago, IL: University of Chicago Press.

    1967. *Studies in Philosophy, Politics and Economics.* Chicago, IL: University of Chicago Press.

    1973. *Law, Legislation and Liberty. Volume I: Rules and Order.* London: Routledge.

    1976. *Law, Legislation and Liberty. Volume II: The Mirage of Social Justice.* London: Routledge.

    1978. *New Studies in Philosophy, Politics, Economics and the History of Ideas.* London: Routledge.

    1979. *Law, Legislation and Liberty.* Volume III: *The Political Order of a Free People.* London: Routledge.

    1984. *The Essence of Hayek*, eds. C. Nishiyama and K. R. Leube. Stanford, CA: Hoover Institution Press.

    1988. *The Fatal Conceit: The Errors of Socialism.* Chicago, IL: University of Chicago Press.

Helfand, Michael S. 1977. "T. H. Huxley's 'Evolution and Ethics': The Politics of Evolution and the Evolution of Politics." *Victorian Studies* 20: 159–177.

Henderson, Lawrence J. 1917. *The Order of Nature.* Cambridge: Harvard University Press.

Henson, Pamela. 2008. "Alfred Edwards Emerson." *Complete Dictionary of Scientific Biography* 20: 389–393.

Herrel, Anthony Huyghe. 2008. "Rapid Large-Scale Evolutionary Divergence in Morphology and Performance Associated with Exploitation of a Different Dietary Resource." *Proceedings of the National Academy of Sciences USA* 105 (12): 4792–4795.

Hinckfuss, I. 1987. *The Moral Society: Its Structure and Effects.* Canberra: Australian National University.

Hitchens, C. 2007. *God Is Not Great: How Religion Poisons Everything.* New York: Twelve.

Hobbes, Thomas. (1651) 2008. *Leviathan.* Oxford: Oxford University Press. Reprint, Gaskin, J. C. A. *Leviathan,* edited with introduction.

Hodge, M. J. S. 2009. "Capitalist Contexts for Darwinian Theory: Land, Finance, Industry and Empire." *Journal of the History of Biology* 42: 399–416.

Hodgson G. M. 1991. "Hayek's Theory of Cultural Evolution: An Evaluation in the Light of Vanberg's Critique." *Economics and Philosophy* 7: 67–82.

1993. *Economics and Evolution: Bringing Life Back into Economics.* Cambridge: Polity Press.

2004. "Social Darwinism in Anglophone Academic Journals: A Contribution to the History of the Term." *Journal of Historical Sociology* 17(4): 428–463.

Hofstadter, Richard. 1944. *Social Darwinism in American Thought.* Boston, MA: Beacon Press.

Huemer, Michael. 2005. *Ethical Intuitionism.* London: Palgrave Macmillan.

2008. "Revisionary Intuitionism." *Social Philosophy and Policy* 25: 368–392.

Hull, David L. 1978. "A Matter of Individuality." *Philosophy of Science* 45 (3): 335–360.

1980. "Individuality and Selection." *Annual Review of Ecology and Systematics* 11: 311–332.

Hume, David. (1739) 2000. *A Treatise of Human Nature.* London: John Noon. Reprint, eds. David Fate Norton and Mary J. Norton. New York: Oxford University Press.

Hurka, T. 2014. *British Ethical Theorists from Sidgwick to Ewing.* Oxford: Oxford University Press.

Hutter, Thiago, Carine Gimbert, Frédéric Bouchard, and François-Joseph Lapointe. 2015. "Being Human Is a Gut Feeling." *Microbiome* 3 (1): 9. doi:10.1186/s40168-015-0076-7.

Huxley, J., ed. 1961. *The Humanist Frame.* New York: Harper & Brothers Publishers.

Huxley, Julian. 1912. *The Individual in the Animal Kingdom.* Cambridge: Cambridge University Press.

(1923) 1970. *Essays of a Biologist.* Freeport, NY: Books for Libraries Press.

1926. "The Biological Basis of Individuality." *Journal of Philosophical Studies* 1 (July): 305–319.

1934. *If I Were Dictator.* New York: Harper and Brothers.

1942. *Evolution: The Modern Synthesis.* London: Allen and Unwin.

1943. *TVA: Adventure in Planning.* London: Scientific Book Club.

1948. *UNESCO: Its Purpose and Its Philosophy.* Washington, DC: Public Affairs Press.

Huxley, Leonard. 1900. *Life and Letters of Thomas Henry Huxley* (2 vols). New York: D. Appleton.

Huxley, Thomas Henry. 1871. "Administrative Nihilism." *Fortnightly Review* 16: 525–543.

(1888) 2011. "The Struggle for Existence in Human Society." In *Collected Essays*, Volume 9: Evolution and Ethics, 195–236. Cambridge: Cambridge Library Collection.

1888. "The Struggle for Existence: A Programme." *Nineteenth Century* 23: 161–180.

(1893) 2011. "Evolution and Ethics." *In: Collected Essays*, Volume 9: Evolution and Ethics, 46–116. Cambridge: Cambridge Library Collection.

1893 *Evolution and Ethics: The Romanes Lecture 1893.* London: Macmillan

1894. *Evolution and Ethics and Other Essays.* London: MacMillan and Co.

Huxley, T. H., and J. S. Huxley. 1947. *Evolution and Ethics 1893–1943*. London: Pilot.

Ichikawa, J., and M. Steup. 2014. "The Analysis of Knowledge." In *The Stanford Encyclopedia of Philosophy* (Spring 2014 Edition), ed. Edward N. Zalta. http://plato.stanford.edu/archives/spr2014/entries/knowledge-analysis/.

Irons, W. 1996. "Morality, Religion, and Human Nature." In *Religion and Science: History, Method, Dialogue*, eds. W. M. Richardson and W. Wildman, 375–399. New York: Routledge.

Jackson, Frank. 1998. *From Metaphysics to Ethics: A Defense of Conceptual Analysis*. Oxford: Clarendon Press.

James, William. 1904. "The Chicago School." *Psychological Bulletin* 1: 1–5.

Janaway, Christopher. 2007. *Beyond Selflessness: Reading Nietzsche's Genealogy*. Oxford: Oxford University Press. Oxford Scholarship Online, 2007. doi: 10.1093/acprof:oso/9780199279692.001.0001.

——— 2012. Nietzsche on Morality, Drives, and Human Greatness. In *Nietzsche, Naturalism, and Normativity*, eds. Christopher Janaway and Simon Robertson. Oxford: Oxford University Press. Oxford Scholarship Online, 2013. doi: 10.1093/acprof:oso/9780199583676.003.0008.

Jardine, Nicholas. 2003. "Whigs and Stories: Herbert Butterfield and the Historiography of Science." *History of Science* 41: 125–140.

Jaworska, Agnieszka, and Julie Tannenbaum. 2013. "The Grounds of Moral Status." In *The Stanford Encyclopedia of Philosophy* (Summer 2013 Edition), ed. Edward N. Zalta. http://plato.stanford.edu/archives/sum2013/entries/grounds-moral-status/.

Jennings, Herbert Spencer. 1906. *Behavior of the Lower Organisms*. New York: Columbia University Press.

——— 1927. "Diverse Doctrines of Evolution, Their Relation to the Practice of Science and Life." *Science* 65: 19–25.

Jensen, Christopher X. J. 2010. "Robert Trivers and Colleagues on Nowak, Tarnita, and Wilson's 'The Evolution of Eusociality.'" Christopherxjjensen.com (October 10). http://www.christopherxjjensen.com/2010/10/13/robert-trivers-and-colleagues-on-nowak-tarnita-and-wilsons-the-evolution-of-eusociality.

Joyce, R. 2001. *The Myth of Morality*. Cambridge: Cambridge University Press.

——— 2005. "Moral Fictionalism." In *Fictionalism in Metaphysics*, ed. M. Kalderon, 287–313. Oxford: Oxford University Press.

——— 2006a. *The Evolution of Morality*. Cambridge, MA: MIT Press.

——— 2006b. "Is Human Morality Innate?" In *The Innate Mind: Culture and Cognition*, eds., P. Carruthers, S. Laurence, and S. Stich, 257–279. Oxford: Oxford University Press.

——— 2008. "Precis of Evolution of Morality and Reply to Critics." *Philosophy and Phenomenological Research* 77: 213–267.

——— 2013. "Irrealism and the Genealogy of Morals." *Ratio* 26: 351–372. doi:10.1111/rati.12027.

——— 2014. "The Evolutionary Debunking of Morality." In *Reason and Responsibility: Readings in Some Basic Problems of Philosophy*, 15th ed., eds. J. Feinberg and R. Shafer-Landau, 527–534. Boston: Wadsworth.

2016. "Reply: Confessions of a Modest Debunker." In *Explanation in Ethics and Mathematics: Debunking and Dispensibility,* eds. U. Leibowitz and N. Sinclair. Oxford: Oxford University Press.

Jumonville, Neil. 2002. "The Cultural Politics of the Sociobiology Debate." *Journal of the History of Biology* 35: 569–593.

Kahane, Guy. 2011. "Evolutionary Debunking Arguments." *Noûs* 45: 103–125.

Kant, I. 1959. *Foundations of the Metaphysics of Morals.* Indianapolis: Bobbs-Merrill.

Katsafanas, Paul. 2013. "Nietzsche's Philosophical Psychology." In *The Oxford Handbook of Nietzsche,* eds. John Richardson and Ken Gemes, 727–757. Oxford: Oxford University Press.

Kelly, Thomas, and Sarah McGrath. 2010. "Is Reflective Equilibrium Enough?" *Philosophical Perspectives* 24: 325–359.

Kettlewell, B. 1973. *The Evolution of Melanism.* Oxford: Clarendon Press.

Kinna, Ruth. 1992. "Kropotkin and Huxley." *Politics* 12: 41–47.

Kitcher, P. 1985. *Vaulting Ambition: Sociobiology and the Quest for Human Nature.* Cambridge, MA: MIT Press.

2005. "Biology and Ethics." In *Oxford Handbook of Ethical Theory,* ed. David Copp, 575–586. New York: Oxford University Press.

2006. "Four Ways of 'Biologicizing' Ethics." In *Conceptual Issues in Evolutionary Biology,* 3rd ed. (Bradford Books), ed. Elliott Sober, 439–450. New York: MIT Press.

2011. *The Ethical Project.* Cambridge, MA: Harvard University Press.

Kley R. 1994. *Hayek's Social and Political Thought.* Oxford: Clarendon Press.

Kozo-Polyansky, Boris. 2010. *Symbiogenesis: A New Principle of Evolution,* trans. Victor Fet, eds. Victor Fet and Lynn Margulis. Cambridge, MA: Harvard University Press.

Krebs, D. 2011. *The Origins of Morality: An Evolutionary Account.* New York: Oxford University Press.

Kropotkin, Pyotr. 1887. "The Scientific Bases of Anarchy." *Nineteenth Century* 21: 238–252.

1890. "Mutual Aid among Animals." *Nineteenth Century* 28: 337–354, 699–719.

1891. "Mutual Aid among Savages." *Nineteenth Century* 29: 538–559.

1896. "Mutual Aid amongst Ourselves." *Nineteenth Century* 39: 914–936.

1902. *Mutual Aid: A Factor in Evolution.* Boston, MA: Extending Horizons Books.

Kvanvig, Jonathan. 1992. *The Intellectual Virtues and the Life of the Mind: On the Place of the Virtues in Contemporary Epistemology.* Savage, MD: Rowman and Littlefield.

Laham, S. 2009. "Expanding the Moral Circle." *Journal of Experimental Social Psychology* 45 (1): 250–253.

Lakatos, I. 1970. "Falsification and the Methodology of Scientific Research Programmes." In *Criticism and the Growth of Knowledge,* eds. I. Lakatos and A. Musgrave, 91–195. Cambridge: Cambridge University Press.

Leach, C., N. Ellemers, and M. Barreto. 2007. Group Virtue: The Importance of Morality (vs. Competence and Sociability) in the Positive Evaluation of In-Groups. *Journal of Personality and Social Psychology* 93 (2): 234–249.

Leibowitz, U. D. 2011. "Scientific Explanation and Moral Explanation." *Noûs* 45: 472–503.

2014. "Explaining Moral Knowledge." *Journal of Moral Philosophy* 11: 35–56.

Leibowitz, U. D., and N. Sinclair, eds. 2016. *Explanation in Ethics and Mathematics: Debunking and Indispensability.* Oxford: Oxford University Press.

Leiter, Brian. 2002. *Nietzsche on Morality.* London: Routledge.

Lekan, Todd. 2003. *Making Morality: Pragmatist Reconstruction in Ethical Theory.* Nashville, TN: Vanderbilt University Press.

Levins, R., and R. C. Lewontin. 1985. *The Dialectical Biologist.* Cambridge, MA: Harvard University Press.

Lévy-Bruhl, L. 1905. *Ethics and Moral Science.* London: Archibald Constable & Co.

Lewis, D. 1973. *Counterfactuals.* Cambridge, MA: Harvard University Press.

Lewontin, Richard. 1976. "Sociobiology: A Caricature of Darwinism." *Proceedings of the Biennial Meeting of the Philosophy of Science Association* 2: 22–23, 30.

Lillehammer, H. 2003. "Debunking Morality: Evolutionary Naturalism and Moral Error Theory." *Biology and Philosophy* 18: 567–581.

2010. "Methods of Ethics and the Descent of Man: Sidgwick and Darwin on Ethics and Evolution." *Biology and Philosophy* 25: 361–378.

2011. "The Epistemology of Ethical Intuitions." *Philosophy* 86: 175–200.

2016. "'An Assumption of Extreme Significance': Moore, Ross and Spencer on Ethics and Evolution." In *Explanation in Ethics and Mathematics: Debunking and Dispensability*, eds. Uri D. Leibowitz and Neil Sinclair, 103–123. Oxford: Oxford University Press.

Linville, Mark. 2012. "The Moral Argument." In *The Blackwell Companion to Natural Theology,* eds. William Lane Craig and J. P. Moreland, 391–448. Oxford: Blackwell.

Lloyd, E., D. Wilson, and E. Sober. 2011. *Evolutionary Mismatch and What To Do about It: A Basic Tutorial.* Wesley Chapel: The Evolution Institute.

Locke, Dustin. "Darwinian Normative Skepticism." In *Challenges to Moral And Religious Belief: Disagreement and Evolution,* eds. Michael Bergmann and Patrick Kain, 220–236, Oxford: Oxford University Press.

Losco, Joseph. 2011. "From Outrage to Orthodoxy? Sociobiology and Political Science at 35." *Politics and the Life Sciences* 30: 80–84.

Machery, E., and R. Mallon. 2010. "The Evolution of Morality." In *The Moral Psychology Handbook,* eds. J. Doris, G. Harman, S. Nichols, J. Prinz, W. Sinnott-Armstrong, and S. Stich, 3–46. Oxford: Oxford University Press.

MacIntyre, A. 1984. *After Virtue.* London: Duckworth.

Mackie, J. 1977. *Ethics: Inventing Right and Wrong.* New York: Penguin.

(1978) 1985. "The Law of the Jungle: Moral Alternatives and Principles of Evolution." Republished in *Persons and Values: Selected Papers*, Volume II, eds. J. Mackie and P. Mackie, 120–131. Oxford: Oxford University Press.

1979. *Hume's Moral Theory.* London: Routledge and Kegan Paul.

1980. "Review of Sociobiology: Sense or Nonsense, by Michael Ruse." *Erkenntnis* 15: 189–194.

1982a. "Co-operation, Competition, and Moral Philosophy." *Erkenntnis* 17: 152–169.

1982b. "Morality and Retributive Emotions." *Erkenntnis* 15: 206–219.

Mackie, J. L., 1982, "Morality and the Retributive Emotions," *Criminal Justice Ethics*, 1: 3–10.

Marciano, A. 2007. "Economists on Darwin's Theory of Social Evolution and Human Behaviour." *European Journal of the History of Economic Thought* 14(4): 681–700.

Margulis, Lynn. 1993. *Symbiosis in Cell Evolution: Microbial Communities in the Archean and Proterozoic Eons.* New York: W. H. Freeman & Co.

Mason, K. 2010. "Debunking Arguments and the Genealogy of Religion and Morality." *Philosophy Compass* 5: 770–778.

Marett, R. R. 1902. "Origin and Validity in Ethics." In *Personal Idealism: Philosophical Essays by Eight Members of the University of Oxford*, ed., 221–287. London: Macmillan.

Mates, B. 1996. *The Skeptic Way: Sextus Empiricus's Outlines of Pyrrhonism.* Oxford: Oxford University Press.

Matheson, C., and J. Dallmann. 2015. "Historicist Theories of Scientific Rationality." In *The Stanford Encyclopedia of Philosophy* (Summer 2015 Edition), ed. Edward N. Zalta. http://plato.stanford.edu/archives/sum2015/entries/rationality-historicist/.

Maund, B. 1995. *Colors: Their Nature and Representation.* Cambridge: Cambridge University Press.

Mayr, Ernst. 1990. "When Is Historiography Whiggish?" *Journal of the History of Ideas* 51: 301–309.

2001. *What Evolution Is.* New York: Basic Books.

Mayr, E. and W. B. Provine, eds. 1980. *The Evolutionary Synthesis.* Cambridge, MA: Harvard University Press.

McDonald, R., B. Newell, and T. Denson. 2014. "Would You Rule Out Going Green? The Effect of Inclusion Versus Exclusion Mindset on Pro-environmental Willingness." *European Journal of Social Psychology* 44 (5): 507–513.

McDowell, J. 1985. "Values and Secondary Properties." In *Morality and Objectivity*, ed. T. Honderich, 110–29. London: Routledge and Kegan Paul.

1988. "Values and Secondary Properties." In *Essays on Moral Realism*, ed. G. Sayre-McCord, 110–129. Ithica, NY: Cornell University Press.

McFarland, D. F. 1998. "Watering Dairy Cattle." *Dairy Feeding Systems Management, Components and Nutrients (NRAES-116).* Ithaca, NY: Natural Resources, Agriculture and Engineering Services.

McGranahan, Lucas. 2011. "William James's Social Evolutionism in Focus." *Pluralist* 6: 80–92.

Melis, A., K. Altricher, A. Schneider, and M. Tomasello. 2013. "Allocation of Resources to Collaborators and Free-Riders by 3-Year-Olds." *Journal of Experimental Child Psychology* 114: 364–370.

Mendelson, Everett. 2008. "Locating 'Fitness' and L. J. Henderson." In *Fitness of the Cosmos for Life: Biochemistry and Fine-Tuning*, ed. John D. Barrow, Simon Conway Morris, Stephen J. Freeland, and Charles L. Harper Jr., 3–96. Cambridge: Cambridge University Press.

Midgley, M. 1978. *Beast and Man: The Roots of Human Nature.* Ithaca, NY: Cornell University Press.

1979. "Gene-juggling." *Philosophy* 54: 439–58.

Mikhail, J. 2011. *Elements of Moral Cognition: Rawls' Linguistic Analogy and the Cognitive Science of Moral and Legal Judgment.* Cambridge: Cambridge University Press.

Mill, John Stuart. 1863. *On Liberty.* Boston, MA: Ticknor and Fields.

Mitman, Gregg. 1992. *The State of Nature: Ecology, Community, and American Social Thought, 1900–1950.* Chicago, IL: University of Chicago Press.

Moore, A.W 1910. *Pragmatism and Its Critics.* Chicago: University of Chicago Press.

Moore, G. E. 1903. *Principia Ethica*, rev. ed., ed. Tom Baldwin. Cambridge: University of Chicago Press.

1912. *Ethics.* Oxford: Oxford University Press.

Moore, Gregory. 2002. *Nietzsche, Biology, and Metaphor.* Cambridge: Cambridge University Press.

Morgan, C. Lloyd. 1923. *Emergent Evolution.* London: Williams and Norgate, Ltd.

Morris, Paul J. 1997. "Louis Agassiz's Arguments against Darwinism in his Additions to the French Translation of the 'Essay on Classification.'" *Journal of the History of Biology* 30: 121–134.

Murphy, J. 1982. *Evolution, Morality, and the Meaning of Life.* Totowa, NJ: Rowman and Littlefield.

Myers, F. W. H. 1881. "George Eliot." *Century Magazine* 23 (November): 57–64.

Nagel, Ernest. 1954. "Naturalism Reconsidered." *Proceedings and Addresses of the American Philosophical Association* 28: 5–17.

Nagel, T. 2012. *Mind and Cosmos: Why the Materialist Neo-Darwinian Conception of Nature is Almost Certainly False.* Oxford: Oxford University Press.

Nanay, B. 2005. "Can Cumulative Selection Explain Adaptation?" *Philosophy of Science* 72: 1099–1112.

Neander, K. 1988. "What Does Natural Selection Explain? Correction to Sober." *Philosophy of Science* 55: 422–426.

Nichols, S. 2004. *Sentimental Rules: On the Natural Foundations of Moral Judgment.* Oxford: Oxford University Press.

Nietzsche, Friedrich. 1954. *The Antichrist*, trans. W. Kaufmann. New York: Viking.

1966. *Beyond Good and Evil*, trans. W. Kaufmann. New York: Vintage Books.

1968. *The Will to Power*, trans. W. Kaufmann and R. J. Hollingdale. New York: Vintage.

1983. *Untimely Meditations*, trans. R. J. Hollingdale. Cambridge: Cambridge University Press.

1986. *Human, All-too-Human*, trans. R. J. Hollingdale. Cambridge: Cambridge University Press.

1997. *Daybreak: Thoughts on the Prejudices of Morality*, eds. Maudemarie Clark and Brian Leiter, trans. R. J. Hollingdale. Cambridge: Cambridge University Press.

1998. *On the Genealogy of Morality*, trans. Maudemarie Clark and Alan Swensen. Indianapolis: Hackett.

2001. *The Gay Science*, trans. Josephine Nauckhoff. Cambridge: Cambridge University Press.

Nolan, D., G. Restall, and C. West. 2005. "Moral Fictionalism Versus the Rest." *Australasian Journal of Philosophy* 83 (3): 307–330.

Nowak, Martina A., Corina E. Tarnita, and Edward O. Wilson. 2010. "The Evolution of Eusociality." *Nature* 446 (August 26): 1057–1062.

Nozick, Robert. 1981. *Philosophical Explanations.* Oxford: Oxford University Press.

Okasha, Samir. 2006. *Evolution and the Levels of Selection.* New York: Oxford University Press.

2010. "Altruism Researchers Must Cooperate." *Nature* 467 (October 7): 653–655.

Olsen, J. 2011. "Getting Real About Moral Fictionalism." *Oxford Studies in Metaethics* 6: 181.

Olson, Jonas. 2011. "In Defence of Moral Error Theory." In *New Waves in Metaethics,* ed. M. Brady, 62–84. Hampshire: Palgrave Macmillan.

2014. *Moral Error Theory: History, Critique, Defence.* Oxford: Oxford University Press.

Osborn, Henry Fairfield. 1894. "The Discussion between Spencer and Weismann." *Psychological Review* 1: 312–315.

Pappas, Gregory Fernando. 2008. *John Dewey's Ethics: Democracy as Experience.* Bloomington: Indiana University Press.

Parascandola, John. 1971. "Organismic and Holistic Concepts in the Thought of L. J. Henderson." *Journal of the History of Biology* 4: 63–113.

Parfit, D. 1984. *Reasons and Persons.* Oxford: Oxford University Press.

2011. *On What Matters,* 2 vols. Oxford: Oxford University Press.

Parker, George Howard. 1924a. "Some Implications of the Evolutionary Hypothesis." *Philosophical Review* 33 (November): 593–603.

1924b. "Organic Determinism." *Science* 59: 517–521.

1938. "Biographical Memoir of William Morton Wheeler." *Biographical Memoirs* 19 (National Academy of Science, 1938): 203–241.

Peacocke, C. 2004. *The Realm of Reason.* Oxford: Oxford University Press.

Pearce, Trevor. 2010a. "From 'Circumstances' to 'Environment': Herbert Spencer and the Origins of the Idea of Organism-Environment Interaction." *Studies in History and Philosophy of Biological and Biomedical Sciences* 41: 241–252.

2010b. "'A Great Complication of Circumstances' – Darwin and the Economy of Nature." *Journal of the History of Biology* 43: 493–528.

2014a. "The Dialectical Biologist, circa 1890: John Dewey and the Oxford Hegelians." *Journal of the History of Philosophy* 52: 747–778.

2014b. "The Origins and Development of the Idea of Organism-Environment Interaction." In *Entangled Life: Organism and Environment in the Biological and Social Sciences,* eds. Gillian Barker, Eric Desjardins, and Trevor Pearce, 13–32. Dordrecht: Springer.

2017. [Review of] Beth Eddy, *Evolutionary Pragmatism and Ethics, Transactions of the Charles S. Peirce Society* 53.

Peterson, Erik L. 2011. "The Excluded Philosophy of Evo-Devo? Revisiting C. H. Waddington's Failed Attempt to Embed Alfred North Whitehead's 'Organicism' in Evolutionary Biology." *History and Philosophy of the Life Sciences* 33: 301–320.

Peterson, Michael. 2012. *Reason and Religious Belief.* Oxford: Oxford University Press.

Peterson, Michael, and Michael Ruse. 2017. *Science, Evolution, and Religion: A Debate about Atheism and Theism.* New York: Oxford University Press.

Philipse, Herman. 2012. *God in the Age of Science: A Critique of Religious Reason.* Oxford: Oxford University Press.

Phillips, Paul T. 2007. "One World, One Faith: The Quest for Unity in Julian Huxley's Religion of Evolutionary Humanism." *Journal of the History of Ideas* 68 (October): 613–633.

Pinker, S. 2009. *How the Mind Works.* New York: W. W. Norton & Co.

2012. *The Better Angels of Our Nature: Why Violence Has Declined.* New York: Penguin Books.

Plantinga, Alvin. 1993. *Warrant and Proper Function.* New York: Oxford University Press.

Plantinga, Alvin, and Daniel Dennett. 2011. *Science and Religion: Are They Compatible?* New York: Oxford University Press.

Pollock, John. 1986. *Contemporary Theories of Knowledge.* Savage, MD: Rowman and Littlefield.

Pollock, John, and Joseph Cruz. 1991. *Contemporary Theories of Knowledge*, 2nd ed. Lanham, MD: Rowman and Littlefield.

Portier. 1918. *Les Symbiotes.* Ann Arbor: University of Michigan Library.

Pradeu, Thomas. 2012. *The Limits of the Self: Immunology and Biological Identity.* Oxford: Oxford University Press.

Prinz, J. 2007. *The Emotional Construction of Morals.* Oxford: Oxford University Press.

2008. "Is Morality Innate?" In *Moral Psychology,* Volume 1: *The Evolution of Morality: Adaptations and Innateness,* ed. W. Sinnott-Armstrong, 367–406. Cambridge, MA: MIT Press.

2009. "Against Moral Nativism." In *Stich and his Critics,* eds. D. Murphy and M. Bishop, 1 Malden: Blackwell.

Pritchard, H. A. 2002. *Moral Writings,* ed. J. MacAdam, Oxford: Oxford University Press.

Putnam, H. 1981. *Reason, Truth and History.* Cambridge: Cambridge University Press.

Quine, W. V. O. 1951. "Two Dogmas of Empiricism." *The Philosophical Review* 60: 20–43.

1969. "Epistemology Naturalized." Reprint, *Ontological Relativity and Other Essays,* 69–90. New York: Columbia University Press.

Quinn, W. 1986. "Truth and Explanation in Ethics." *Ethics* 96: 524–544.

Railton, P. 2003. *Facts, Values and Norms: Essays Toward a Morality of Consequence.* Cambridge: Cambridge University Press.

Rakison D., and J. Derringer. 2008. "Do Infants Possess an Evolved Spider-Detection Mechanism?" *Cognition* 107: 381–393.

Rashdall, H. 1907. *The Theory of Good and Evil,* 2 vols. Oxford: Oxford University Press.

Ratnieks, Francis L. W., Kevin R. Foster, and Tom Wenseleers. 2011. "Darwin's Special Difficulty: The Evolution of 'Neuter Insects' and Current Theory." *Behavioral Ecology and Sociobiology* 65: 481–492.

Rawls, J. 1971. *A Theory of Justice.* Cambridge: Harvard University Press.

Rée, Paul. 2003. *Basic Writings,* ed. and trans. Robin Small. Urbana: University of Illinois Press.

Richards, R. A. 2005. "Evolutionary Naturalism and the Logical Structure of Valuation." *Cosmos and History: The Journal of Natural and Social Philosophy* 1(2): 270–294.

Richards, Robert J. 1986. "A Defense of Evolutionary Ethics." *Biology and Philosophy* 1: 265–293.

    1987. *Darwin and the Emergence of Evolutionary Theories of Mind and Behavior.* Chicago, IL: University of Chicago Press.

    1993. *The Meaning of Evolution: The Morphological Construction and Ideological Reconstruction of Darwin's Theory.* Chicago: University of Chicago Press.

    2013. *Was Hitler a Darwinian? Disputed Questions in the History of Evolutionary Theory.* Chicago, IL: University of Chicago Press.

Richards, Robert J., and Michael Ruse. 2016. *Debating Darwin.* Chicago, IL: University of Chicago Press.

Richardson, John. 2004. *Nietzsche's New Darwinism.* Oxford: Oxford University Press.

Richerson, P., and J. Henrich. 2012. "Tribal Social Instincts and the Cultural Evolution of Institutions to Solve Collective Action Problems." *Cliodynamics* 3 (1): 38–80.

Richerson, Peter J., and Robert Boyd. 2005. *Not by Genes Alone : How Culture Transformed Human Evolution.* Chicago, IL: University of Chicago Press.

Robinson, S. J. 2008. "How to Be an Error Theorist about Morality." *Polish Journal of Philosophy* 2: 107–125.

Rosenberg, Eugene, and Ilana Zilber-Rosenberg. 2011. "Symbiosis and Development: The Hologenome Concept." *Birth Defects Research Part C: Embryo Today: Reviews* 93 (1): 56–66. doi:10.1002/bdrc.20196.

Ross, W. D. 1930. *The Right and the Good.* Oxford: Oxford University Press.

    1939. *The Foundations of Ethics.* Oxford: Oxford University Press.

    1995. *Aristotle,* 6th ed. London: Routledge.

Rottschaefer, W. A., and D. Martinsen. 1995. "Really Taking Darwin Seriously: An Alternative to Michael Ruse's Darwinian Metaethics." In *Issues in Evolutionary Ethics,* ed. P. Thompson, 375–408. Albany, NY: State University of New York Press.

Rousseau, Jean-Jacques. 1762. "Du Contrat Social." In *Collected Writings,* 13 vols., eds. Roger Masters and Christopher Kelly. Dartmouth: University Press of New England, 1990–2010.

Royce, Josiah. 1885. *The Religious Aspect of Philosophy.* Boston, MA: Houghton, Mifflin.

Ruse, Michael. 1979a. *The Darwinian Revolution: Science Red in Tooth and Claw.* Chicago: University of Chicago Press.

    1979b. *Sociobiology: Sense or Nonsense?.* London: D. Reidel.

    1986. *Taking Darwin Seriously.* Oxford: Blackwell.

    1989. *The Darwinian Paradigm.* London: Routledge.

    1993. "The New Evolutionary Ethics." In *Evolutionary Ethics,* eds. M. H. Nitecki and D. V. Nitecki, 133–162. Albany, NY: Suny Press.

    1995. *Evolutionary Naturalism.* New York: Routledge.

    1998. *Taking Darwin Seriously. A Naturalistic Approach to Philosophy.* Amherst, NY: Prometheus Books.

    2006. "Is Darwinian Metaethics Possible (And If It Is, Is It Well Taken)?" In *Giovanni Boniolo an Gabriele de Anna, Evolutionary Ethics and Contemporary Biology,* 13–26. Cambridge: Cambridge University Press.

2009a. "Evolution and Ethics: The Sociobiological Approach." In *Philosophy After Darwin,* ed. Michael Ruse, 489–511. Princeton, NJ: Princeton University Press.

2009b. "Introduction." In *Evolution and Ethics,* Thomas Henry Huxley, vii–xxxvi. Princeton, NJ: Princeton University Press.

2011. "Harvard Holism." *Chronicle of Higher Education* (March 30). http://chronicle.com/blogs/brainstorm/harvard-holism/33763.

2012. *The Philosophy of Human Evolution.* New York: Cambridge University Press.

2013. *The Gaia Hypothesis: Science on a Pagan Planet.* Chicago: University of Chicago Press.

2017. *Darwinism as Religion: What Literature Tells Us about Evolution.* Oxford: Oxford University Press.

Ruse, Michael, and E. O. Wilson. 1985. "The Evolution of Morality." *New Scientist* 1478: 108–128.

1986. "Moral Philosophy as Applied Science." *Philosophy* 61 (236): 173–192.

1989. "The Evolution of Ethics." *New Scientist* 17: 108–128.

Russell, B. (1912) 1988. *The Problems of Philosophy. Great Books in Philosophy.* Buffalo). New York: Prometheus Books.

(1950) 2009. *Unpopular Essays.* London: Routledge.

Russell, Doug. 2015. "Toward a Pragmatist Epistemology: Arthur O. Lovejoy's and H. S. Jennings's Biophilosophical Responses to Neovitalism, 1909–1914." *Journal of the History of Biology* 48: 37–66.

Russett, C. E. 1976. *Darwin in America: The Intellectual Response. 1865-1912.* San Francisco: Freeman.

Rutland, Adam, Melanie Killen, and Dominic Abrams. 2010. "A New Social-Cognitive Developmental Perspective on Prejudice." *Perspectives on Psychological Science* 5(3): 279–291.

Sagan, C., and A. Druyan. 1993. *Shadows of Forgotten Ancestors.* New York: Random House.

Sapp, Jan. 1994. *Evolution by Association: A History of Symbiosis.* New York: Oxford University Press.

Sarkissian, H., J. Parks, D. Tien, J. Wright, and J. Knobe. 2011. "Folk Moral Relativism." *Mind and Language* 26 (4): 482–505.

Sayre-McCord, G. 1986. "The Many Moral Realisms." *Southern Journal of Philosophy* 12 (1): 1–22.

ed. 1988. *Essays on Moral Realism.* Ithaca, NY: Cornell University Press.

2015. "Moral Realism." *The Stanford Encyclopedia of Philosophy* (Spring 2015 Edition), ed. Edward N. Zalta. Available at https://plato.stanford.edu/entries/moral-realism/.

Schafer, Karl. 2010. "Evolution and Normative Skepticism." *Australasian Journal of Philosophy* 88 (3): 471–488.

Schermer, M. 2015. *The Moral Arc: How Science and Reason Lead Humanity toward Truth, Justice, and Freedom.* New York: Henry Holt.

Schloss, J., and M. Murray, eds. 2009. *The Believing Primate: Scientific, Philosophical and Theological Reflections on the Origin of Religion.* Oxford: Oxford University Press.

Schmidt, M., H. Rakoczy, and M. Tomasello. 2013. "Young Children Understand and Defend the Entitlements of Others." *Journal of Experimental Child Psychology* 116(4): 930–944.

Schneewind, J. 2003. *Moral Philosophy from Montaigne to Kant.* Cambridge: Cambridge University Press.

Schweber, S. S. 1980. "Darwin and the Political Economists: Divergence of Character." *Journal of the History of Biology* 13(2): 195–289.

Sclafani, Anthony. 2004. "Oral and Postoral Determinants of Food Reward." *Physiology & Behavior, Proceedings from the 2003 Meeting of the Society for the Study of Ingestive Behavior (SSIB)* 81 (5): 773–779. doi:10.1016/j.physbeh.2004.04.031.

Seabright, P. 2010. *The Company of Strangers: A Natural History of Economic Life.* Princeton, NJ: Princeton University Press.

Searle, John R. 1964. "How to Derive 'Ought' from 'Is.'" *Philosophical Review* 73 (1): 43–58.

Segerstrale, Ullica. 2000. *Defenders of the Truth: the Battle for Science in the Sociobiology Debate and Beyond.* Oxford: University Press.

Seigfried, Charlene Haddock. 1996. *Pragmatism and Feminism: Reweaving the Social Fabric.* Chicago, IL: University of Chicago Press.

1999. "Socializing Democracy: Jane Addams and John Dewey." *Philosophy of the Social Sciences* 29: 207–230.

Sellars, Roy Wood. 1922. *Evolutionary Naturalism.* Chicago: Open Court.

Shafer-Landau, R. 2003. *Moral Realism: A Defense.* Oxford: Oxford University Press.

2007. "Moral and Theological Realism: The Explanatory Argument." *Journal of Moral Philosophy* 4 (3): 311–329.

2012. "Evolutionary Debunking, Moral Realism and Moral Knowledge." *Journal of Ethics & Social Philosophy* 7 (1): 1–37.

Sidgwick, H. 1876. "The Theory of Evolution in its Application to Practice." *Mind* 1: 52–67.

1907. *The Methods of Ethics*, 7th ed. London: Macmillan.

Simion, F., L. Regolin, and H. Bulf . 2008. "A Predisposition for Biological Motion in the Newborn Baby." *Proceedings of the National Academy of Sciences* 105: 809–813.

Simpson, B., A. Harrell, and R. Willer. 2013. "Hidden Paths from Morality to Cooperation: Moral Judgments Promote Trust and Trustworthiness." *Social Forces* 91: 1529–1548.

Simpson, G. G. 1949. *The Meaning of Evolution.* New Haven, CT: Yale University Press.

1964. *This View of Life.* New York: Harcourt, Brace, and World.

Sinclair, N. 2012. "Metaethics, Teleosemantics and the Function of Moral Judgements." *Biology and Philosophy* 27: 639–662.

Singer, P. 1972. Famine, Affluence and Morality. *Philosophy and Public Affairs* 1: 229–43.

1981. *The Expanding Circle: Ethics and Sociobiology.* New York: Farrar, Straus, and Giroux.

1995. *How Are We to Live?* New York: Prometheus Books.

2005. "Ethics and Intuitions." *Journal of Ethics* 9: 331–352.

2011. *The Expanding Circle: Ethics, Evolution and Moral Progress.* Princeton, NJ: Princeton University Press.

Singer, P. and K. de Lazari-Radek. 2012. "The Objectivity of Ethics and the Unity of Practical Reason." *Ethics* 123: 9–31.

Sinnott-Armstrong, Walter. 2006. *Moral Skepticisms.* Oxford: Oxford University Press.

Skarsaune, Knut. 2011. "Darwin and Moral Realism: Survival of the Iffiest." *Philosophical Studies* 152: 229–243.

Slater, A., and P. Quinn. 2001. "Face Recognition in the Newborn Infant." *Infant and Child Development* 10: 21–24.

Small, D., G. Loewenstein, and P. Slovic. 2007. "Sympathy and Callousness: The Impact of Deliberative Thought on Donations to Identifiable and Statistical Victims." *Organizational Behavior and Human Decision Processes* 102(2): 143–153.

Small, Robin. 2005. *Nietzsche and Rée: A Star Friendship.* Oxford: Oxford University Press. Oxford Scholarship Online, 2005. doi: 10.1093/0199278075.001.0001.

2007. "Nietzsche's Evolutionary Ethics." In *Nietzsche and Ethics*, ed. Gudrun von Tevenar, 119–135. Bern: Peter Lang.

Smocovitis, Betty. 1992. "Unifying Biology: The Evolutionary Synthesis and Evolutionary Biology." *Journal of the History of Biology* 25: 1–65.

1996. *Unifying Biology: The Evolutionary Synthesis and Evolutionary Biology.* Princeton, NJ: Princeton University Press.

2008. "The Unifying Vision: Julian Huxley, Evolutionary Humanism, and the Evolutionary Synthesis." In *Pursuing the Unity of Science: Ideology and Scientific Practice between the Great War and the Cold War.* Ashgate Press.

Smuts, Jan C. 1926. *Holism and Evolution.* New York: MacMillan Company.

Sober, Elliott. 1984. *The Nature of Selection: Evolutionary Theory in Philosophical Focus.* Chicago, IL: University of Chicago Press.

1994. *"Prospects for an Evolutionary Ethics." From a Biological Point of View: Essays in Evolutionary Philosophy.* Cambridge: Cambridge University Press.

Sober, Elliott, and David Sloan Wilson. 1998. *Unto Others: The Evolution and Psychology of Unselfish Behavior.* Cambridge, MA: Harvard University Press.

Sommers, Tamler, and Alex Rosenberg. 2003. "Darwin's Nihilistic Idea: Evolution and the Meaningless of Life." *Biology and Philosophy* 18 (5): 653–668.

Sosa, Ernest. 1999. "How to Defeat Opposition to Moore." *Noûs* (Supplement: Philosophical Perspectives *13, Epistemology*) 33: 141–153.

2002. "Reliability and the A Priori." In *Conceivability and Possibility*, eds. Tamar Szabo Gendler and John Hawthorne, 369–384. Oxford: Oxford University Press.

Spaulding, Edward Gleason. 1918. *The New Rationalism.* New York: Henry Holt.

Spencer, Herbert. 1851. *Social Statics: Or, the Conditions Essential to Human Happiness Specified, and the First of them Developed.* London: John Chapman.

1852. "A Theory of Population, Deduced from the General Law of Animal Fertility." *Westminster Review* 1: 468–501.

1853. "The Universal Postulate." *Westminster Review* 60: 513–550.

1855. *The Principles of Psychology.* London: Longman, Brown, Green, and Longmans.

1857. "Progress: Its Law and Cause." *Westminster Review* 67 (April): 244–267.

1860. "The Social Organism." *Westminster Review* XVII: 90–121.

1864. *The Principles of Biology,* Volume 1. London: Williams and Norgate.

1871. "Specialized Administration." *Fortnightly Review* 16: 628–654.

1879. *The Data of Ethics.* London: Williams and Norgate.

1881. *The Principles of Sociology I.* New York: Appleton and Co.

1884. *The Man versus the State.* London: Williams and Norgate.

1892. *The Principles of Ethics.* London: Williams and Northgate.

Sterelny, K. 2007. *Dawkins vs Gould: Survival of the Fittest.* Thriplow: Totem Books.

2014a. "A Paleolithic Reciprocation Crisis: Symbols, Signals and Norms." *Biological Theory* 9 (1): 65–77.

2014b. Cooperation, Culture and Conflict. *British Journal for the Philosophy of Science.* doi: 10.1093/bjps/axu024.

Stich, S. Forthcoming. "The Moral Domain." In *The Atlas of Moral Psychology: Mapping Good and Evil in the Mind,* eds. K. Gray and J. Graham. New York: Guilford.

Strauss, David Friedrich. 1872. *The Old Faith and the New,* trans. Mathilde Blind. New York: Prometheus Books.

Street, Sharon. 2006. "A Darwinian Dilemma for Realist Theories of Value." *Philosophical Studies* 127 (1): 109–166.

2008a. "Reply to Copp: Naturalism, Normativity, and the Varieties of Realism Worth Worrying about." *Philosophical Issues* 18: 207–228.

2008b. "Constructivism about Reasons." In *Oxford Studies in Mathematics,* ed. Russ Shafer-Landau, 207–45. Oxford: Oxford University Press.

2009. "In Defense of Future Tuesday Indifference: Ideally Coherent Eccentrics and the Contingency of What Matters." *Philosophical Issues* 19: 273–298.

2011. "Mind-Independence without the Mystery: Why Quasi-Realists Can't Have It Both Ways." In *Oxford Studies in Metaethics,* Volume 6, ed. Russ Shafer-Landau, 1–32. Oxford: Oxford University Press.

2016. "Objectivity and Truth: You'd Better Rethink It." In *Oxford Studies in Metaethics,* Volume 11, ed. Russ Shafer-Landau. Oxford University Press.

Streumer, B. 2013. "Can We Believe the Error Theory?" *Journal of Philosophy* 110: 194–212.

Stroud, Scott R. 2011. *John Dewey and the Artful Life: Pragmatism, Aesthetics, and Morality.* University Park: Pennsylvania State University Press.

Sturgeon, Nicholas. 1985. "Moral Explanations." In *Morality, Reason, and Truth: New Essays on the Foundations of Ethics,* eds. David Copp and David Zimmerman, 49–78. Totowa, NJ: Rowman and Allanheld.

1988. "Moral Explanations." In *Essays on Moral Realism,* ed. G. Sayre-McCord, 229–255. Ithaca, NY: Cornell University Press.

2002. "Ethical Intuitionism and Ethical Naturalism." In *Ethical Intuitionism: Re-Evaluations,* ed. Philip Stratton-Lake, 184–211. Oxford: Oxford University Press.

Sullivan, Shannon. 2001. *Living Across and Through Skins: Transactional Bodies, Pragmatism, and Feminism.* Bloomington: Indiana University Press.

Tangney, J., and K. Fischer. 1995. *Self-Conscious Emotions: The Psychology of Shame, Guilt, Embarrassment, and Pride.* New York: Guilford Press.

Teehan, John. 2002. "Evolution and Ethics: The Huxley/Dewey Exchange." *Journal of Speculative Philosophy* 16: 225–238.

Tenbrunsel, A. E., and K. Smith-Crowe. 2008. "Ethical Decision Making: Where We've Been and Where We're Going." *Academy of Management Annals* 2: 545–607.

Thompson, E. 2000. "Comparative Color Vision: Quality Space and Visual Ecology." In *Color Perception: Philosophical, Psychological, Artistic and Computational Perspectives*, ed. S. Davis. Oxford University Press.

Thompson, R. Paul 2011. *Agro-Technology: A Philosophical Introduction.* Cambridge: Cambridge University Press.

Thornhill, R., and C. T. Palmer. 2001. *A Natural History of Rape: Biological Bases of Sexual Coercion.* Cambridge, MA: MIT Press.

Todes, Daniel P. 1989. *Darwin without Malthus: The Struggle for Existence in Russian Evolutionary Thought.* Oxford: Oxford University Press.

Tomasello, M., M. Carpenter, J. Call, T. Behne, and H. Moll. 2005. "Understanding and Sharing Intentions: The Origins of Cultural Cognition." *Behavioral and Brain Sciences* 28: 675–691.

Tomasello, M., and A. Vaish. 2013. "Origins of Human Cooperation and Morality." *Annual Review of Psychology* 64: 231–255.

Toner, C. 2010. "Evolution, Naturalism, and the Worthwhile: A Critique of Richard Joyce's Evolutionary Debunking of Morality." *Metaphilosophy* 42: 520–546.

Trachtenberg, Alan. 1982. *The Incorporation of America: Culture and Society in the Gilded Age.* New York: Hill and Wang.

Tremaroli, Valentina, and Fredrik Bäckhed. 2012. "Functional Interactions between the Gut Microbiota and Host Metabolism." *Nature* 489 (7415): 242–249. doi:10.1038/nature11552.

Trivers, R. 1971. "The Evolution of Reciprocal Altruism." *Quarterly Review of Biology* 46 (1): 35–57.

Tufts, James H. 1895. "[Review of] History and Natural Science. W. Windelband. Inaugural Address as Rector. Strassburg, May, 1894." *Psychological Review* 2: 96–97.

Turnbaugh, Peter J., Micah Hamady, Tanya Yatsunenko, Brandi L. Cantarel, Alexis Duncan, Ruth E. Ley, Mitchell L. Sogin, et al. 2009. "A Core Gut Microbiome in Obese and Lean Twins." *Nature* 457: 480–484.

Turner, J. Scott. 2000. *The Extended Organism the Physiology of Animal-Built Structures.* Cambridge, MA: Harvard University Press.

2004. "Extended Phenotypes and Extended Organisms." *Biology and Philosophy* 19(3): 327–352.

2013. "Superorganisms and Superindividuality: The Emergence of Individuality in a Social Insect Assemblage." In *From Groups to Individuals Evolution and Emerging Individuality*, , eds. Frédéric Bouchard and Philippe Huneman, 219–241. Cambridge, MA: MIT Press.

Tversky, Amos and Daniel Kahneman. 1981. "The Framing of Decisions and the Psychology of Choice." *Science* 211(4481): 453–458.

Van der Linden, S. 2011. "Charitable Intent: A Moral or Social Construct? A Revised Theory of Planned Behavior Model." *Current Psychology* 30: 355–374.

Vavova, Katia. 2014. "Debunking Evolutionary Debunking." In *Oxford Studies in Metaethics*, 76–101. Oxford: Oxford University Press.

2015. "Evolutionary Debunking of Moral Realism." *Philosophy Compass* 10: 104–116.

n.d. "A Dilemma for the Darwinian Debunker."

Vlerick, Michael. 2012. "How Can Human Beings Transgress Their Biologically Based Views?" *South African Journal of Philosophy* 31 (4): 717–735.

2014. "Biologising Putnam: Saving the Realism in Internal Realism." *South African Journal of Philosophy* 33 (3): 271–283.

2016. "Explaining Universal Social Institutions: A Game-Theoretic Approach." *Topoi* 35: 291–300.

Wainwright, William. 2010. "In Defense of Non-Natural Theistic Realism: A Response to Wielenberg." *Faith and Philosophy* 27 (4): 457–463.

Wallin, Ivan E. 1927. *Symbionticism and the Origin of Species*. Baltimore: Williams & Wilkins Company.

Wallis Budge, E. A. 1901. *Book of the Dead*. London: Kegan Paul, Trench, Trübner, and Co.

Walls, Ramona L., John Deck, Robert Guralnick, Steve Baskauf, Reed Beaman, Stanley Blum, Shawn Bowers, et al. 2014. "Semantics in Support of Biodiversity Knowledge Discovery: An Introduction to the Biological Collections Ontology and Related Ontologies." *PLoS ONE* 9 (3): e89606. doi:10.1371/journal.pone.0089606.

Warneken, F., K. Lohse, A. Melis, and M. Tomasello. 2011. "Young Children Share the Spoils after Collaboration." *Psychological Science* 22: 267–273.

Watson, J. D and F. H. Crick. April 1953. "Molecular Structure of Nucleic Acids; a Structure for Deoxyribose Nucleic Acid (PDF)." *Nature* 171(4356): 737–738.

Welchman, Jennifer. 1995. *Dewey's Ethical Thought*. Ithaca, NY: Cornell University Press.

Wharton, E. 1905. *The House of Mirth*. New York: Charles Scribner's Sons.

Wheeler, William Morton. 1910. *Ants: Their Structure, Development and Behavior*. New York: Columbia University Press.

1911. "The Ant Colony as an Organism." *Journal of Morphology* 22: 307–326.

1918. "A Study of Some Ant Larvae, with a Consideration of the Origin and Meaning of the Social Habit among Insects." *Proceedings of the American Philosophical Society* 57: 293–343.

1923. "Social Life among the Insects." *Scientific Monthly* 16: 160–177.

1926. "Emergent Evolution and the Social." *Science* 64(1662): 433–440.

1927. *Emergent Evolution and the Social*. London: Kegan Paul.

1928. *Emergent Evolution and the Development of Societies*. New York: Norton and Company.

White, R. 2010. "You Just Believe That Because...." *Philosophical Perspectives* 24: 573–615.

Whitehead, Alfred North. 1925. *Science and the Modern World*. New York: MacMillan.

Whitman, Charles Otis. 1891. "Specialization and Organization." In *Biological Lectures Delivered at Marine Biological Laboratory*. Boston: Ginn and Company.

1894. "The Inadequacy of the Cell-Theory of Development." In *Biological Lectures Delivered at the Marine Biological Laboratory*. Boston: Ginn and Company.

Wielenberg, Eric. 2005. "In Defense of Non-Natural Non-Theistic Moral Realism." *Faith and Philosophy* 26 (1): 23–41.

2009. "A Defense of Moral Realism." *Faith and Philosophy* 26 (1): 23–41.

2010. "On the Evolutionary Debunking of Morality." *Ethics* 120: 441–464.

2014. *Robust Ethics: The Metaphysics and Epistemology of Godless Normative Realism.* Oxford: Oxford University Press.

2016. "Evolutionary Debunking Arguments in Religion and Morality." In *Explanation in Ethics and Mathematics: Debunking and Dispensibility,* eds. U. Leibowitz and N. Sinclair. Oxford: Oxford University Press. DOI: 10.1093/acprof:oso/9780198778592.001.0001

Williams, B. 1985. *Ethics and the Limits of Philosophy.* London: Fontana.

Williams, G. C. 1966. *Adaptation and Natural Selection: A Critique of Some Current Evolutionary Thought.* Princeton, NJ: Princeton University Press.

Williamson, T. 2009. "Reply to Alvin Goldman." In *Williamson on Knowledge,* eds. P. Greenough and D. Pritchard, 305–312. Oxford: Oxford University Press.

Wilson, D. 2002. *Darwin's Cathedral: Evolution, Religion and the Nature of Society.* Chicago, IL: University of Chicago Press.

2004. "The New Fable of the Bees: Multilevel selection, Adaptive Societies, and the Concept of Self Interest." Ed. R. Koppl, 201–220. Oxford: Elsevier.

Wilson, D. S., E. Dietrich, and A. B. Clark. 2003. "On the Inappropriate Use of the Naturalistic Fallacy in Evolutionary Psychology." *Biology and Philosophy* 18: 669–682.

Wilson, David Sloan, and Edward O. Wilson. 2007. "Rethinking the Theoretical Foundations of Sociobiology." *Quarterly Review of Biology* 82: 327–348.

2008. "Evolution for the Good of the Group." *American Scientist* 96: 380–389.

Wilson, D. S, and E. Sober. 1989. "Reviving the Superorganism." *Journal of Theoretical Biology* 136 (3): 337–356.

Wilson, Edward O. 1953. "The Origin and Evolution of Polymorphism in Ants." *Quarterly Review of Biology* 28: 136–156.

1971. *The Insect Societies.* Cambridge: Belknap Press.

1975. *Sociobiology: The New Synthesis.* Cambridge: Belknap Press.

1976. "Academic Vigilantism and the Political Significance of Sociobiology." *BioScience* 26: 187–190.

1978. *On Human Nature: With a New Preface,* rev. ed. Cambridge, MA: Harvard University Press.

1992. *The Diversity of Life.* Cambridge, MA: Harvard University Press.

1994. *Naturalist.* Washington, DC: Island Press.

2005. "Kin Selection as the Key to Altruism: Its Rise and Fall." *Social Research* 72: 159–166.

2012. *The Social Conquest of Earth.* New York: Norton and Co.

Wilson, H. V. 1907. Migration and Rearrangement of Cells Within Sponges.

2014. *The Meaning of Human Existence.* Liveright: NYC.

Wilson, Jack. 1999. *Biological Individuality the Identity and Persistence of Living Entities.* Cambridge, NY: Cambridge University Press.

Wilson, Robert A. 2004. *Genes and the Agents of Life: The Individual in the Fragile Sciences Biology.* Cambridge, NY: Cambridge University Press.

Wittig, M., K. Jensen, and M. Tomasello. 2013. Five-Year-Olds Understand Fair as Equal in a Mini-Ultimatum Game. *Journal of Experimental Child Psychology* 116(2): 324–337.

Wolfe, Elaine Claire Daughetee. 1975. "Acceptance of the Theory of Evolution in America: Louis Agassiz vs. Asa Gray." *American Biology Teacher* 37: 244–247.

Wright, J., P. Grandjean, and C. McWhite. 2013. "The Meta-Ethical Grounding of our Moral Beliefs: Evidence for Meta-Ethical Pluralism." *Philosophical Psychology* 26 (3): 336–361.

Wright, J. C., C. B. McWhite, and P. T. Grandjean. 2014. "The Cognitive Mechanisms of Intolerance: Do Our Meta-Ethical Commitments Matter." In *Oxford Studies in Experimental Philosophy*, Volume 1, eds. T. Lombrozo, J. Knobe, and S. Nichols, 28–61. Oxford: Oxford University Press.

Wright, L. 1976. *Teleological Explanations*. Berkeley: University of California Press.

Wright, R. 1994. *The Moral Animal: Why We Are The Way We Are*. New York: Vintage Books.

Wynne-Edwards, V. C. 1962. *Animal Dispersion in Relation to Social Behaviour*. New York: Hafner Publishing.

Young, L., and A. Durwin. 2013. "Moral Realism as Moral Motivation: The Impact of Meta-Ethics on Everyday Decision-making." *Journal of Experimental Social Psychology* 49 (2): 302–306.

Index